Gas-Phase Chemistry in Space

From elementary particles to complex organic molecules

AAS | IOP Astronomy

AAS Editor in Chief

About the program:

AAS-IOP Astronomy ebooks is the official book program of the American Astronomical Society (AAS), and aims to share in depth the most fascinating areas of astronomy, astrophysics, solar physics and planetary science. The program includes publications in the following topics:

- GALAXIES AND COSMOLOGY
- INTERSTELLAR MATTER AND THE LOCAL UNIVERSE
- STARS AND STELLAR PHYSICS
- EDUCATION, OUTREACH AND HERITAGE
- HIGH-ENERGY PHENOMENA AND FUNDAMENTAL PHYSICS
- THE SUN AND THE HELIOSPHERE
- THE SOLAR SYSTEM, EXOPLANETS, AND ASTROBIOLOGY
- INSTRUMENTATION, SOFTWARE, LABORATORY ASTROPHYSICS AND DATA

Books in the program range in level from short introductory texts on fast-moving areas, graduate and upper-level undergraduate textbooks, research monographs and practical handbooks.

For a complete list of published and forthcoming titles, please visit iopscience.org/books/aas.

About the American Astronomical Society

The American Astronomical Society (aas.org), established 1899, is the major organization of professional astronomers in North America. The membership (~7,000) also includes physicists, mathematicians, geologists, engineers and others whose research interests lie within the broad spectrum of subjects now comprising the contemporary astronomical sciences. The mission of the Society is to enhance and share humanity's scientific understanding of the universe.

Gas-Phase Chemistry in Space

From elementary particles to complex organic molecules

François Lique
Universite Le Havre Normandie, France

Alexandre Faure
Universite Grenoble Alpes, France

IOP Publishing, Bristol, UK

ISBN 978-0-7503-1425-1 (ebook)
ISBN 978-0-7503-1426-8 (print)
ISBN 978-0-7503-1427-5 (mobi)

DOI 10.1088/2514-3433/aae1b5

Version: 20190201

AAS–IOP Astronomy
ISSN 2514-3433 (online)
ISSN 2515-141X (print)

British Library Cataloguing-in-Publication Data: A catalogue record for this book is available from the British Library.

Published by IOP Publishing, wholly owned by The Institute of Physics, London

IOP Publishing, Temple Circus, Temple Way, Bristol, BS1 6HG, UK

US Office: IOP Publishing, Inc., 190 North Independence Mall West, Suite 601, Philadelphia, PA 19106, USA

Contents

Preface

The study of molecules in space, astrochemistry, is a rapidly growing field in astrophysics. Molecules are found everywhere, from our Solar System to high redshift galaxies. For half a century, spectral exploration at radio, infrared, visible and ultra-violet wavelengths has led to the discovery of about 50 different molecules in comets and more than 200 molecules in the interstellar medium. In meteorites, high-resolution mass spectrometry has even revealed tens of thousands of organic compounds. Molecules are also being found in protoplanetary disks and in the atmosphere of exoplanets, with ever greater complexity. Many of these molecules are invaluable thermometers and barometers to probe local physical conditions but some also play a direct role in the dynamics of protostellar collapse and star formation.

The chapters in this book collectively address the physics and chemistry of astrophysical molecules with a focus on gas-phase processes. They introduce essential concepts that govern the formation, excitation and destruction of molecules at a postgraduate level. They cover a broad range of topics, from early Universe chemistry (Chapter 1) and stellar nucleosynthesis (Chapter 2) to the study of bimolecular reaction kinetics (Chapter 3). A special attention has been given to describe specific radiative and electron collision processes (Chapters 4 and 5) as well as the spectroscopy and excitation of astrophysical molecules (Chapters 6 and 7). Recent examples of the interplay between observational and laboratory astrophysics are finally provided in the last chapter (Chapter 8). With more than 100 figures, the volume nicely illustrates the novel and exciting advances in studies devoted to the interaction of molecules with photons, electrons, atoms and other molecules, both from theoretical and experimental perspectives.

The purpose, then, of this monograph is to review the recent research in the multidisciplinary field of astrochemistry, with emphasis on small gas-phase molecules. We believe this volume is very timely, with the advent of new powerful telescopes such as the Atacama Large Millimeter/submillimeter Array (ALMA) and the James Webb Space Telescope (JWST). Major astrophysical advances will surely come from these facilities, but maximizing a broad scientific return depends on our understanding of the many molecular processes that control the observed spectra. This is precisely the object of this book to present the state-of-art of research in gas-phase astrochemistry.

François Lique & Alexandre Faure

"Astrochemistry has made extraordinary advances in the past decade, driven mostly by observations with sophisticated instrumentation of enhanced sensitivity and improved spectral and spatial resolution over an extended range of frequencies but with crucial support from theoretical models and laboratory investigations."

Alex Dalgarno, introductory lecture "The growth of molecular complexity in the Universe" (Faraday Discussions 2006, 133, 9)

Acknowledgements

We wish to warmly thank the authors for their contributions and the referees of the submitted papers for their helpful comments—in particular our colleagues Pierre Hily-Blant (University of Grenoble Alpes), Laurent Margulès (University of Lille), Denis Puy (University of Montpellier), Ioan Schneider (University of Le Havre), Jean Christophe Loison (University of Bordeaux), Corinne Charbonnel (Genova University) and Thierry Stoecklin (University of Bordeaux). This work was supported by the Programme National "Physique et Chimie du Milieu Interstellaire" (PCMI) of CNRS/INSU with INC/INP co-funded by CEA and CNES.

About the Editors

François Lique

François Lique (born in 1980, in France) studied physics at the Université Pierre et Marie Curie, Paris (France) and received his PhD in 2006, from the same university, under the supervision of Nicole Feautrier and Annie Spielfiedel. His dissertation focused on the collisional excitation of interstellar species, from the theoretical modeling to astrophysical applications. During his PhD, he spent several months in Madrid (Spain) working on astrophysical modeling. As a postdoc, he joined Millard Alexander's group at the University of Maryland (USA) to work on inelastic and reactive collisions implying open-shell molecules. In 2008, he obtained a lecturer position at the Laboratoire Ondes et Milieux Complexes of the University of Le Havre (France). In 2010, he defended his Habilitation. He was promoted to a full professor position in 2017. The same year, He was elected fellow of the Institut Universitaire de France. His research focuses on the modeling of physical and chemical processes of astrophysical interest and its applications. The objective is to determine accurate data on for processes (radiative, collisional, or reactive) in which interstellar molecules are involved. The determination of these molecular data also requires the development of theoretical methods, which is another main research field on its own. He has scientific expertise in three main topics: the use of quantum chemistry methods to model the interaction between particles, the study of quantum dynamics of nuclei, and astrophysical modeling. His research allows significant advances in the field of molecular astrophysics, such as observations of new molecules or solving problems linked to molecules abundances. In 2014, he was awarded the Young Researcher Prize of the French Astronomical Society, and in 2016, he obtained the Early Career Prize of the Laboratory Astrophysics Division of the American Astronomical Society.

Alexandre Faure

Alexandre Faure (born in 1972, in France) studied Physics at the Université Joseph Fourier, Grenoble (France) and received his PhD in 1999 under the supervision of Pierre Valiron. His dissertation focused on the chemistry of interstellar radicals at very low temperature. In the period 2000–2003, he held various postdoctoral positions; in particular, he spent two years at University College London (UCL) in the group of Prof. J. Tennyson, as an individual Marie-Curie fellow, to work on the electron-impact excitation of astrophysical molecules. Dr. Faure obtained a permanent academic position at CNRS in Grenoble in 2003. His research focuses on molecular processes of astrophysical relevance and mainly concerns quantum aspects (state-resolved collisions). He has made contributions in scattering theory and astrochemistry, and is the author of about 150 publications.

Contributors

Alfredo Aguado
Unidad Asociada de Química-Física Aplicada CSIC-UAM
Facultad de Ciencias,
Universidad Autónoma de Madrid
Spain
alfredo.aguado@uam.es

Ludovic Biennier
Institut de Physique de Rennes.
UMR 6251 Université de Rennes 1 - CNRS
Bat 11 C - Campus de Beaulieu,
263 Av. du Général Leclerc, 35042 Rennes Cedex, France
ludovic.biennier@univ-rennes1.fr

Carla Maria Coppola
Dipartimento di Chimica.
Universitá degli Studi "Aldo Moro" di Bari
Via Orabona 4, I-70126, Bari, Italy
carla.coppola@uniba.it

Paul J. Dagdigian
Department of Chemistry
Johns Hopkins University
Baltimore, MD 21218, USA
pjdagdigian@jhu.edu

Alexandre Faure
UJF-Grenoble 1/CNRS, Institut de Planétologie et d'Astrophysique de Grenoble - UMR 5274
Grenoble F-38041, France
alexandre.faure@univ-grenoble-alpes.fr

Daniele Galli
INAF - Osservatorio Astrofisico di Arcetri,
Largo E. Fermi 5, I-50125, Firenze, Italy
galli@arcetri.astro.it

Susana Gómez-Carrasco
Unidad Asociada CSIC-USAL, Plz. de los Caídos s/n,
Facultad de Química,
Universidad de Salamanca
Plz. de los Caídos s/n, 37008, Salamanca, Spain
susana.gomez@usal.es

Hua Guo
Department of Chemistry and Chemical Biology,
University of New Mexico, Albuquerque,
New Mexico 87131, USA
hguo@unm.edu

Sébastien Le Picard
Institut de Physique de Rennes.
UMR 6251 Université de Rennes 1 - CNRS
Bat 11 C - Campus de Beaulieu, 263 Av. du Général Leclerc, 35042 Rennes Cedex, France
sebastien.le-picard@univ-rennes1.fr

Maurice Monnerville
Laboratoire de Physique des Lasers, Atomes et Molécules,
UMR 8523 du CNRS, Centre d'Études et de Recherches Lasers et Applications,
Université Lille I, Bât. P5, 59655 Villeneuve d'Ascq Cedex, France
maurice.monnerville@univ-lille1.fr

Nikos Prantzos
Institut d'Astrophysique de Paris
98bis Boulevard Arago, F-75014 Paris, France
prantzos@iap.fr

Octavio Roncero
Instituto de Física Fundamental (IFF-CSIC), C.S.I.C.,
Serrano 123, 28006 Madrid, Spain
octavio.roncero@csic.es

Evelyne Roueff
Sorbonne Universit, Observatoire de Paris, Universit PSL,CNRS, LERMA
5 Place J. Janssen, 92190 Meudon, France
evelyne.roueff@obspm.fr

Stephan Schlemmer
Physikalisches Institut,
Universität zu Köln
Zülpicher Straße 77, 50937 Kln
schlemmer@ph1.uni-koeln.de

Jonathan Tennyson
Department of Physics and Astronomy,
University College London
London WC1E 6BT, UK
j.tennyson@ucl.ac.uk

Chapter 1

The Chemistry of the Early Universe

Carla Maria Coppola and Daniele Galli

1.1 Cosmological Background

The standard cosmological model (SCM) is a theoretical framework that successfully accounts for the main observational characteristics of our universe. For reasons that will be explained at the end of this section, it is specifically referred to as the Λ-cold dark matter model (ΛCDM). The SCM is rooted on Big Bang cosmology, and was formulated during the second half of the 20th century on the basis of two important observational evidences: (i) the recession of galaxies, expressed by the famous "law" formulated by Hubble in 1929, $v(d) = H_0 d$, where v is the galaxy's velocity away from the Milky Way, d its distance, and H_0 is the Hubble constant; (ii) the existence of the cosmic microwave background (CMB), discovered at microwave wavelengths by Penzias and Wilson in 1965. Both discoveries lend credence to the Big Bang model of the origin of the universe. It was eventually found that H_0 is the present value, known today to be equal to 67.74 ± 0.46 km s^{-1} Mpc^{-1} (see Table 1.1)[1], of a time-dependent variable $H(t)$, called the Hubble parameter, related to the geometry of the universe and its evolution through the scale factor $a(t)$. The CMB discovered in 1965 was interpreted as the emission coming from the epoch of matter and radiation decoupling about 400,000 yr after the Big Bang. At this time, the universe became optically transparent, the photons mean free path being longer than the radius of the universe itself (this epoch defines the "surface of last scattering").[2] The spectrum of these relic photons corresponds to a blackbody with temperature equal to 2.7255 ± 0.0006 K (see Table 1.1). Later on, experiments such as the *Wilkinson Microwave Anisotropy Probe* (*WMAP*), Boomerang, and Planck revealed tiny spatial temperature fluctuations in the CMB, known as CMB anisotropies, representing the expected signatures of the formation of the first cosmological structures.

[1] Galactic distances are expressed in Megaparsec (1 Mpc = 10^6 pc = 3.086×10^{22} m). The parsec (pc) is a unit of astronomical distance corresponding to the distance at which the semimajor axis of the Earth's orbit subtends an angle of one second of arc.

[2] For a detailed insight on General Relativity and Cosmology, see Gasperini (2007, 2016).

doi:10.1088/2514-3433/aae1b5ch1

Table 1.1. Cosmological Parameters, from Planck and *WMAP* Data (Planck Collaboration 2016; Fixsen 2009; Komatsu et al. 2011)

Parameter	Numerical value
H_0	67.74 ± 0.46 km s^{-1} Mpc^{-1}
ρ_{cr}	$(8.62 \pm 0.12) \times 10^{-27}$ kg m^{-3}
T_0	2.7255 ± 0.0006 K
Ω_m	0.3089 ± 0.0062
Ω_b	0.0486 ± 0.0010
Ω_{dm}	0.2589 ± 0.0057
Ω_Λ	0.6911 ± 0.0062
z_{eq}	3361 ± 27
η_{10}	6.16 ± 0.15

The SCM is based on the following main assumptions:

1. On cosmological distances, the gravitational interaction is described by Einstein's equation of general relativity,

$$G_{\mu\nu} \equiv R_{\mu\nu} - \frac{1}{2}g_{\mu\nu}R = 8\pi G T_{\mu\nu}, \tag{1.1}$$

   where $G_{\mu\nu}$ is the Einstein tensor, $T_{\mu\nu}$ is the energy–momentum tensor, $g_{\mu\nu}$ is the metric tensor, and G is the universal gravitational constant. Here, $T_{\mu\nu}$ represents the effect of all the sources gravitationally coupled to the metric.
2. On large scales of distance, the universe can be considered homogeneous and isotropic.
3. On cosmological scales, the sources of the gravitational field can be represented by barotropic fluids, in which the fluid's pressure p and energy density ρ_e are related by the relation $p = w\rho_e$, with w held constant;[3]
4. The spectrum of the radiation fluid is in thermodynamic equilibrium at a temperature T_r and is therefore described by a Planck spectral distribution.

Several fundamental consequences can be derived from these assumptions:

1. The energy–momentum tensor is maximally symmetric in tridimensional space (i.e., the pressure is isotropic). For a perfect fluid in thermodynamic equilibrium, it takes the form

$$T^{\nu}_{\mu} = \left(\rho_e + \frac{p}{c^2}\right)u_\mu u^\nu - p\delta^{\nu}_{\mu}, \tag{1.2}$$

   where u^μ is the quadrivelocity with respect to a class of special observers that are at rest with respect to the spacetime geometry, and δ^{ν}_{μ} is the Kronecker tensor.

[3] As shown later, three fluids are considered in the ΛCDM model: radiation, with $w = 1/3$; matter, with $w = 0$; and dark energy, responsible for the accelerated expansion of the universe, with $w = -1$.

2. The metric tensor is maximally symmetric in its spatial component. For this reason, the fundamental spacetime interval can be expressed as:

$$ds^2 = g_{\mu\nu}dx^\mu dx^\nu = dt^2 - a^2(t)\left[\frac{dr^2}{1 - Kr^2} + r^2(d\theta^2 + \sin^2\theta d\varphi^2)\right], \quad (1.3)$$

where K is the spatial curvature of the spacetime Riemann manifold, and $a(t)$ is a time-dependent function known as a scale factor. This is called the Friedmann–Robertson–Walker (FRW) metric.

It is possible to separate the spatial and temporal components of Equation (1.1):

$$\frac{\ddot{a}}{a} + 2\left(H^2 + \frac{K}{a^2}\right) = 4\pi G(\rho_e - p), \quad (1.4)$$

$$\frac{\ddot{a}}{a} = -\frac{4\pi G}{3}(\rho_e + 3p), \quad (1.5)$$

where $H(t) = \dot{a}/a$ (the dot indicates the derivative with respect to cosmic time) determines the rate of expansion of the universe.

From Equations (1.4) and (1.5), the following fundamental cosmological equations can be derived:

$$H^2 + \frac{K}{a^2} = \frac{8\pi G}{3}\rho_e, \quad (1.6)$$

$$\dot{\rho}_e + 3H(\rho_e + p) = 0. \quad (1.7)$$

Equation (1.6), known as the Friedmann equation, determines the evolution of the geometry of the universe; Equation (1.7), the cosmological equivalent for the continuity equation, states how the fluid evolves in time. However, rather than time, the quantity commonly used in cosmology is the redshift z, defined as

$$1 + z = \frac{a_0}{a(t)}, \quad (1.8)$$

where t_0 is the present time and $a_0 = a(t_0)$. This definition expresses the kinematical properties of the FRW metric. It asserts that, even in the absence of relative motion between the emitting source and the observer (i.e., in absence of Doppler effect), a wavelength shift of the signal is present. If λ_{obs} is the observed wavelength, λ_{em} the wavelength of the emitted signal, and t_{obs} and t_{em} the observation time and emission time of the periodic signal, respectively, Equation (1.8) implies

$$\frac{\lambda_{obs}}{\lambda_{em}} = \frac{a(t_{obs})}{a(t_{em})} = \frac{1 + z_{em}}{1 + z_{obs}}. \quad (1.9)$$

Thus, the higher the redshift, the earlier the corresponding cosmological time. Using the equations of state for each fluid, Equation (1.7) can be solved for each component, using the corresponding equation of state, assuming that matter and radiation evolve without exchanging energy. This gives the time-dependent expressions

$$\rho_r \sim (1+z)^4, \qquad \rho_m \sim (1+z)^3, \tag{1.10}$$

for the evolution of radiation and matter, respectively. They indicate the existence at early times in the evolution of the universe of an epoch dominated by radiation, followed by an epoch dominated by matter (the present epoch). The densities of radiation and matter were the same at a redshift z_{eq} (defined below) corresponding to a time $\sim 5 \times 10^4$ yr after the Big Bang, when the temperature (of both matter and radiation) was $T_m \approx T_r \approx 9700$ K.

In addition to matter and radiation, the total density ρ_e includes the contribution of a third component, introduced to account for the observational evidence, coming both from the Hubble diagram of Type Ia supernovae and the analysis of CMB anisotropies, that our universe is experiencing an accelerated expansion ($\ddot{a} < 0$) (Riess et al. 1998; Perlmutter et al. 1999). To satisfy this condition, it is necessary to introduce in the model a third fluid component characterized by a negative pressure resulting in a repulsive effect on the matter evolution. This is usually referred to as "dark energy." The simplest dark energy model is based on the introduction of vacuum in the Einstein equation, the so-called "cosmological constant" Λ, and the corresponding density parameter is denoted Ω_Λ. For this reason, this cosmological model is usually referred to as the Λ-cold dark matter model (ΛCDM).

Defining the present-day critical density

$$\rho_{cr} = \frac{3H_0^2}{8\pi G}, \tag{1.11}$$

Equation (1.6) can be written as

$$\Omega_m + \Omega_r + \Omega_\Lambda + \Omega_K = 1, \tag{1.12}$$

where Ω_m, Ω_r, and Ω_Λ are the densities of the three components in units of the critical density and

$$\Omega_K = -\frac{K}{a_0^2 H_0^2} \tag{1.13}$$

is a "curvature density," constrained by observations to be consistent with zero, $|\Omega_K| \lesssim 10^{-2}$. In addition, Ω_m includes both the baryonic and dark matter contribution to the matter fluid

$$\Omega_m = \Omega_b + \Omega_{dm}. \tag{1.14}$$

For example, the baryon density at redshift z is

$$\rho_b = \Omega_b \rho_{cr} (1+z)^3. \tag{1.15}$$

The density of radiation Ω_r can be related to Ω_m as $\Omega_r = \Omega_m/(1 + z_{eq})$, where z_{eq} represents the redshift at which $\rho_r = \rho_m$, defined by

$$1 + z_{eq} = \frac{3(cH_0)^2}{8\pi G(1 + f_\nu)U}\Omega_m, \tag{1.16}$$

with f_ν being the neutrino contribution to the energy density in relativistic species (approximately equal to 0.68 for massless neutrino models), and U the photon energy density.

Using the definition of redshift together with the dependence of radiation and matter densities on the scale factor, the Friedmann Equation (1.6) can be rearranged as

$$\frac{dt}{dz} = -\frac{1}{(1 + z)H(z)}, \tag{1.17}$$

where

$$H(z) = H_0\sqrt{\Omega_r(1 + z)^4 + \Omega_m(1 + z)^3 + \Omega_K(1 + z)^2 + \Omega_\Lambda}. \tag{1.18}$$

According to this definition, the dependence of the redshift on cosmic time strictly depends on the adopted cosmological model; in particular, the redshift increases with decreasing cosmic time, the divergence at $z \to \infty$ corresponding to the Big Bang singularity. Updated values of the cosmological parameters can be found in Planck Collaboration (2016) or derived by data therein; they are listed in Table 1.1.

1.1.1 Energy Exchanges: Radiation and Matter Temperatures

The temperatures of the various components of the universe are functions of the scale factor $a(t)$, controlling the overall expansion. As long as the fluids evolve adiabatically, these temperatures can be derived from the corresponding densities, expressed by Equation (1.10). For the matter or radiation fluids, the first law of thermodynamics $dE = TdS - pdV$ implies that

$$d(\rho V) = TdS - pdV \Rightarrow dS = \frac{1}{T}[d(\rho V) + pdV], \tag{1.19}$$

where S is the entropy of the system, and V and T its volume and temperature, respectively. If these are considered as independent variables, then $Tdp = (\rho + p)dT$. With the help of this relation, direct integration of Equation (1.19) gives the relation between density and temperature for a fluid in thermal equilibrium:

$$T \sim \rho^{\gamma/(1+\gamma)}, \tag{1.20}$$

where γ is the adiabatic index ($\gamma = 1/3$ and $\gamma = 2/3$ for radiation and matter, respectively). On the other hand, the density of matter and radiation constituting our universe as a function of cosmic time, deduced from the continuity Equation (1.5), are given by Equation (1.10). Combining these results, Equation (1.20) gives

$$T \sim a(t)^{-3\gamma}, \tag{1.21}$$

which, for the radiation and (nonrelativistic) matter fluids, becomes

$$T_r \sim 1 + z. \qquad T_m \sim (1 + z)^2. \tag{1.22}$$

The proportionality constant is fixed by the present value of the radiation temperature $T_0 = 2.7255$ K (see Table 1.1), such that the radiation temperature is

$$T_r = T_0(1 + z). \tag{1.23}$$

However, energy exchanges between matter particles and the radiation field play an important role, especially at high redshifts. This is shown by the equation describing the time evolution of the matter temperature T_m,

$$\frac{dT_m}{dt} = -2H(t)T_m + \frac{8\sigma_T a T_r^4 (T_r - T_m)x_e}{3m_e c} + (\Gamma - \Lambda)_{mol}, \tag{1.24}$$

where σ_T is the Thomson cross section, a is the radiation constant, m_e is the electron mass, and x_e is the ionization fraction. The first term in Equation (1.24) represents the adiabatic cooling of the universe due to its expansion; the second represents the net energy transfer between the CMB and the free electrons due to Compton scattering; the last term is the heating/cooling contribution due to molecular processes $(\Gamma - \Lambda)_{mol}$, namely: bremsstrahlung (or free–free cooling), photoionization heating, photorecombination cooling, line cooling, collisional ionization cooling, and collisional recombination heating. This last term, negligible in the expansion phase of the universe, becomes crucial for the thermal evolution of the gas during the collapse phase for the formation of the first gravitationally bound structures.

Figure 1.1 shows the evolution of the matter and radiation temperatures T_m and T_r as a function of the redshift z, described by Equations (1.23) and (1.24); it makes evident the decoupling between gas and radiation, i.e., the moment in which the energy exchange between the two fluids became inefficient and the universe became transparent to radiation.

1.2 Big Bang Nucleosynthesis

According to the standard Big Bang model, the first nuclei were formed by thermonuclear reactions when the universe cooled from about 10 to 1 billion degrees K. This phase of primordial nucleosynthesis lasted a very short time, from about 1 s to a few minutes after the Big Bang. As a result, ~25% in mass of the baryonic component of the universe, consisting of protons and neutrons, was converted into He nuclei; trace amounts of D, ^{3}He, and ^{6,7}Li were also formed, while the rest was left in the form of protons (see, e.g., Coc 2013 and Cyburt et al. 2016 for recent reviews).

Two of the three parameters controlling the production of nuclei in the standard Big Bang model (the number of neutrino families and the lifetime of the neutron) are experimentally well-determined and the residual uncertainties only marginally affect the results. The remaining parameter is the value of the density of baryons in the

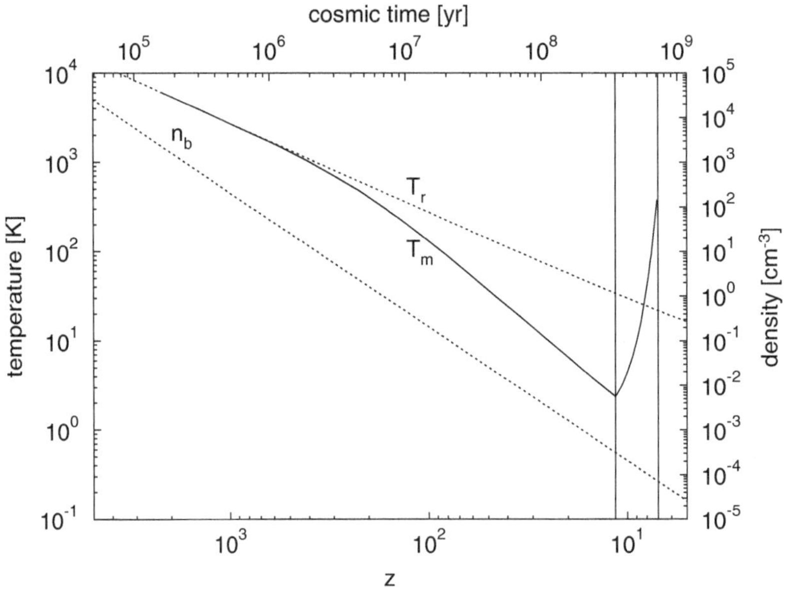

Figure 1.1. Temperatures of matter and radiation T_m and T_r as a function of redshift z. The phase of plasma-radiation energy coupling and the following adiabatic phase, where the two fluids evolve independently, are evident here. The increase of the gas temperature at lower redshifts (shown here qualitatively), is due to the formation of the first luminous sources and is responsible for the reionization of the gas. The baryon density n_b of the expanding universe is also shown.

universe, combined with the Hubble constant. It is usually expressed in terms of the number of photons per baryon η, which is independent on redshift:

$$\eta \equiv \frac{n_\gamma}{n_b} = 2.739 \times 10^{-8}\Omega_b h^2 \tag{1.25}$$

where h is the present value of the Hubble constant H_0 in units of 100 km s^{-1} Mpc^{-1} (see Table 1.1).

Observational constraints on η are obtained from the analysis of the CMB power spectrum, specifically from the first-to-second peak ratio relative to the first-to-third peak ratio, now well-determined by the *WMAP* satellite. The seven-year *WMAP* data give $\Omega_b h^2 = 0.02255 \pm 0.00054$ (Komatsu et al. 2011), resulting in $\eta_{10} = 6.16 \pm 0.15$, where $\eta_{10} = 10^{10}\eta$. This value is consistent, within the errors, with measurements of the abundance of D and He in low-metallicity galaxies (Noterdaeme et al. 2012; Aver et al. 2012). The situation for Li remains less clear and potentially critical, owing to a persistent discrepancy of a factor of ~3 between the predictions of the standard Big Bang model and measurements of Li in metal-poor halo dwarfs (e.g., Bonifacio et al. 2012). However, the detection of a fractional abundance of Li in the interstellar medium of the Small Magellanic Cloud, in excellent agreement with the cosmological predictions (Howk et al. 2012), seems to indicate that the "lithium problem" might not signal a crisis for the standard cosmological model. This discrepancy between standard Big Bang nucleosynthesis

Table 1.2. Primordial Abundances Relative to H, According to Standard Big Bang Nucleosynthesis (from Coc et al. 2012)

^{2}H	2.59×10^{-5}	^{3}He	1.04×10^{-5}	^{4}He	8.23×10^{-2}	^{6}Li	1.23×10^{-14}
^{7}Li	5.24×10^{-10}	^{9}Be	9.60×10^{-19}	^{10}B	3.00×10^{-21}	^{11}B	3.05×10^{-16}
^{12}C	5.34×10^{-16}	^{13}C	1.41×10^{-16}	^{14}C	1.62×10^{-21}	^{14}N	6.76×10^{-17}
^{15}N	2.25×10^{-20}	^{16}O	9.13×10^{-20}				

and observations has also been justified introducing a non-Boltzmann-distribution for the velocities of nucleons (Hou et al. 2017).

Because of the absence of stable nuclei with five or eight nucleons, Big Bang nucleosynthesis produced only traces of elements heavier than He ("metals," in astrophysical jargon). The rapid cooling and density decrease due to the cosmological expansion prevented the synthesis of C, N, and O nuclei by the three-body reactions that are activated in long-lived stars. In addition to ^{3,4} He, D, and ^{6,7} Li, only tiny amounts of "metals" (^{9}Be, ^{11}B, and CNO elements) were produced by cosmological nucleosynthesis, and their subsequent chemical evolution is generally ignored (for an exception, see Vonlanthen et al. 2009). Their production could also be a result of possible fluctuations in the primordial baryon abundances (Jedamzik et al. 1994). Table 1.2 summarizes the composition of the baryonic component of the universe according to Coc et al. (2012) for $\eta_{10} = 6.16$.

1.3 The Recombination Era

The expansion of universe allowed the primordial plasma of H and He nuclei, radiation, and electrons to cool down enough to form neutral species. The chemical reactions through which the neutralization proceeded are called recombination processes; for this reason, the epoch between $z \approx 3000$ and $z \approx 1000$ in which neutralization of H and He took place is usually referred to as the recombination era (Kolb & Turner 1990; Padmanabhan 2002; Peebles 1968, 1993). This epoch is crucial for the evolution of the universe: the birth of the first gravitationally bound systems, like galaxies, critically depends on when and how universe became neutral. Here, we examine the case of H recombination; He recombination follows a similar dynamics (see Figure 1.2).

The first models of the recombination kinetics of H in the pregalactic plasma were made in the 1960s by Peebles (1968) and Zel'dovich et al. (1968). These early calculations were based on a three-level hydrogen atom model (ground state + first excited state + continuum). Direct recombination to the ground state,

$$H^+ + e^- \rightarrow H(1s) + \gamma, \tag{1.26}$$

is inefficient because the recombination process is followed by the emission of energetic photons (~13.6 eV), which can immediately ionize H atoms newly formed

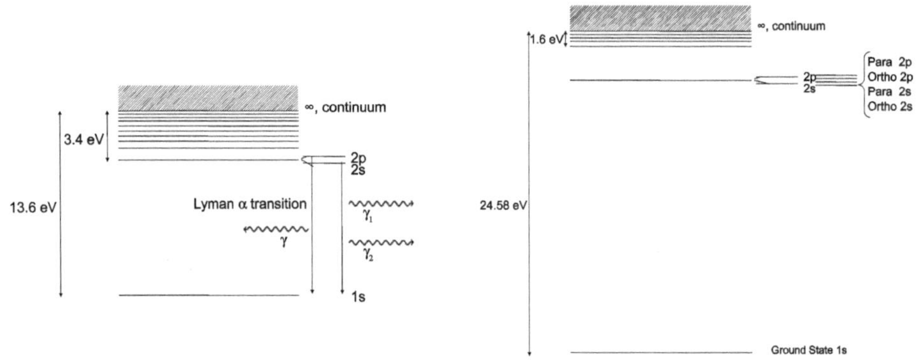

Figure 1.2. Energy levels of H and He, ordered by increasing energy. In the case of He, the energies of the first excited angular momentum states for the ortho- and para-nuclear states (corresponding to the total nuclear spins equal to 1 and 0, respectively) are also shown.

by recombination. An alternative pathway through which recombination could take place involves intermediaries excited states. In the case of Lyα transition,

$$H^+ + e^- \rightarrow H(n = 2, l = 1) \rightarrow H(1s) + \gamma, \tag{1.27}$$

as in the case of direct free–bound recombination, the rate of recombination is much larger than the rate of expansion of the universe; therefore, emitted photons are eventually recaptured by other H atoms, photoionizing them and slowing down the process of recombination. Peebles, Zel'dovich, and coworkers independently found that H recombination can overcome this "bottleneck" by two-photon decay,

$$H^+ + e^- \rightarrow H(2s) \rightarrow H(1s) + \gamma_1 + \gamma_2, \tag{1.28}$$

in which the two emitted photons γ_1 and γ_2 are not able to ionize hydrogen, having energies below the ionization threshold. Two-photon decay occurs at a rate higher than the reverse ionization process, resulting in a non-equilibrium situation. A proper account of the cascade from higher excited states is then essential to characterize the ionization history. Moreover, because of the expansion of the universe, the photons emitted in the recombination processes are redshifted, becoming progressively less energetic. This would eventually cause Lyα photons to be inefficient in ionizing hydrogen.

Taking all these considerations into account, the reactions included in the three-level model are

$$\begin{aligned} H^+ + e^- &\leftrightarrow H(2p, 2s) + \gamma, \\ H(2s) &\leftrightarrow H(1s) + 2\gamma, \\ H(2p) &\leftrightarrow H(1s) + \gamma. \end{aligned} \tag{1.29}$$

Furthermore, due to computational limitations, a number of simplifications were adopted: (i) the populations of excited states follow a Boltzmann distribution; (ii) the temperatures of matter and radiation remain the same throughout the whole recombination era; (iii) bound–bound and bound–free radiative transitions, as well as photons that are redshifted to the Lyα resonance, do not affect the thermal spectrum of the radiation field; (iv) the population of l substates of the same principal quantum number, $N_{nl} = N_n(2l + 1)/n^2$, is determined by statistical equilibrium; (v) the rate of the inverse of the two-photon absorption process is given by the thermal rate. Under these simplifying assumptions, it can be shown (Padmanabhan 2002) that the ionization fraction $x_e = n_e/n_H$ is given by the recombination equation:

$$\frac{dx_e}{dz} = \frac{n_H \alpha_B x_e^2 - \beta_B (1 - x_e) e^{-E_{2s}/k_B T_m}}{H(z)(1 + z)} \times \frac{1 + K\Lambda_{2s1s} n_H (1 - x_e)}{1 + K\Lambda_{2s1s} n_H (1 - x_e) + K\beta_B n_H (1 - x_e)} \tag{1.30}$$

where α_B is the recombination coefficient, which takes into account the recombination due to higher atomic energy levels (Pequignot et al. 1991; Hummer 1994); β_B is the total photoionization rate,

$$\beta_B = \alpha_B \left(\frac{2m_e k_B T_m}{h^2} \right)^{3/2} e^{-(E_i - E_{2s})/k_B T_m}, \tag{1.31}$$

with $E_{2s} = 10.2$ eV, $E_i = 13.6$ eV energy of the $2s$ level and ionization energy, respectively; $\Lambda_{2s1s} = 8.22458$ s^{-1} is the two-photon rate for the transition $2s$–$1s$ (Zeldovich et al. 1968); and

$$K = \frac{\lambda_\alpha^3}{8\pi H(z)}, \tag{1.32}$$

where λ_α represents the Lyα rest wavelength ($2p$–$1s$ transition).

The network (1.29) takes into account only H recombination. Clearly, an accurate calculation of the electron fraction x_e requires the simultaneous solution of the H^+, He^{2+}, and He^+ recombination processes. Thanks to the power of present-day computers, recent models of multi-level H and He recombination have largely removed all the limitations of earlier studies. The freely available code RECFAST (Seager et al. 1999, 2000), for example, includes 300 energy levels of H and all bound–bound and bound–free transitions originating from them, taking into account the He recombination history and the evolution of the matter temperature with redshift. The code COSMOREC (Switzer & Hirata 2008; Chluba et al. 2010; Grin & Hirata 2010; Rubiño-Martín et al. 2010; Chluba & Thomas 2011) introduces additional features and degrees of detail in the treatment of the recombination processes, relaxing the assumption (iv) of statistical equilibrium (Chluba & Sunyaev 2006; Rubiño-Martín et al. 2006) and including recombination from highly excited states ($n > 100$) (Ali-Haïmoud & Hirata 2010), thus bypassing the computational problem of explicitly solving the differential equations for such levels. The resulting ionization fraction as a function of redshift obtained using COSMOREC is shown in Figure 1.3.

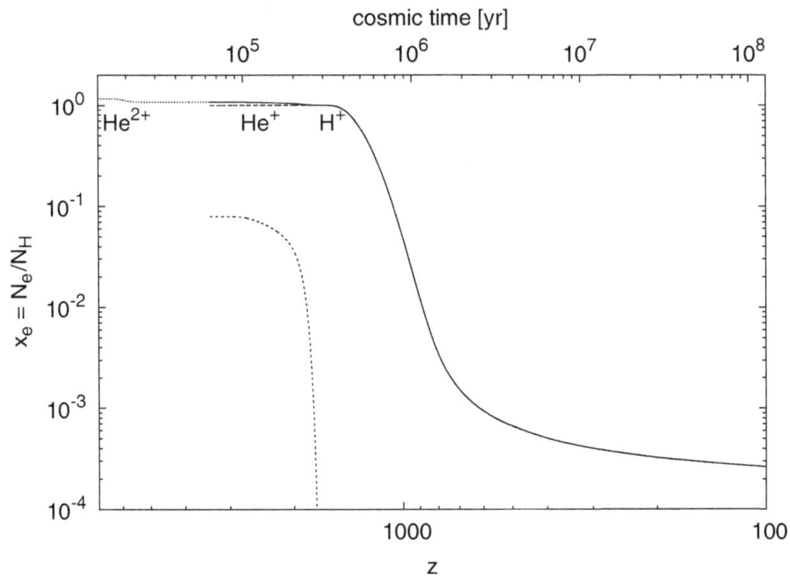

Figure 1.3. Total residual ionization fraction. Residual ionization fraction obtained with COSMOREC. Both H and He contributions are included: from left to right, the first step represents the recombination of He^{2+} to He^{+}, the second the recombination of He^{+} to He, and the last the recombination of H^{+} to H. For $z > 3400$, the output from RECFAST has been adopted.

1.4 Chemistry

1.4.1 Chemical Models

There are important differences between chemical kinetics in the early universe and in other contexts. First, the volume in which the reactions are occurring is expanding and therefore the gas density decreases in time, eventually stopping reactions between colliding particles and "freezing" the abundances of several species; second, T_m and T_r steadily decrease and differ from each other after the decoupling epoch, implying that reaction rates (including photoreactions) must be known over a large temperature range. A cartoon depicting the basic ideas of chemical kinetics in the expanding universe is shown in Figure 1.4.

The first step needed to describe how the abundances evolve in time is to identify the chemical species of the network and to list all the possible formation and destruction reactions for each of them. In computational terms, if N chemical species are inserted in the model and R processes are identified in the network, this corresponds to writing a set of N coupled ordinary differential equations (ODEs), one for the fractional abundance of each species. Each differential equation contains terms corresponding to the R processes, as sources or sink terms, according to the effect that the process has on each species. These ODEs can be written in terms of the abundances x_i and of the reaction rates k of the processes (also called rate

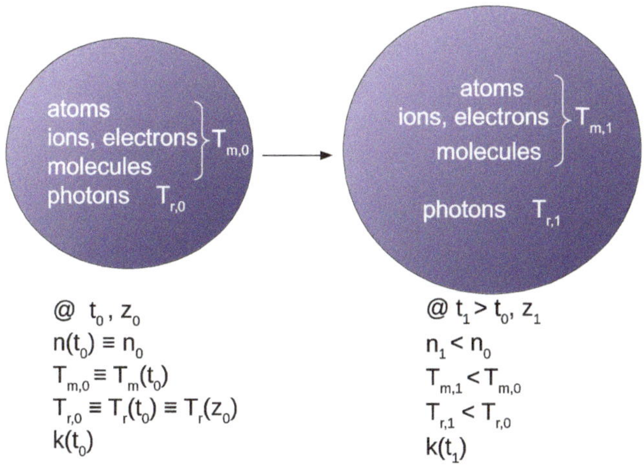

Figure 1.4. Basic scheme of chemical kinetics in the expanding universe.

coefficients or specific rate constants), functions of T_m or T_r. Specifically, each ODE takes the form

$$\begin{aligned}\frac{dx_i}{dt} = n_\mathrm{b} \sum_{l,m=1}^{N} k_{l,\,m}(T_\mathrm{m})x_l x_m + \sum_{\mathrm{ph}} \sum_{l \neq i} k_{l,\,\mathrm{ph}}(T_\mathrm{r})x_l \\ - n_\mathrm{b} x_i \sum_{l=1}^{N} k_{i,\,l}(T_\mathrm{m})x_l - x_i \sum_{\mathrm{ph}} k_{i,\,\mathrm{ph}}(T_\mathrm{r}),\end{aligned} \tag{1.33}$$

where n_b is the baryon number density (decreasing with time as the universe expands), and the index i covers all the N chemical species. The subscript "ph" stands for "photoprocess" (photodissociation, photoionization, photodetachment, etc.). Each ODE contains collisional and radiative terms. The first term describes the formation of the ith species as a result of the collision between the lth and mth species, whose reaction rate $k_{l,\,m}(T_\mathrm{m})$ is obtained integrating the specific cross section over the collision energy. The second term represents the formation of the ith species as a result of photoprocesses: the two sums are performed over the number of photoprocesses relevant to the formation of the ith species and the chemical species involved, respectively. The third and fourth terms represent the loss of the ith species: the former is due to collisional processes involving the ith species and other atomic/molecular partners, the latter to photodestruction processes. The resulting fractional abundances are usually computed as a function of z rather than time using the relation (1.17) between cosmic time and redshift. Notice that, in Equation (1.33), three-body reactions have been neglected: the probability of a collision among three or more partners is negligible in the early universe, owing to its low density. However, three-body reactions are crucial for models of collapsing clouds, where the typical densities are orders of magnitude higher than in the background gas.

The numerical solution of the resulting system of ODEs, generally "stiff," gives the evolution with redshift of the chemical species; alternatively, if the characteristic timescale for chemical reactions is much shorter than the expansion timescale at any given redshift, one can assume that chemical equilibrium holds at each time step. In this last case, the fractional abundances can be derived by imposing that the time derivatives in the ODEs are identically equal to zero and numerically solving the resulting system of nonlinear algebraic equations. Such an assumption is usually referred to as a "steady state" model, although the abundances change with time following the evolution of the matter and radiation temperature.

The first studies describing the possible routes of formation and destruction of early atoms and molecules date back to the 1960s and have mainly addressed the formation of H_2 in the early universe. In particular, Saslaw & Zipoy (1967) and Peebles (1968) realized the importance of the so-called H^- and H_2^+ channels for the formation of H_2. The first chemical networks in the early universe were numerically implemented in the late 1980s–90s (Lepp et al. 1984; Dalgarno & Lepp 1987; Puy et al. 1993; Galli & Palla 1998; Lepp et al. 1998; Dalgarno et al. 1998; Stancil et al. 1998; Puy & Signore 1999; Puy et al. 1999) and fully developed into a complete chemo-physical description of the chemistry of the early universe in the subsequent decade (Puy & Signore 2002; Lepp et al. 2002; Schleicher et al. 2008).

1.4.2 Formation and Destruction Processes

Figure 1.5 shows the most relevant pathways for the chemistry of the early universe. For the majority of them, only the forward mechanics are shown, except for the most important processes. Each blue rectangle contains a reactant, and a reaction partner is shown next to each arrow leading to the formation of the species shown at the corresponding arrowhead. The most important channels for each chemical species are indicated by green arrows. In the following, we summarize the main routes for the formation and destruction of the most abundant chemical species.

HeH^+

The first molecular ion ever to appear in the universe is HeH^+, formed after the recombination of He nuclei with free electrons in the primordial plasma. The dominant formation mechanism is radiative association,

$$He^+ + H \rightarrow HeH^+ + \gamma. \tag{1.34}$$

and the main destruction channel is via the reverse reaction, photodissociation.

H_2^+

Formation of H_2^+ can occur in two ways:

$$H^+ + H \rightarrow H_2^+ + \gamma, \tag{1.35}$$

and

$$HeH^+ + H \rightarrow H_2^+ + He. \tag{1.36}$$

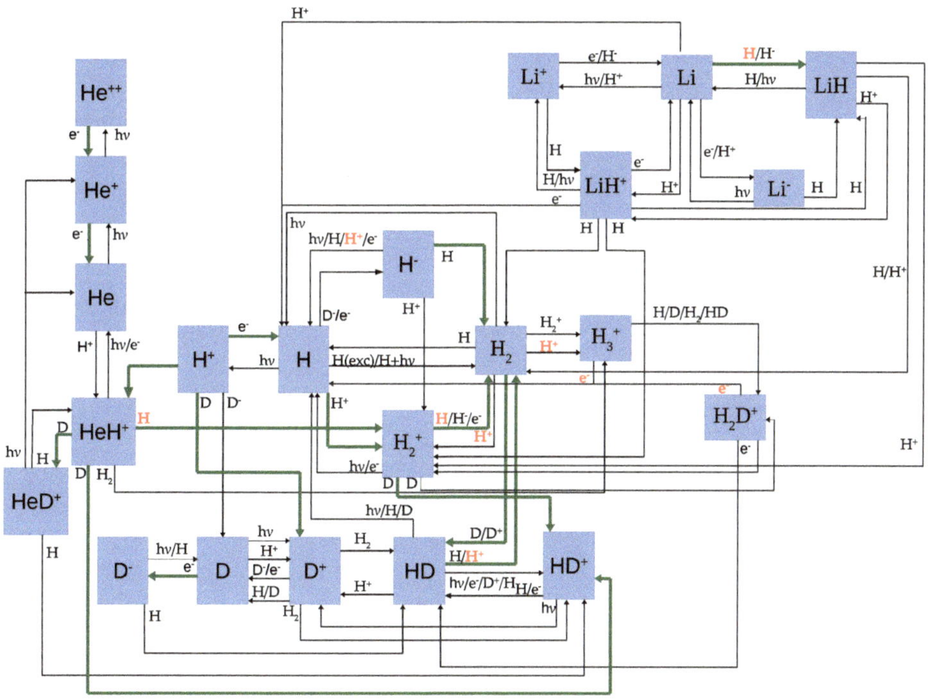

Figure 1.5. Chemical network in the early universe. The most relevant channels for the formation and destruction of chemical species are shown. The green arrows indicate the main pathways. Species in red indicate reactions for which updated state-to-state rates are needed.

Radiative association of H^+ and H is a relatively simple process. Nevertheless, state-to-state reaction rates have been made available only recently and are obtained by detailed balance applied to the available photodissociation data (Babb 2015; Vujčič et al. 2015). The reverse of process (1.35) represents the main destruction channel for H_2^+. Process (1.36) has been investigated by means of semiclassical and quantum methods, leading to a full set of state-to-state data (Esposito et al. 2015).

H_2

As anticipated, the two main pathways for the formation of H_2 in the early universe have been recognized already in the 1960s (Saslaw & Zipoy 1967; Peebles 1968). They are known as the H_2^+ and H^- channels:

$$H_2^+ + H \rightarrow H_2 + H^+, \tag{1.37}$$

and

$$H^- + H \rightarrow H_2 + e^-. \tag{1.38}$$

the former being the charge transfer of H_2^+ and the latter the associative detachment of H^- in collisions with H atoms. These two processes reach the maximum efficiency at the peak of formation of H_2^+ ($z \approx 300$) and H^- ($z \approx 50$), respectively. Reaction (1.38) has been extensively investigated by means of theoretical and experimental tools; in particular, see Kreckel et al. (2010). State-to-state experimental data are also available (Cízek et al. 1998). Conversely, process (1.37) has not been fully analyzed in terms of rovibrationally resolved data, the main obstacle being the complex dynamics involving non-adiabatic coupling between different potential energy surfaces.

HD

The kinetics of HD is strictly connected to the kinetics of its isotopic variant, H_2. It is formed by charge transfer,

$$D^+ + H_2 \rightarrow HD + H^+, \tag{1.39}$$

deuteron exchange with H_2,

$$D + H_2 \rightarrow HD + H, \tag{1.40}$$

and to a much lesser extent, by radiative association

$$H + D \rightarrow HD + \gamma. \tag{1.41}$$

State-resolved cross sections are available for specific reactions. In particular, reactions (1.39) and (1.40) have been investigated by González-Lezana et al. (2013) and Cheikh Sid Ely et al. (2017), respectively. Nevertheless, state-to-state data are still required for a detailed description of the kinetics of the rovibrational manifold. In particular, inelastic collisional cross sections are needed to describe the cooling due to HD, which is very important to cool the system to temperatures lower than 100 K. In fact, due to its mass asymmetry, HD has a dipole moment that allows larger cross sections (compared to H_2) for transitions to lower energy levels (Combes & Pineau Des Forets 2000).

1.4.3 Recent Developments in Chemical Modeling

Over the last decades, considerable progress has been made in the modeling of the chemistry of the primordial gas: starting from the early works, limited to a few dominant species and a restricted number of reactions (Lepp 1984; Black 1990), several improvements and refinements have been made by many researchers (Puy et al. 1993; Lepp et al. 1998, Galli & Palla 1998, 2002; Stancil et al. 1998; Vonlanthen et al. 2009; Signore & Puy 2009; Gay et al. 2011; Coppola et al. 2011; Longo et al. 2011). The subject was recently reviewed by Galli & Palla (2013). This progress was made possible by the availability of accurate theoretical and/or experimental determinations of reaction rates relevant for the chemistry of the early universe. Current networks of primordial chemistry include about 250 reactions for about 30 species (e.g., Gay et al. 2011). For particular species, the interested reader may consult the following studies: Latter & Black (1991), Lepp et al. (2002), Coppola et al. (2011), and Longo et al. (2011) for H; Lepp et al. (2002), Glover & Abel (2008) and Gay et al. (2011) for D; Schleicher et al. (2008) and Bovino et al. (2011b) for He; and Dalgarno et al. (1996), Stancil et al. (1996), Bougleux & Galli (1997), and Bovino et al. (2011a, 2012) for Li. Figure 1.6 shows the typical output of a primordial chemistry code: one can clearly see the rapid rise of atomic and molecular abundances, followed by a "freeze-out" phase, or a gentle decline with

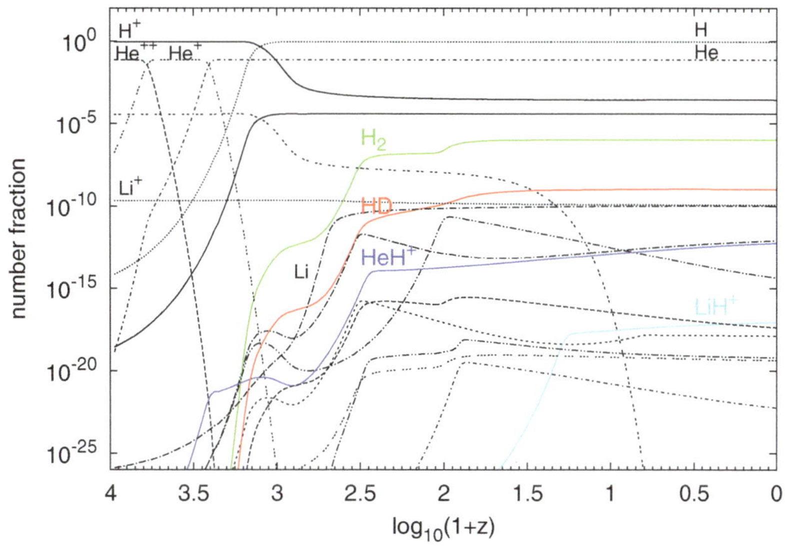

Figure 1.6. Typical output of a primordial chemistry code (Galli & Palla 2013), showing atomic and molecular abundances as function of redshift. The colored curves identify the abundances of representative species, like H_2 (green), HD (red), HeH^+ (blue), and LiH^+ (cyan). The fractional abundances of H, He and Li, in their neutral and atomic forms, are also reported to follow the recombination processes.

redshift. Below $z \approx 10$, the chemical composition of the early universe is deeply affected by the formation of the first structures, and the curves shown in Figure 1.6 cannot be trusted. The colored curves identify the abundances of representative species, like H_2 (green), HD (red), HeH^+ (blue), and LiH^+ (cyan). At the highest redshifts, it is easy to recognize the recombination of He (from He^{2+} to He^+ to He), the recombination of H, leaving a residual electron fraction of $\sim 2 \times 10^{-4}$, and the recombination of D (in the latter case, one can notice how residual D^+ is promptly removed at low redshifts via exothermic charge exchange with ambient H atoms). Recombination of Li, on the other hand, is never completed, owing to its low ionization potential: at $z = 10$, Li^+ is still 98% ionized.

State-to-State Approach and Non-LTE

The traditional approach to the study of the chemistry of the early universe rests on the assumption that the velocities and internal states of atoms and molecules, as well as the spectrum of the radiation field, can be described by equilibrium distributions, i.e., Maxwell–Boltzmann for atomic- and molecular-level populations and Planck for CMB photons. Equilibrium distributions are also assumed when averaging state-to-state cross sections to calculate reaction rates. Although a priori this hypothesis might be considered valid because of the density and gas temperature regimes at high redshift, whether or not the distributions are thermalized can only be determined by a state-resolved kinetic model. This requires solving a complex chemical network where each species is split in a number of internal states. In this approach, the main problem concerns the availability of reaction rates for each

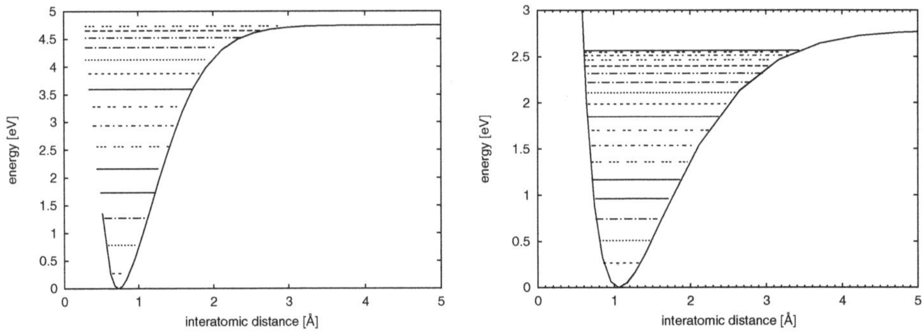

Figure 1.7. Potential energy curve and vibrational energy levels of H_2 (left panel) and H_2^+ (right panel). Data are from Kolos & Wolniewicz (1968) and Shaw et al. (2005) for H_2, and Fábri et al. (2009) for H_2^+. In both cases, energies are relative to the dissociation limit.

state-resolved channel, either calculated or experimentally determined. The need for accurate data is more stringent for the most relevant channels for the formation and destruction, together with the channels that allow for the internal redistribution of level population (radiative and collisional transitions). The presence of possible departures from equilibrium in the level population is especially important to evaluate accurate cooling rates at low temperatures.

The first molecule for which the distribution of the level population has been evaluated in the context of the early universe in a complete formation/destruction chemical scheme is H_2^+ (Hirata & Padmanabhan 2006; see the procedure for calculating the level population adopting a steady-state approach in Martin et al. 1996, Tiné et al. 1998, and Flower & Le Bourlot 2000). In particular, Martin et al. (1996) computed the distribution of rovibrational levels, showing that departures from the Maxwell–Boltzmann equilibrium are enhanced at low densities due to the lower frequency of collisions. However, no formation or destruction pathways were included. Recently, Coppola et al. (2011) and Longo et al. (2011) have focused on the non-equilibrium kinetics of H_2 and H_2^+, specifically developing a time-dependent network for the vibrational states (see Figure 1.7 for the vibrational energy levels of H_2 and H_2^+). The main finding of these works is that vibrational distributions can be considered as thermalized only in a small range of redshifts around $z \approx 1000$; at lower redshifts, when the expansion of the universe reduces the matter density and therefore the frequency of collisions, the internal level population of molecules cannot be described by a Boltzmann distribution. On the contrary, strong deviations from equilibrium are present, as shown in the top panel of Figure 1.8. Recently, the state-resolved approach has been extended also to the rotational levels of HD (Walker 2015, 2018).

Distortion Photons

At any redshift z, the radiation field in which the primordial gas is embedded, the CMB, is generally described by a Planck distribution with temperature $T_r(z)$. At the present time, the spectrum of the CMB is consistent within the errors with a perfect

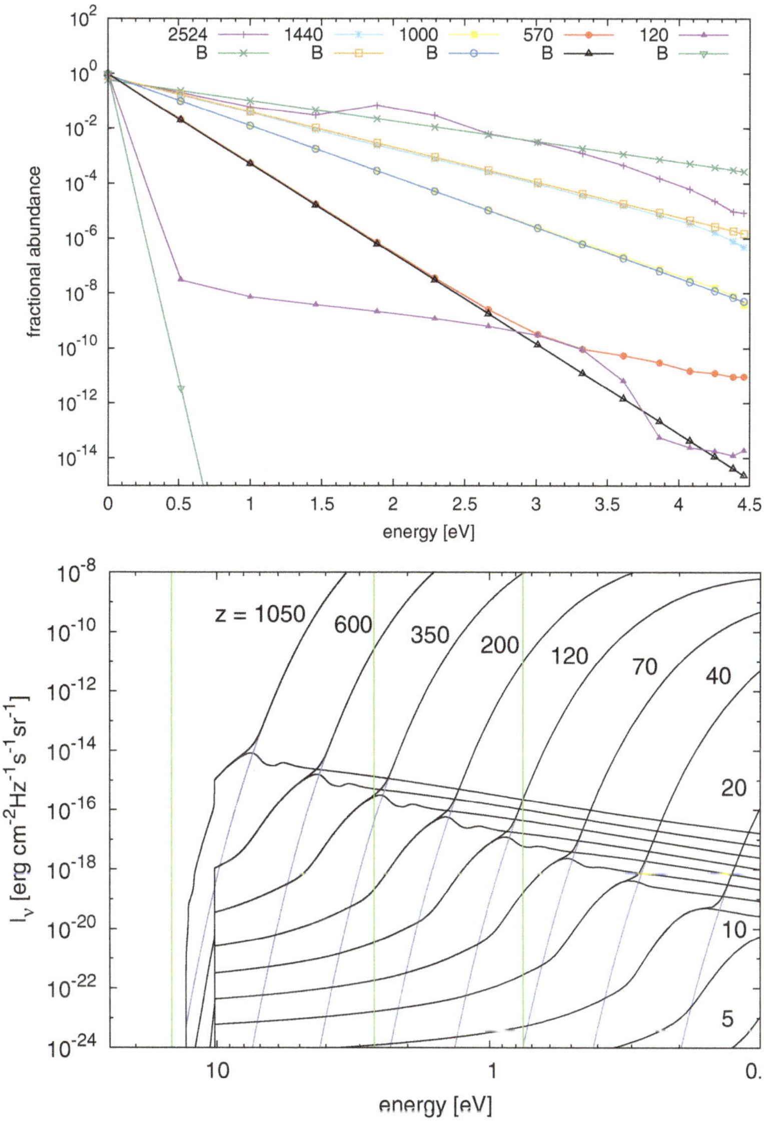

Figure 1.8. The imprint of non-equilibrium processes in the chemical and thermal history of the universe. Top panel: each curve shows the vibrational distribution of H_2 at the redshifts indicated at the top (numbers) compared with the corresponding Boltzmann distributions (B) (Coppola et al. 2011; Longo et al. 2011). Bottom panel: radiation spectra for various values of the redshift z: thermal (Planck) and non-thermal (recombination) contribution. As a reference, the energy thresholds for H_2 $(v = 0, j = 0)$ and H_2^+ $(v = 0, j = 0)$ photodissociation, together with the threshold for H^- photodetachment, are shown by green lines (from higher to lower energies).

blackbody with temperature $T_r(0) = 2.7255 \pm 0.0006$ (Fixsen 2009; see Table 1.1). However, deviations from a Planck profile are expected to be generated by any process that affects the thermodynamic equilibrium between photons and baryons, as first pointed out by Weymann, Zel'dovich, Sunyaev, and coworkers (Weymann

1966; Zeldovich & Sunyaev 1969; Sunyaev & Zeldovich 1970; Illarionov & Sunyaev 1974; Sunyaev 1979). These include: matter/antimatter annihilation, decaying particles, matter–radiation interaction, and atomic and molecular processes. The spectral distortions produced by photons emitted during the process of H and He recombination have been calculated (Chluba & Sunyaev 2006; Switzer & Hirata 2008; Ali-Haïmoud & Hirata 2010; Chluba et al. 2010; Grin & Hirata 2010; Rubiño-Martín et al. 2010; Chluba & Thomas 2011) and recently updated (Chluba & Ali-Hamoud 2016). The bottom panel of Figure 1.8 shows the spectra of thermal (CMB) and non-thermal (recombination) photons for different values of the redshift.

Spectral signatures in the CMB can also be produced by scattering processes of photons with the molecules themselves. It is therefore important to evaluate the redshift-integrated optical depth of each molecular species (Dubrovich 1994). In addition to the abundance, the relevant parameter for the optical depth is represented by the dipole moment of the species. The molecules that have been considered in this context are LiH (Bougleux & Galli 1997) and HeH^+ (Schleicher et al. 2008), with the latter being the most promising.

1.4.4 Non-standard Kinetics

Several models have been developed to take into account some "non-standard" effects, like modified Big Bang nucleosynthesis, and phase transition in the early universe. Although typically ignored in the standard framework, they may provide additional insights and details of the chemical and physical conditions at high redshifts.

Heavy Elements from Primordial Nucleosynthesis

Although the commonly accepted framework in primordial nucleosynthesis predicts that only light nuclei are synthesized during the first three minutes after the Big Bang (see Section 1.2), several works have developed kinetic models for heavier nuclei. In these models, molecules such as CH and OH can be formed (Vonlanthen et al. 2009), with abundances that might be comparable to those of hydrogen-bearing and hydrogen-isotope-bearing molecules such as HD^+. The heavier hydride foreseen in this framework is HF (Puy et al. 2007).

Hydrogen Phase Transition

The modeling of the evolution of the early universe described so far has taken into account only the existence of one phase of matter, namely the gas phase. However, particular conditions in the temperature and density inhomogeneities can lead to the formation of solid and liquid phases, especially at redshifts at which the first stars are expected to form. In particular, at low temperatures, H_2 can be described as a two-phase (or even three-phase) fluid (Füglistaler & Pfenniger 2016). Hydrogen shows a peculiar phase diagram: its triple point occurs for temperature $T \approx 10$ K and pressure $p \approx 10^{-1}$ bar. Some models have introduced H ice at the end of the cosmological dark ages (Pfenniger & Puy 2003); these solid structures are labeled as "quantum crystals" because of the importance of quantum effects in their dynamics.

This introduction is relevant both for the correct thermal description of the coupling of matter and radiation and for the chemistry that would be driven by gas–grain processes as well.

At lower redshifts, when contamination of the pristine gas with elements synthesized by the first stars has occurred, the physical and chemical complexity increases: magnetohydrodynamics as well as dusty plasma processes must be taken into account to describe the chemical evolution of the universe. A proper modeling of these situations requires appropriate kinetic approaches (see, e.g., Caselli et al. 1998). These include a modification of the rate equations for reactions on grain surfaces, alternative to the Monte Carlo procedures that are typically adopted because of the discrete nature of gas–grain physical processes.

1.4.5 Cooling Functions

One of the main goals of studying the chemistry of the primordial gas is finding applications to models for the formation of the first gravitationally bound structures in the universe. This is achieved by calculating the fractional abundances of species as a function of the gas density and temperature, and the energy content and exchanges among the matter and radiation components (cooling rates). At low redshifts, a dominant role in determining the energy losses of the gas is played by rovibrational molecular transitions, which are able to efficiently radiate the energy stored in the internal degrees of freedom of molecules following a collisional excitation.

Consider the collisions of a molecule m with number density n_m with particles of a species s with number density n_s, and consider two energy levels i and j of the molecule, with $E_i < E_j$. The cooling function Λ is given by the sum of all collisional excitations followed by radiative decay

$$n_m n_s \Lambda = \sum_j \sum_{i<j} n_i C_{ij} P^r_{ji}(E_j - E_i), \tag{1.42}$$

whereas the heating function Γ is given by the sum of all radiative excitations followed by collisional de-excitation

$$n_m n_s \Gamma = \sum_j \sum_{i<j} n_i R_{ij} P^c_{ji}(E_j - E_i), \tag{1.43}$$

where the probabilities of radiative and collisional decay are:

$$P^r = \frac{R_{ji}}{R_{ji} + C_{ji}}, \qquad P^c = \frac{C_{ji}}{R_{ji} + C_{ji}}. \tag{1.44}$$

The collisional excitation and de-excitation rates are

$$C_{ij} = n_s \gamma_{ij}, \tag{1.45}$$

where γ_{ij} is the collisional coefficient. The radiative excitation and de-excitation rates are

$$R_{ij} = \mathrm{B}_{ij} J_\nu^{ij}(T_r), \qquad R_{ji} = \mathrm{A}_{ji} + \mathrm{B}_{ji} J_\nu^{ij}(T_r), \tag{1.46}$$

respectively. Here A_{ji}, B_{ij} and B_{ji} are the Einstein coefficient and $J_\nu^{ij}(T_r)$ is the Planck function at frequency $\nu_{ij} = (E_j - E_i)/h$ and radiation temperature T_r,

$$\mathrm{B}_{ij} = \frac{g_j}{g_i} \mathrm{B}_{ji}, \qquad \mathrm{B}_{ji} = \frac{c^2}{2h\nu^3} \mathrm{A}_{ji}, \tag{1.47}$$

and

$$J_\nu^{ji}(T_r) = \left(\frac{2h\nu^3}{c^2} \right) \frac{1}{e^{h\nu_{ij}/kT_r} - 1}. \tag{1.48}$$

It follows that

$$R_{ji} = \eta_{ij} \mathrm{A}_{ji}, \qquad R_{ij} = \frac{g_j}{g_i} \eta_{ij} \mathrm{A}_{ji}, \tag{1.49}$$

where $\eta_{ij} = (e^{h\nu_{ij}/kT_r} - 1)^{-1}$ is the photon occupation number. The net heat transfer function $\Phi = \Gamma - \Lambda$ is then

$$n_m n_s \Phi = \sum_j \sum_{i<j} n_i (C_{ij} P_{ji}^r - R_{ij} P_{ji}^r)(E_j - E_i). \tag{1.50}$$

In the early universe, the function Φ expresses the transfer of energy from the radiation to the gas and vice versa via the transitions of the molecule m. If collisions with more than a species s are considered, Φ must be summed over all possible collision partners s. With the additional hypothesis of statistical equilibrium, the condition of detailed balance between any couple of levels i and j,

$$n_j (R_{ji} + C_{ji}) = n_i (R_{ij} + C_{ij}), \tag{1.51}$$

can be used to simplify the expression of Φ, obtaining

$$n_m n_s \Phi = \sum_j \sum_{i<j} (n_i C_{ij} - n_j C_{ji})(E_j - E_i), \tag{1.52}$$

a form often adopted in cosmological studies. In the absence of a radiation field, $R_{ij} = 0$ and $R_{ji} = \mathrm{A}_{ji}$, and the heat transfer function reduce to a pure cooling rate:

$$n_m n_s \Lambda = \sum_j \sum_{i<j} n_j \mathrm{A}_{ji} (E_j - E_i). \tag{1.53}$$

This is the form in which the (equilibrium) cooling rate is usually given in the astrophysical literature, e.g., Tiné et al. (1998), Flower & Le Bourlot (2000), and Lipovka et al. (2005).

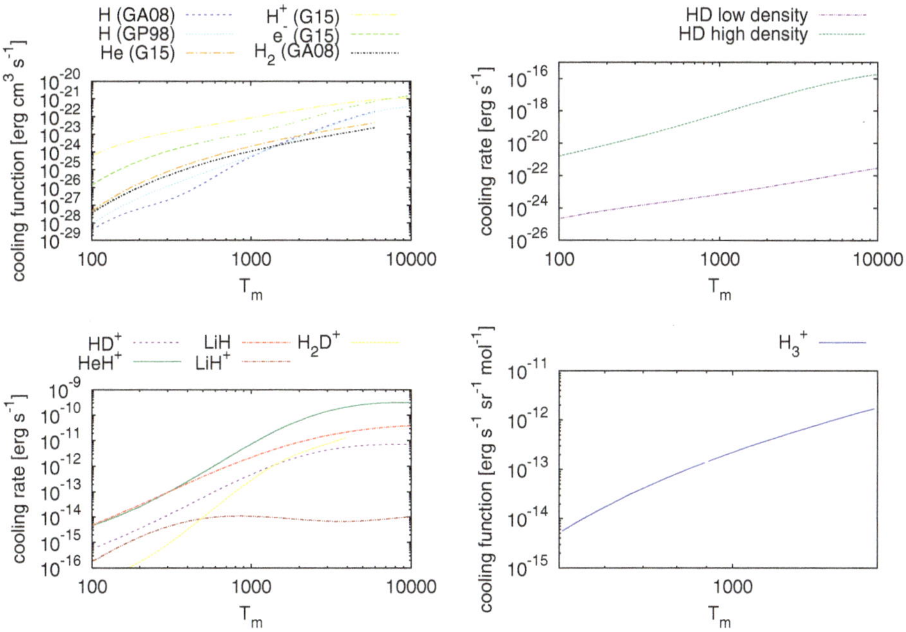

Figure 1.9. Molecular cooling rates as a function of the gas temperature. Top panel, left: cooling functions for collisions of H_2 with different partners (H, H^+, He, H_2, and electrons). The fits have been calculated by Galli & Palla (1998; GP98), Glover & Abel (2008; GA08) and Glover (2015; G15). Top panel, right: HD for collisions with H. The shown fits have been calculated in the low-density regime by Lipovka et al. (2005) and in the high-density limit by Lipovka et al. (2005) and Lipovka et al. (2005). Bottom panel, left: cooling functions of other primordial molecules at high density limit. The fits for HeH^+, LiH, LiH^+, HD^+ are provided by Coppola et al. (2011). Bottom panel, right: H_3^+ cooling rate as calculated and fitted by Miller et al. (2010) and Neale et al. (1996).

Figure 1.9 shows the cooling rates of the dominant species in the early universe as a function of the gas temperature. The figure shows the cooling rate per molecule: the actual cooling must be multiplied by the fractional abundance of each species. Recent calculations of the H_2 cooling function (C. M. Coppola et al. 2018, in preparation) based on the new collisional coefficients calculated by Lique (2015) and on the FRIGUS code (M. V. Kazandjian & C. M. Coppola 2018, in preparation) showed that, as expected, the inclusion of the effect of nuclei exchange at low temperatures increases the cooling by almost one order of magnitude compared to previous calculations.

1.5 Conclusions

In this chapter, we have introduced the basic concepts that underlie the formation of the first bound chemical systems, i.e., atoms and molecules. The presence of molecules, even a trace abundance of H_2 and HD, affects the cooling properties of the primordial gas, which would be otherwise an extremely poor radiator. Because the evolution of primordial density fluctuation is largely controlled by the ability of the gas to cool down to low temperatures, an accurate determination of its chemical composition, not limited to the formation of H_2, is crucial to address the problem of the formation of the first stars, black holes, and galaxies. The basic approach is to start from a certain value of redshift z and follow the chemical, thermal, and dynamical properties of an overdense region in the expanding universe, from the linear phase to the phase of gravitational instability. In a way, this represents the reverse situation of the chemistry in the expanding universe. Low-mass primordial clouds are less efficient than their high-mass counterparts in dissipating their energy and cooling. Thus, there exists a critical mass above which cooling can occur faster than the characteristic time of expansion of the universe, whereas less massive clouds will merely remain in pressure equilibrium and never collapse. To compute this critical mass and thus predict the properties of the first luminous objects is one of the main objectives of current research in the field (for an overview of the subject, see Bromm 2013).

The chemical evolution of the universe sketched in this chapter takes place during the Dark Ages, in a redshift interval spanning from $z \approx 1000$ to $z \approx 100$. Thus, the imprint left by atomic and molecular transitions on the CMB in this range of redshift offers, in principle, a way to probe the conditions of the universe during this unknown epoch of the evolution of the universe. Unfortunately, the low abundances of the relevant chemical species and the rapid expansion of the universe make any signal from the Dark Ages too small to be detectable with current instrumentation (Schleicher et al. 2008). Much of the effort on the observational side is focused toward lower redshifts, say, around the epoch of galaxy formation and the subsequent reionization of the intergalactic gas at $z \approx 10$ or slightly above. The spin-flip transition in neutral hydrogen, also known as the HI 21 cm line, is considered to be the most promising tool to probe the epoch of reionization. This will be accomplished by measuring the line intensity at frequencies below $\sim$100 MHz, where a transition from emission to absorption is expected as one progressively accesses the epoch when the universe was essentially neutral (e.g., Furlanetto et al. 2006). In the near future, it is also expected that H_2 line cooling in primordial dark matter halos will likely be the primary tracer to study the Dark Ages at $z > 20$ with next-generation facilities like SPICA and *JWST* (Mizusawa et al. 2005) or the the Far-Infrared Surveyor (Meixner et al. 2016).

References

Ali-Haïmoud, Y., & Hirata, C. M. 2010, PhRvD, 82, 063521

Aver, E., Olive, K. A., & Skillman, E. D. 2012, JCAP, 4, 004

Babb, J. F. 2015, ApJS, 216, 21

Black, J. H. 1990, in Molecular Astrophysics, ed. T. Hartquist (Cambridge: Cambridge Univ. Press), 473

Bonifacio, P., Sbordone, L., Caffau, E., et al. 2012, A&A, 542, A87

Bougleux, E., & Galli, D. 1997, MNRAS, 288, 638

Bovino, S., Tacconi, M., Gianturco, F. A., & Galli, D. 2011b, A&A, 529, A140

Bovino, S., Tacconi, M., Gianturco, F. A., Galli, D., & Palla, F. 2011a, ApJ, 731, 107

Bovino, S., Čuřík, R., Galli, D., Tacconi, M., & Gianturco, F. A. 2012, ApJ, 752, 19

Bromm, V. 2013, RPPh, 76, 112901
Caselli, P., Hasegawa, T. I., & Herbst, E. 1998, ApJ, 495, 309
Cheikh Sid Ely, S., Coppola, C. M., & Lique, F. 2017, MNRAS, 466, 2175
Chluba, J., & Sunyaev, R. A. 2006, A&A, 458, L29
Chluba, J., & Thomas, R. M. 2011, MNRAS, 412, 748
Chluba, J., Vasil, G. M., & Dursi, L. J. 2010, MNRAS, 407, 599
Chluba, J., & Ali-Hamoud, Y. 2016, MNRAS, 456, 3494
Cízek, M., Horácek, J., & Domcke, W. 1998, JPhB, 31, 2571
Coc, A. 2013, JPhCS, 420, 012136
Coc, A., Goriely, S., Xu, Y., Saimpert, M., & Vangioni, E. 2012, ApJ, 744, 158
Combes, F., & Pineau Des Forets, G. (ed) 2000, Molecular Hydrogen in Space (Cambridge: Cambridge Univ. Press)
Coppola, C. M., Lodi, L., & Tennyson, J. 2011, MNRAS, 415, 487
Coppola, C. M., Longo, S., Capitelli, M., Palla, F., & Galli, D. 2011, ApJS, 193, 7
Cyburt, R. H., Fields, B. D., Olive, K. A., & Yeh, T.-H. 2016, RvMP, 88, 015004
Dalgarno, A., Kirby, K., & Stancil, P. C. 1996, ApJ, 458, 397
Dalgarno, A., & Lepp, S. 1987, in IAU Symp. 120, Astrochemistry, ed. M. S. Vardya, & S. P. Tarafdar (Dordrecht: Springer), 109
Dalgarno, A., Stancil, P. C., & Lepp, S. 1998, in Stellar Evolution, Stellar Explosions and Galactic Chemical Evolution, ed. A. Mezzacappa (Bristol: Institute of Physics Publishing), 137
Dubrovich, V. K. 1994, A&AT, 5, 57
Esposito, F., Coppola, C. M., & De Fazio, D. 2015, JPCA, 119, 12615
Fábri, C., Czakó, G., Tasi, G., & Császár, A. G. 2009, JChPh, 130, 134314
Fixsen, D. J. 2009, ApJ, 707, 916
Flower, D. R., Le Bourlot, J., Pineau des Forêts, G., & Roueff, E. 2000, MNRAS, 314, 753
Füglistaler, A., & Pfenniger, D. 2016, A&A, 591, A100
Furlanetto, S. R., Oh, S. P., & Briggs, F. H. 2006, PhR, 433, 181
Galli, D., & Palla, F. 1998, A&A, 335, 403
Galli, D., & Palla, F. 2002, P&SS, 50, 1197
Galli, D., & Palla, F. 2013, ARA&A, 51, 163
Gasperini, M. 2007, Elements of String Cosmology (Cambridge: Cambridge Univ. Press)
Gasperini, M. 2016, Theory of Gravitational Interactions (Berlin: Springer)
Gay, C. D., Stancil, P. C., Lepp, S., & Dalgarno, A. 2011, ApJ, 737, 44
Glover, S. C. O., & Abel, T. 2008, MNRAS, 388, 1627
Glover, S. C. O. 2015, MNRAS, 451, 2082
González-Lezana, T., Honvault, P., & Scribano, Y. 2013, JChPh, 139, 054301
Grin, D., & Hirata, C. M. 2010, PhRvD, 81, 083005
Hirata, C. M., & Padmanabhan, N. 2006, MNRAS, 372, 1175
Hou, S. Q., He, J. J., Parikh, A., et al. 2017, ApJ, 834, 165
Howk, J. C., Lehner, N., Fields, B. D., & Mathews, G. J. 2012, Natur, 489, 121
Hummer, D. G. 1994, MNRAS, 268, 109
Illarionov, A. F., & Sunyaev, R. A. 1974, AZh, 51, 1162
Jedamzik, K., Fuller, G. M., Mathews, G. J., & Kajino, T. 1994, ApJ, 422, 423
Kolb, E. W., & Turner, M. S. 1990, The Early Universe (Boulder, Co: Westview Press)
Kolos, W., & Wolniewicz, L. 1968, JChPh, 49, 404

Komatsu, E., Smith, K. M., Dunkley, J., et al. 2011, ApJS, 192, 18
Kreckel, H., Bruhns, H., Čížek, M., et al. 2010, Sci, 329, 69
Latter, W. B., & Black, J. H. 1991, ApJ, 372, 161
Lepp, S., Dalgarno, A., & Shull, J. M. 1984, BAAS, 16, 1015
Lepp, S., & Shull, J. M. 1984, ApJ, 280, 465
Lepp, S., Stancil, P. C., & Dalgarno, A. 1998, MmSAI, 69, 331
Lepp, S., Stancil, P. C., & Dalgarno, A. 2002, JPhB, 35, R57
Lipovka, A., Núñez-López, R., & Avila-Reese, V. 2005, MNRAS, 361, 850
Lique, F. 2015, MNRAS, 453, 810
Longo, S., Coppola, C. M., Galli, D., Palla, F., & Capitelli, M. 2011, Rendiconti Lincei, 22, 119
Martin, P. G., Schwarz, D. H., & Mandy, M. E. 1996, ApJ, 461, 265
Meixner, M., Cooray, A., Carter, R., et al. 2016, Proc. SPIE, 9904, 99040K
Miller, S., Stallard, T., Melin, H., & Tennyson, J. 2010, FaDi, 147, 283
Mizusawa, H., Omukai, K., & Nishi, R. 2005, PASJ, 57, 951
Neale, L., Miller, S., & Tennyson, J. 1996, ApJ, 464, 516
Noterdaeme, P., López, S., Dumont, V., et al. 2012, A&A, 542, L33
Padmanabhan, T. 2002, Theoretical Astrophysics—Volume 3, Galaxies and Cosmology (Cambridge: Cambridge Univ. Press)
Peebles, P. J. E. 1993, Principles of Physical Cosmology (Princeton: Princeton Univ. Press)
Peebles, P. J. E., & Dicke, R. H. 1968, ApJ, 154, 891
Pequignot, D., Petitjean, P., & Boisson, C. 1991, A&A, 251, 680
Perlmutter, S., Aldering, G., Goldhaber, G., et al. 1999, ApJ, 517, 565
Pfenniger, D., & Puy, D. 2003, A&A, 398, 447
Planck Collaboration 2016, A&A, 594, A13
Puy, D., Alecian, G., Le Bourlot, J., Leorat, J., & Pineau Des Forets, G. 1993, A&A, 267, 337
Puy, D., Dubrovich, V., Lipovka, A., Talbi, D., & Vonlanthen, P. 2007, A&A, 476, 685
Puy, D., & Signore, M. 1999, NewAR, 43, 223
Puy, D., & Signore, M. 2002, NewAR, 46, 709
Riess, A. G., Filippenko, A. V., Challis, P., et al. 1998, AJ, 116, 1009
Rubiño-Martín, J. A., Chluba, J., Fendt, W. A., & Wandelt, B. D. 2010, MNRAS, 403, 439
Rubiño-Martín, J. A., Chluba, J., & Sunyaev, R. A. 2006, MNRAS, 371, 1939
Saslaw, W. C., & Zipoy, D. 1967, Natur, 216, 976
Schleicher, D. R. G., Galli, D., Palla, F., et al. 2008, A&A, 490, 521
Seager, S., Sasselov, D. D., & Scott, D. 1999, ApJL, 523, L1
Seager, S., Sasselov, D. D., & Scott, D. 2000, ApJS, 128, 407
Shaw, G., Ferland, G. J., Abel, N. P., Stancil, P. C., & van Hoof, P. A. M. 2005, ApJ, 624, 794
Signore, M., & Puy, D. 2009, EPJC, 59, 117
Srećković, V. A., Jevremović, D., Vujčić, V., et al. 2017, IAUT, 12, 393
Stancil, P. C., Lepp, S., & Dalgarno, A. 1996, ApJ, 458, 401
Stancil, P. C., Lepp, S., & Dalgarno, A. 1998, ApJ, 509, 1
Sunyaev, R. A. 1979, NASA STI/Recon Technical Report, n. 79, 1
Sunyaev, R. A., & Zeldovich, Y. B. 1970, Ap&SS, 7, 20
Switzer, E. R., & Hirata, C. M. 2008, PhRvD, 77, 083006
Tiné, S., Lepp, S., & Dalgarno, A. 1998, MmSAI, 69, 345
Vonlanthen, P., Rauscher, T., Winteler, C., et al. 2009, A&A, 503, 47

Vujčič, V., Jevremović, D., Mihajlov, A. A., et al. 2015, JApA, 36, 693
Walker, K. M. 2015, PhD thesis, Univ. Georgia
Walker, K. M., Porter, R. L., & Stancil, P. C. 2018, ApJ, 867, 152
Weymann, R. 1966, ApJ, 145, 560
Zeldovich, Y. B., Kurt, V. G., & Syunyaev, R. A. 1968, ZhETF, 55, 278
Zeldovich, Y. B., & Sunyaev, R. A. 1969, Ap&SS, 4, 301

AAS | IOP Astronomy

Gas-Phase Chemistry in Space
From elementary particles to complex organic molecules
François Lique and Alexandre Faure

Chapter 2

Nucleosynthesis: The Origin of the Chemical Elements

Nikos Prantzos

2.1 Introduction

The theory of nucleosynthesis emerged around the middle of the 20th century as a result of rapid progress in our understanding of three different fields:

(1) the composition of the Sun and the solar system,
(2) the physical conditions prevailing in the interiors of stars during their various evolutionary stages, and
(3) the systematic properties of nuclei and nuclear reactions.

The idea that all nuclei are synthesized in the hot stellar interiors was promoted by the British astronomer Fred Hoyle in the 1940s. On the other hand, at about the same period, the Russian physicist George Gamow argued that all nuclei were produced in the hot primordial universe[1] via successive neutron captures. However, it was rapidly shown that, in those conditions, it was extremely difficult to synthesize anything heavier than ^4He. Moreover, observations in the 1950s showed that, although all stars have similar amounts of the light (and most abundant) elements H and He, they may differ considerably in their heavy element content. It was clear then that heavy nuclei had to be produced *after* the Big Bang, in successive stellar generations that progressively enriched the galaxies.

In the following, a brief account is presented of the basic ideas underlying nucleosynthesis. Section 2.2 describes the cosmic abundances and the nuclear stability of the nuclei, because those two properties (macroscopic and microscopic) are correlated to a large extent. Section 2.3 describes the primordial nucleosynthesis,

[1] The name *Big Bang* was ironically given to the theory of the hot early universe by Fred Hoyle, who promoted instead the *steady-state theory* for the universe. Hoyle was proven to be wrong in his cosmological views, but correct as to the origin of the elements; the opposite happened with Gamow.

doi:10.1088/2514-3433/aae1b5ch2

producing in the hot, early universe the light nuclides H, D, He-3, He-4, and Li7. Section 2.4 and 2.5 present the early and advanced stages of stellar evolution, respectively, with the accompanying nucleosynthesis of nuclei up to the Fe peak. Explosive nucleosynthesis in supernovae is presented in Section 2.6, while the synthesis of heavier than Fe elements through neutron captures is discussed in Section 2.7.

2.2 Nuclei in the Cosmos

2.2.1 Solar and Cosmic Abundances

According to our current understanding, the material of the protosolar nebula had a remarkably homogeneous composition, as a result of high temperatures (which caused the melting of the quasi-totality of dust grains) and thorough mixing. This composition characterizes the present-day surface layers of the Sun, which remain unaffected by nuclear reactions occurring in the solar interior (with a few exceptions, e.g., the fragile D and Li). The abundances of most elements in the solar photosphere are now established to a fairly good precision.

Once various physico-chemical effects are taken into account,[2] it appears that the elemental composition of the Earth and meteorites matches the solar photospheric composition extremely well.[3] On the other hand, Earth and meteoritic materials provide an opportunity to measure their isotopic composition with extreme accuracy in the laboratory, while such measurements are, in general, impossible in the case of the Sun.[4]

A combination of solar and meteoritic measurements allows us to establish the solar composition (Figure 2.1), presumably reflecting the one of the protosolar nebula 4.5 Gyr ago. An early attempt to obtain such a curve was made by Goldsmith in 1938. In 1956, Suess and Urey provided the first relatively complete and precise data set, on which the founding works of nucleosynthesis (Burbidge et al. 1957; Cameron 1957) were based. The most recent major compilations are those of Lodders et al. (2009) and Asplund et al. (2009).

In the early 1950s, it was realized that the composition of stars in the Milky Way presents both striking similarities to and considerable differences from the solar composition. The universal predominance of H (90% by number) and He (9% by number) and the relative abundances of "metals" (elements heavier than He, with O, C, N, Ne, and Fe being systematically more abundant than the other species) is the most important similarity. On the other hand, the fraction of metals ("metallicity") appears to vary considerably (Figure 2.2), either within the solar vicinity (where the oldest stars have a metallicity of 0.1 solar), across the Milky Way disk (with stars in

[2] For instance, the low gravity of the Earth and smaller bodies of the solar system was insufficient to retain the light H and He of the proto-solar mixture; despite their high initial abundances, these elements are absent from those bodies (the H combined to chemically reactive O in the form of water was added to the Earth after its formation, during a later accretion period).

[3] Some long-standing discrepancies between meteoritic and solar measurements concerning the abundances of several key elements were solved in the 1990s (for Fe) and early 2000s (for O and C).

[4] Nuclear spectroscopy, through the detection of characteristic γ-ray lines from the de-excitation of nuclei in solar flares, offers a (limited) possibility to determine the Sun's isotopic composition.

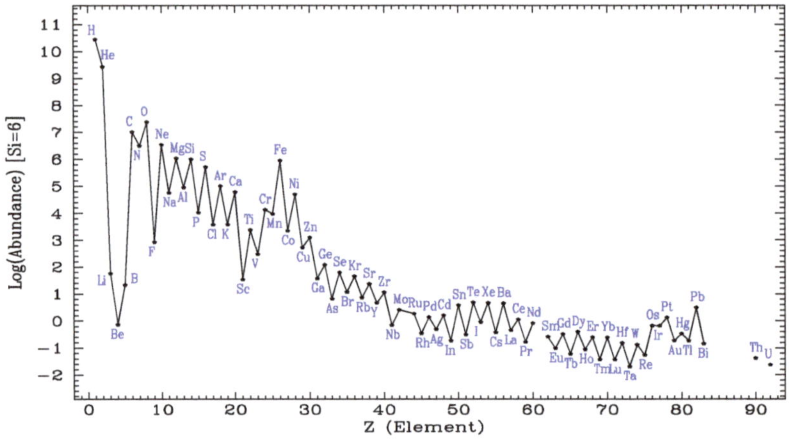

Figure 2.1. Solar system abundances by number, in a scale where log(Si)=6, as a function of the charge number Z of the element. H and He are by far the most abundant elements (90% and 9%, respectively, by number, or 70% and 28% by mass). Note also the extremely low abundances of Li, Be, and B with regard to the neighboring C, N, and O; the peak in the abundance curve around Fe; the "saw-tooth" overall pattern; and the high abundance of Pb (abundances from Lodders et al. 2009).

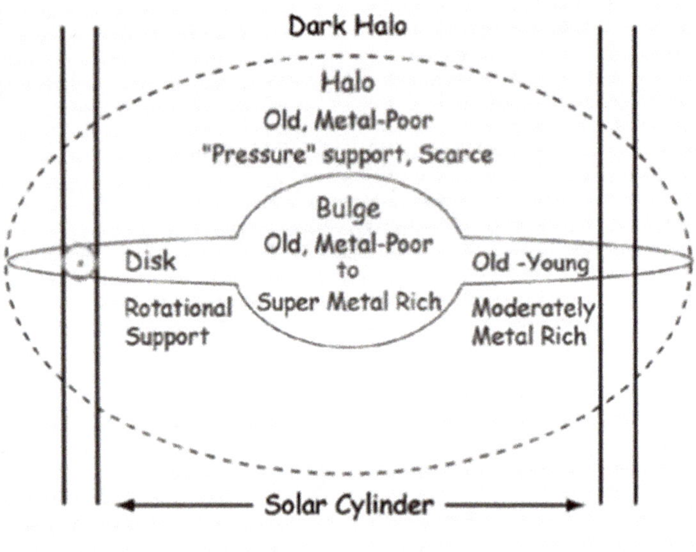

Figure 2.2. Schematic cross-section of the Milky Way, with the Sun at a distance of 25,000 light years from the galactic center and the various galactic components (halo, bulge, disk) displayed, along with the main properties of the corresponding stellar populations.

the inner Galaxy having three times more metals than the Sun) or in the galactic halo (with stellar metallicities ranging from 0.1 to 0.00001 solar).

These variations in composition are extremely important for understanding the "chemical evolution" of the Milky Way. Indeed, they reflect the progressive enrichment of the various components of the Galaxy (halo, bulge, disk) with metals produced and ejected by successive stellar generations. The first generation was presumably formed from gas of primordial composition, i.e., H and He (with a trace amount of Li) resulting from the early hot universe of the Big Bang. However, the large diversity of the metallicity of stars (also observed in the interstellar medium of the Milky Way and other galaxies) should not mask the important fact of uniformity in the basic pattern, namely the predominance of H and He and the quasi-uniformity (within a factor of a few) in the abundance ratios between metals. It is precisely that uniformity that calls for an explanation involving nuclear reactions in appropriate astrophysical sites.

2.2.2 Cosmic Abundances versus Nuclear Properties

The solar or cosmic abundances of the various nuclear species (Figure 2.3, upper panel) constitute an important macroscopic property of (baryonic) matter. It was realized that this property is closely related to a microscopic one, namely the "binding energy per nucleon" (BEN, see lower panel in Figure 2.3). This quantity represents the energy per nucleon required to break a nucleus into its constituent particles, and is a measure of the nuclear stability. In the framework of the "liquid drop model" of the nucleus (composed of A nucleons), it is described approximately by the formula of Weizsäcker (1935):

$$B(A, Z) = f_V - f_S A^{-1/3} - f_E Z(Z-1)A^{-4/3} - f_{\rm SM}(A-2Z)^2/A^2. \qquad (2.1)$$

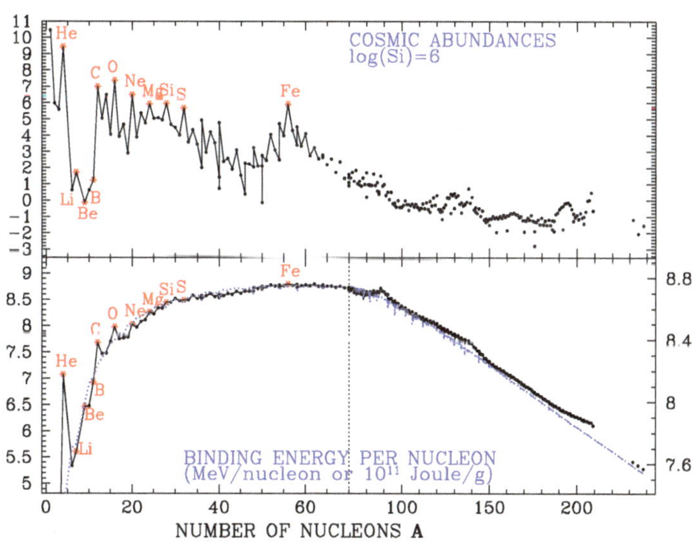

Figure 2.3. Cosmic abundances (top panel) and nuclear binding energy per nucleon (bottom panel) as a function of nuclear mass number A. Strongly bound nuclei ("α nuclei," like ^{4}He, ^{12}C, ^{16}O, ^{20}Ne, ^{24}Mg, ^{28}Si, or Fe-peak nuclei) are more abundant than their neighbors. In the lower panel, the continuous curve connects experimental data points, while the dotted curve is a straight application of the liquid drop model (see text), without quantum-mechanical corrections. Note the change in the horizontal scale at $A = 70$, as well as the different vertical scales in the right and left parts of the lower panel.

According to that formula, the BEN results from the competition (or synergy) of several factors:

- The short-range nuclear attraction between neighboring nucleons, contributing a constant term f_V (where V stands for volume, the corresponding contribution to the total binding energy of the nucleus Be = BEN × A being proportional to the number of nucleons A, which occupy a volume V);
- A symmetry term (with coefficient f_{SM}), arising in part from the Pauli exclusion principle and in part from symmetry effects in the nucleon–nucleon interaction, which favors equal numbers of protons and neutrons, as well as even rather than odd nuclei;
- The long-range electrostatic repulsion between protons (with coefficient f_E), which favors an increasing fraction of neutrons when the mass number A increases;
- A surface term (with coefficient f_S), representing a reduced contribution in the binding energy from nucleons at the "surface" of the nucleus; the importance of this term decreases with increasing A (i.e., with decreasing surface/volume ratio).

A simple calculation of the BEN along those lines satisfactorily reproduces the gross features of the measured curve (see Figure 2.3): BEN increases steadily up to the Fe peak (due to decreasingly important negative contributions from the surface term) and then declines slowly (due to increasingly important negative contributions from the electrostatic and symmetry terms).

However, reproducing several key features of the BEN curve requires a full quantum-mechanical treatment. This is done in the framework of the "nuclear shell model," which assumes that each nucleon is moving in the potential created by all the other nucleons. That treatment leads to quantized energy levels (Figure 2.4) and accounts for the propensity of identical nucleons to form pairs with opposite spins, which introduces a supplementary (positive) contribution to the BEN curve; it also accounts for the exceptional stability of α-nuclei (with nucleon numbers that are multiples of four, like ^{4}He, ^{12}C, ^{16}O, ^{20}Ne, ^{24}Mg, etc.) and for the stability of nuclei with "magic" nucleon numbers 2, 8, 20, 28, 82, and 126 (corresponding to filled nuclear shells in Figure 2.4).

The key features of the BEN curve are obviously reflected in the cosmic abundance curve, albeit at a local level only: more stable nuclei are more abundant than their neighbors (e.g., the α-nuclei or the Fe-peak nuclei), while the fragile Li, Be, and B isotopes are extremely underabundant. However, at a global level, the light H and He are overwhelmingly more abundant than the more strongly bound C, N, and O, which in turn are more abundant than the even more stable Fe peak nuclei.

The (local) correlation between cosmic abundances and nuclear stability suggests that nuclear reactions have shaped the abundances of elements in the universe. The fact that the correlation is only local and does not hold at a global level tells us that nuclear processes have affected only a small fraction of the baryonic matter in the universe (less than a few percent). This is good news, because the overabundant H and He nuclei constitute the main fuel of stars (see Section 2.3). The history of stars and of the associated nuclear transmutations is far from its end yet.

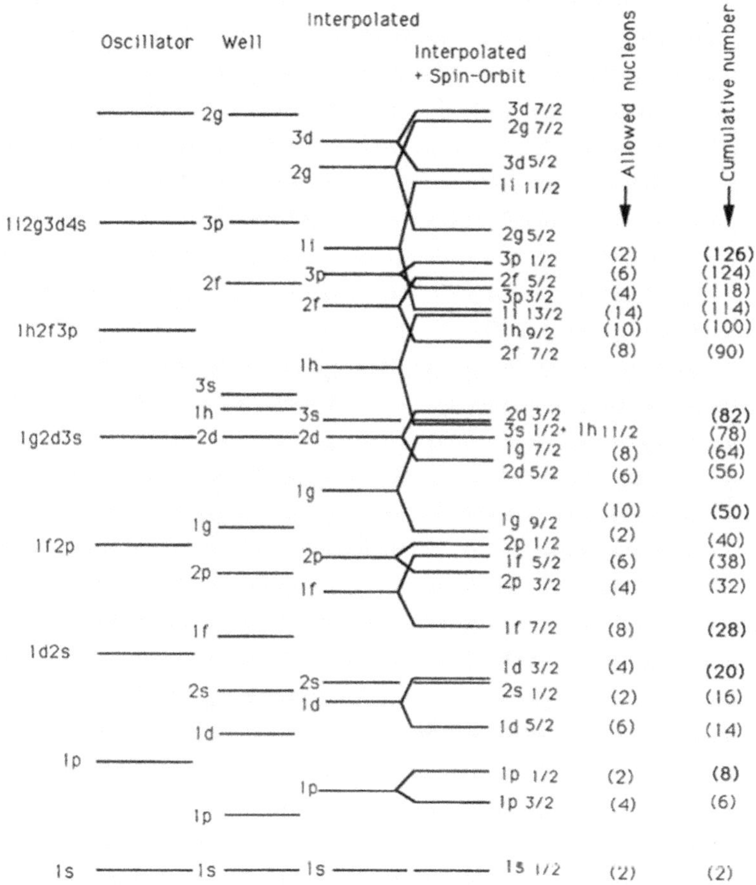

Figure 2.4. Nuclear level scheme for various nuclear potentials. The spin–orbit interaction (fourth column) was of fundamental importance for a correct understanding of nuclear properties. Nuclei with filled shells (with 2, 8, 20, 28, 50, 82, and 126 nucleons, last column) are considerably more stable than their neighbors, a property also reflected in the cosmic abundance curve.

2.2.3 Overview of Nucleosynthesis

Based on a rapidly growing body of empirical data, both astronomical (abundances in stars and meteorites) and nuclear (binding energies and nuclear reaction rates), as well as on an elementary understanding of stellar structure and evolution, Burbidge et al. (1957)[5] and Cameron (1957) identified in two landmark papers the main nucleosynthetic processes in nature. These processes have been thoroughly studied throughout the second half of the 20th century. It is now well-established that:

[5] One of the authors, W. A. Fowler, received the 1984 Nobel Prize in Physics for his contribution to our understanding of the origin of the elements.

- The light isotopes of H and He, along with 10% of the fragile ^{7}Li, have been produced in the hot early universe by thermonuclear reactions between neutrons and protons (Section 2.3).[6]
- All the elements between C and the Fe peak have been produced by thermonuclear reactions inside stars, either during their quiescent evolutionary stages or during the violent explosions (supernovae) that mark the deaths of some stars (Sections 2.4, 2.5, 2.6).
- Elements heavier than those of the Fe peak have been produced by neutron captures in stars, either in low neutron densities and long timescales (s-elements) or in high neutron densities and short timescales (r-elements); a minor fraction of those heavy elements has been produced by photodisintegration of the heavy isotopes in supernova explosions (p-isotopes; Section 2.7).
- Finally, the light and fragile isotopes of Li, Be, and B are not produced in stellar interiors (rather, they are destroyed at high temperatures), but by *spallation reactions*, with high-energy cosmic-ray particles removing nucleons from the abundant C, N, and O nuclei of the interstellar medium. We shall not discuss those elements in this chapter.

2.3 Primordial Nucleosynthesis: from H to He

The standard, hot Big Bang model provides a very successful and economical description of the evolution of the (observable) universe, from temperatures as high as $T \sim 10^{12}$ K ($t \sim 10^{-4}$ s after the "bang") until the present epoch ($T \sim 3$ K, $t \sim 13.8\ 10^9$ yr). The observational evidence testifying to the validity of the model is threefold:

- The *universal expansion*, discovered by Hubble in 1929: all galaxies, except those of the local group, are receding from us (and each other) with velocities v proportional to their distances r : $v = H\ r$. The *Hubble parameter* $H(t)$ measures the rate of universal expansion; its current value is $H_0 \sim 70$ km s^{-1} Mpc^{-1} (1 Mpc = 3.26 10^6 light years), and its inverse $t_H \sim H_0^{-1} \sim 14 \times 10^9$ yr (*Hubble time*) is a measure of the age of the universe.
- The *cosmic* (microwave) *background radiation* (CBR in the following), discovered in 1965 by A. Penzias and R. Wilson (who obtained the Nobel Prize in 1978): its spectrum fits with astonishing precision a blackbody of temperature $T_0 = 2.735$ K, and its angular uniformity ($\Delta T/T < 10^{-5}$) combined with its presumed homogeneity (cosmological principle) strongly argue for a hot and homogeneous early universe, where matter and radiation were in equilibrium. The "last scattering surface," i.e., the epoch when matter and radiation "decoupled," is situated at a temperature $T \sim 3000$ K, corresponding to an age of a few 10^5 years after the "bang."

[6] Note that nuclei with mass numbers $A = 5$ and 8 are unstable and do not exist in nature. This fact has important implications for primordial nucleosynthesis, because two-body reactions could not bridge the gap between ^{4}He and ^{12}C. Also, because of the rapid decrease of density and temperature in the hot early universe, the $3\alpha \longrightarrow {}^{12}$C reaction had no time to operate either; this became possible only much later, inside red giant stars.

- The *cosmic abundances of D*, ^{3}He, ^{4},He, ^{7}Li: these light nuclides are predicted to be synthesized in the hot early universe, at temperatures $T \sim 10^9$–10^8 K ($t \sim 10^2$–10^3 s). The successful comparison of their predicted abundances with inferred determinations of their primordial ones is a real triumph of the standard model; at the same time, this is one of the most stringent tests of that model, constraining its parameters. Notice that the epoch of Big Bang nucleosynthesis (BBN), being the earliest period in the life of the universe from which we have "relics" from a well-understood process (i.e., the abundances of the light elements), it is a gateway to the very early universe; the importance of continuous scrutiny of the BBN predictions is manifest, e.g., Coc (2016) and references therein.

2.3.1 Thermodynamics of the Early Universe

The two fundamental assumptions underlying BBN are: (1) that general relativity offers a valid description of gravity; and (2) that the universe once was hotter than $\sim 10^{11}$ K, such that statistical equilibrium was established between all of its components. The basic cosmological equation, describing the expansion of the universe under the gravitational attraction of its material content, is *Friedmann's equation*:

$$\left(\frac{\dot{R}}{R}\right)^2 - \frac{8\pi}{3}G\rho = -\frac{k}{R^2} \tag{2.2}$$

where ρ is the average density; k is a constant related to spacetime curvature; a value of $c = 1$ is assumed for light speed and of $\Lambda = 0$ for the cosmological constant (being negligible in the Big Bang era); and $R(t)$ is a dimensionless *scale factor* (distances in the expanding universe increase proportionally to it, i.e., $r = Rr_0$, where r_0 is a distance at some reference epoch) and $\dot{R}(t)$ is its time derivative. (Note: Although Equation (2.2) is a general relativistic equation, it can also be obtained in the framework of Newtonian mechanics by requiring energy conservation for a test particle on a sphere of radius R and mass $M = 4\pi/3\ \rho R^3$, expanding adiabatically with velocity $v = \dot{R}$, i.e., $E = v^2/2 - GM/R =$ const.; the difference with general relativity is in the interpretation of ρ (matter density in classical mechanics, matter + energy density in relativity) and k (a simple constant in classical mechanics)). Equation (2.2) can be cast in the form:

$$\frac{k}{H^2R^2} = \frac{\rho}{3H^2/8\pi G} - 1 = \frac{\rho}{\rho_c} - 1 = \Omega - 1 \tag{2.3}$$

where $H = \dot{R}/R$, and we have defined the *critical density* $\rho_c = 3H^2/8\pi G$, and the *density parameter* $\Omega = \rho/\rho_c$. For k (>, =, <)0, we have Ω (>, =, <)1 and ρ (>, =, <) ρ_c, corresponding to a universe that is closed, flat, or open, respectively.

The density of the universe has not always been dominated by matter, as it is today: indeed, number densities of all species (i.e., baryons (n_B), dark matter particles (n_{DM}), photons (n_γ), etc.) are inversely proportional to the expanding volume (i.e., $n \propto R^{-3}$) and the same is true for the corresponding *mass densities*, e.g., $\rho_B = n_m m_B \propto R^{-3}$ (m_B being the mass of a baryon); on the other hand, the equivalent energy *density of radiation* $\rho_\gamma = n_\gamma hc/\lambda \propto R^{-4}$ (because the wavelength of a photon $\lambda \propto R^{-1}$ is being "stretched" by the universal expansion). This means that, at some time in the past (which turns out to be shortly before the matter–radiation decoupling period, at $T\sim$ a few 10^4 K), the universe was *radiation-dominated.* Notice that the *baryon/photon ratio* $\eta = n_B/n_\gamma \sim 10^{-9}$ remains approximately constant throughout the evolution of the universe (i.e., from just before the BBN period until now, because the contribution of stellar radiation turns out to be negligible); this makes it a very useful parameter, connecting the early universe of BBN with the observable one.

The present-day universe is very close to being flat, and it can be shown that it was even more so in the past. For $k = 0$ and with an appropriate equation of state, the evolution of the thermodynamic variables of the universe can be deduced from Equation (2.2) and the assumption of adiabatic expansion (see Equation (2.8) below). Notice that *time* is not a good variable because we do not know if we are allowed to extrapolate our physical laws to arbitrarily high temperatures, i.e., back to $t = 0$; *temperature* is used instead, which fixes the *number of relativistic degrees of freedom* of the system:

$$g^*(T) = \Sigma g_B \left(\frac{T_B}{T}\right)^4 + \frac{7}{8}\Sigma g_F \left(\frac{T_F}{T}\right)^4 \tag{2.4}$$

where B and F stand for bosons and fermions, respectively (the factor 7/8 being due to fermion statistics) and the possibility of components with different temperatures (i.e., thermally decoupled from the cosmic plasma) is taken into account. In equilibrium, the *total density* is essentially determined by the number of relativistic species:

$$\rho \sim \rho_R \sim 1/2\, g^*(T)\, \rho_\gamma = 1/2\, g^*(T)\, \alpha T^4 \tag{2.5}$$

(α is a constant), because the density of non-relativistic species of mass m in equilibrium is severely hindered by the Boltzmann factor ($\rho_m = m\, n_m \sim m \exp(-mc^2/kT)$). Total density fixes then the *expansion rate* through Equation (2.2)

$$H = \dot{R}\Big/R \sim \left(\frac{8\pi}{3}G\rho\right)^{1/2} \sim 2(g^*)^{1/2} G^{1/2}\, T^2 \tag{2.6}$$

which also measures the rate of change of the thermodynamic properties of the system (notice the important (and counter-intuitive!) result of a faster expansion for a larger density). On the other hand, the rates of the various reactions (weak, strong, electromagnetic) in the cosmic plasma depend generally on temperature:

$$\Gamma \sim n(T)\, \sigma(T)\, v(T) \sim T^\nu \tag{2.7}$$

where n is the particle density, σ the reaction cross-section, and v the average particle velocity; since $n \propto R^{-3} \propto T^3$, $\nu > 3$. Obviously then, $\Gamma/H \propto T^\mu$ ($\mu > 1$) decreases during the expansion: as long as $\Gamma > H$ (i.e., the reactions are faster than the expansion), equilibrium is maintained, redistributing energy among the various components of the cosmic plasma; when $\Gamma < H$, the reactions can no longer keep in pace with the expansion and the affected species drop out of equilibrium (*decoupling*).

At temperatures greater than the rest mass of some particle S ($kT > m_S c^2$), pair creation by the background radiation maintains particle densities $n_S = n_{\bar{S}} \sim n_\gamma$. When temperature drops below that threshold ($kT < m_S c^2$), particle–antiparticle annihilation is no longer balanced by pair creation, and in principle, $n_S = n_{\bar{S}} < 10^{-17}$ n_γ. However, no antimatter has been found in the observable universe ($n_B \gg n_{\bar{B}}$ for baryons), except in cosmic rays, where it comes as a secondary product of high-energy interactions; moreover, $(n_B - n_{\bar{B}})/n_\gamma = \eta \sim 10^{-9} \gg 10^{-17}$, as we saw before. This observed *excess of matter over antimatter* is unexplained in standard Big Bang models, where it is assumed as an initial condition.

2.3.2 BBN: Results and Comparison to Observations

For our purposes, we start the description of the primordial universe at temperatures $T \sim 10^{12}$ K ($\sim$100 MeV), with a mixture of one non-relativistic baryon (p and n) for every $\sim 10^9$ relativistic species (γ, e^-, e^+, ν, $\bar{\nu}$), all of them in statistical equilibrium. Because the leptonic degrees of freedom dominate the total mass–energy density, we are in the *Leptonic Era*. The assumption of adiabatic expansion ($dE + pdV = 0$) gives:

$$\frac{d}{dt}(\rho R^3) + pd(R^3) = 0 \tag{2.8}$$

which, combined with Equations (2.2) and (2.4) and the equation of state for radiation (i.e., $p = 1/3\rho$), leads to a useful formula for the time dependence of temperature during the whole radiation-dominated epoch:

$$T(K) \sim 1.2 \times 10^{10}\,(g^*)^{-1/4} t(\mathrm{s})^{-1/2}. \tag{2.9}$$

We see that, at $T \sim 10^{12}$ K, $t \sim 10^{-4}$ s; this timescale is larger than all the characteristic timescales of the various interactions (weak, strong, or electromagnetic) at that temperature, which means that the assumption of statistical equilibrium (the starting hypothesis for the standard BBN scenario) is fully justified.

At $t \sim 0.1$ s ($T \sim 3 \times 10^{10}$ K $\sim$4 MeV), neutral current weak interactions (i.e., $e^+ + e^- \leftrightarrow \nu + \bar{\nu}$) become slower than the expansion ($\Gamma_{\mathrm{WEAK}} \propto T^5$, whereas $H \propto T^2$). Neutrinos then decouple from the thermal plasma and expand adiabatically thereafter, with a neutrino temperature $T_\nu \propto R^{-1}$ (this is the "last scattering" epoch for neutrinos; detection of the cosmic neutrino seas would allow us to directly "view" the universe as it was $\sim$0.1 s after the "bang"). However, charged current weak interactions (i.e., $n + e^+ \leftrightarrow p + \bar{\nu}$, $e^- + p \leftrightarrow n + \nu$) proceeding at a rate $\Gamma \propto \tau_n^{-1}$ ($\tau_n \sim 890$ s being neutron's lifetime) are still sufficiently fast to maintain the neutron/proton ratio at its equilibrium value n/p $\sim \exp(-Q/kT)$, where $Q = (m_n - m_p)c^2 \sim 1.28$ MeV; at those temperatures, $(\mathrm{n/p})_{\mathrm{equilibrium}} \sim 1$.

At $t \sim 1$ s ($T \sim 10^{10}$ K $\sim$1 MeV), charged current weak interactions become slower than the expansion; the n/p ratio then essentially "freezes" at its equilibrium value at that *decoupling temperature* T_*: $(\text{n/p})_* \sim \exp(-Q/kT_*) \sim 0.18$. Neutron decay ($n \rightarrow p + e^- + \bar{\nu}$) still continues to operate, slowly modifying the "freeze-out" ratio: $\text{n/p} = (\text{n/p})_* \exp(-t/\tau_n)$.

At $t \sim 10$ s ($T \sim 3 \times 10^9$ K $\sim$0.5 MeV), photons are no longer sufficiently energetic to create $e^- - e^+$ pairs, which disappear, leaving behind 1 e^- for every $\sim 10^9$ photons, such that charge conservation with regard to protons is respected; the photon bath is slightly "heated" by the transfer of the annihilation entropy, but not the neutrino seas, which are decoupled, so henceforth the corresponding temperatures are related by $T_\nu = (4/11)^{1/3} T_\gamma$ (leading to a current neutrino temperature $T_{\nu,0} \sim 2$ K, too low for them to be detected). The disappearance of e^- - e^+ marks the end of the Leptonic and the beginning of the *Radiation Era*, which will end $\sim 10^5$ years later.

During all that time, nuclear reactions could not form any composite nuclei, because of the "deuterium bottleneck:" because of its relatively small binding energy ($\text{BE}_D \sim 2.2$ MeV), deuterium is photodisintegrated as soon as it is formed by $p + n \rightarrow D$, its abundance being kept to tiny equilibrium amounts $n_D/n_B \sim \eta \exp(-\text{BE}_D/kT)$. Even when $kT < \text{BE}_D$, the large value of $\eta^{-1} \sim 10^9$ allows for a large photon population in the high-energy tail of the Planck spectrum (i.e., $E_\gamma > \text{BE}_D$), enough to keep destroying deuterium; further, because three-particle interactions are impossible at those densities ($\rho_B \sim 10$ g cm^{-3}), no heavier nuclei can be formed either. Only when temperature drops down to $\sim 9 \times 10^8$ K ($\sim$0.1 MeV, at $t_D \sim 200$ s) does the deuterium bottleneck "break," and substantial amounts of D start being formed. The n/p ratio at that time is: $(\text{n/p})_0 \sim (\text{n/p})_* \exp(-t_D/\tau_n) \sim 0.15$, not very different from the "freeze-out" value.

This is the starting point of *primordial nucleosynthesis*: several other reactions become operative, i.e., D+p $\leftrightarrow$ ^{3}He+γ, D+n $\leftrightarrow$ ^{3}H+γ, etc. (Figure 2.5), bringing nuclei in statistical equilibrium; because ^{4}He is the most tightly bound nucleus in that region, almost all the neutrons present at that time are finally incorporated in ^{4}He, which allows an easy evaluation of its final mass fraction X_4. Before ^{4}He formation, the (number *and* mass fraction X) abundances of neutrons and protons were related by $(\text{n/p})_0 = (X_n/X_p)_0 \sim 0.15$; because, by definition, $X_{n,0} + X_{p,0} = 1$, we find $X_{p,0} \sim 0.87$ and $X_{n,0} \sim 0.13$. Neutrons combine then with an equal amount of protons to form ^{4}He, so after BBN, we are left with $X_p = X_{p,0} - X_{n,0} \sim 0.74$ for protons and $X_4 \sim 1 - X_p \sim 0.26$ for ^{4}He, i.e., a quarter of the baryonic mass of the universe is transformed in ^{4}He. The result of this rough calculation compares fairly well to the observed amount of ^{4}He in the universe ($X_{4,\text{obs}} \sim 25\%$), which cannot be accounted for by stellar nucleosynthesis.

The nuclear flow essentially stops at ^{4}He because there are no stable nuclei with mass $A = 5$ or $A = 8$. It can be seen that combining the most abundant nuclei, protons, and ^{4}He via two-particle interactions always leads to unstable $A = 5$; and even if ^{4}He combines with rarer nuclei, like ^{3}H or ^{3}He, we still get only to a nucleus with $A = 7$, which, when hit by (abundant) protons or (rare) neutrons yields mass $A = 8$. Nucleosynthesis stops when temperature drops to $\sim 3 \times 10^8$ K ($t \sim 10^3$ s) and

Coulomb barriers can no longer be penetrated. Eventually, ^3H decays to ^3He and any A = 7 nucleus to ^7Li, while all the protons remain as hydrogen (Figure 2.6). Thus, BBN transforms the composition of the primordial universe into (mass fractions): ~75 % H, ~25% ^4He, and traces of D (~10^{-4}), ^3He (~10^{-4}), and ^7Li (~10^{-9}). For standard homogeneous BBN, all other chemical elements are formed later in stars and related processes.

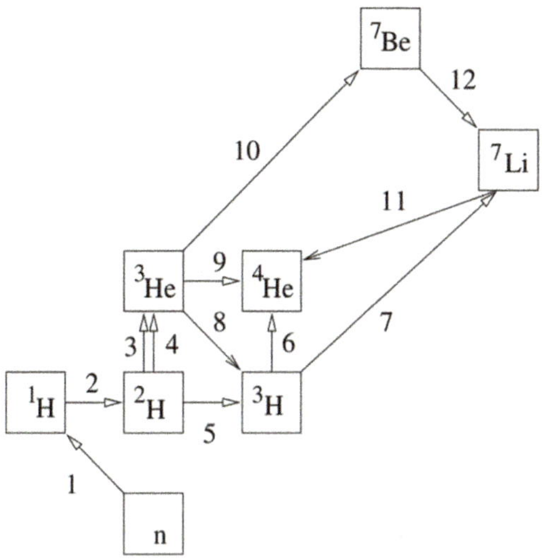

1. $p \longleftrightarrow n$
2. $p(n, \gamma)d$
3. $d(p, \gamma)^3$He
4. $d(d, n)^3$He
5. $d(d, p)t$
6. $t(d, n)^4$He
7. $t(\alpha, \gamma)^7$Li
8. ^{3}He$(n, p)t$
9. ^{3}He$(d, p)^4$He
10. ^{3}He$(\alpha, \gamma)^7$Be
11. ^{7}Li$(p, \alpha)^4$He
12. ^{7}Be$(n, p)^7$Li

Figure 2.5. Minimal network of standard Big Bang nucleosynthesis (SBBN), with the relevant reactions on the right.

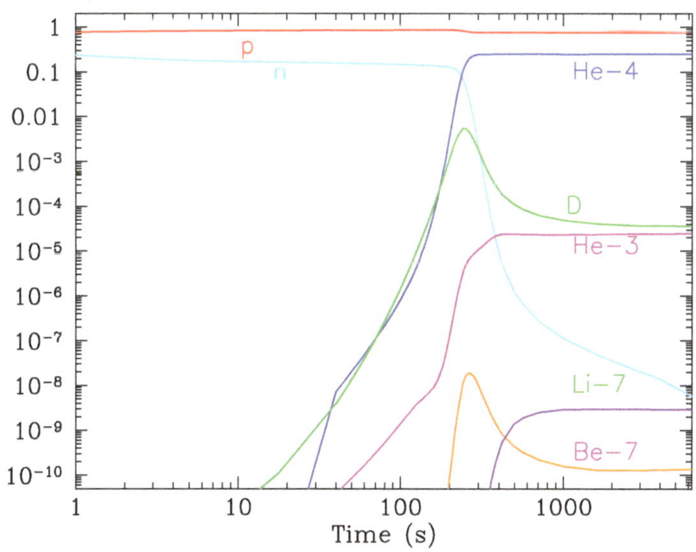

Figure 2.6. Evolution of light element abundances in standard Big Bang nucleosynthesis calculations, as a function of time t. Nucleosynthesis really starts at t ~ 200 s, with the breaking of the "deuterium bottleneck" (see text). ^{7}Be eventually decays to ^{7}Li, with a lifetime of ~54 days.

The BBN yields are essentially determined by the competition between the various reaction rates and the expansion rate; the latter depends on the (unknown) total density ρ, fixed by the number of relativistic degrees of freedom g^*, while the former depend on the (also unknown) densities of the various interacting species (e.g., $\eta = n_B/n_\gamma$ for baryons), and of course, on the rates of the relevant two-body reactions.

The primordial ^{4}He yield (Y_p in the following) depends upon η, N_ν (which parameterizes the number of "light" species, i.e., $mc^2 < 1$ MeV, other than γ and e^-, e^+), and τ_n (the neutron lifetime, which determines the rates for all the weak processes that interconvert neutrons and protons); it is almost insensitive to the rates of the thermonuclear reactions because, for $\eta > 3 \times 10^{-11}$, all of the available neutrons at "freeze-out" are converted in ^{4}He. The yield Y_p is a monotonically increasing function of η, N_ν, and τ_ν. The dependence on η is a weak one: larger η means that the D bottleneck "breaks" earlier, with more neutrons available (because they had less time to decay after the "freeze-out") to form a larger Y_p. More light species (i.e., larger N_ν) means a faster expansion because, from Equation (2.6): $H \sim 2.4 g^{1/2} G^{1/2} T^2$, with $g_{\text{LeptonicEra}} = g_\gamma + 7/8\,(g_e + N_\nu g_\nu) = 43/4$ for a mixture of photons with $g_\gamma = 2$ (2 spin states), electron–positron pairs with $g_e = 4$ (2 spin states), and $N_\nu = 3$ neutrino–antineutrino species with $g_\nu = 2$ (*only one helicity state*); and faster expansion implies an earlier decoupling with a larger abundance of neutrons to form ^{4}He. Increase in τ_n has the same effect because the slower rate of weak interactions ($\Gamma \propto \tau_n^{-1}$) leads to an earlier "freeze-out" and larger amounts of neutrons available.

The primordial yields of D and ^{3}He are particularly sensitive to the density of interacting nuclei during BBN, i.e., to η. Their abundances are found to decrease with increasing η because higher densities favor their larger destruction to ^{4}He. A larger expansion rate H, on the other hand, leads to a larger yield of D and ^{3}He because the reactions that destroy them stop earlier in that case.

The situation is somewhat different for ^{7}Li: its primordial abundance as a function of η has a minimum situated (by accident?) in the range of interest for cosmology, i.e., $10^{-10} < \eta < 10^{-9}$. This is due to its two-fold production mechanism, through ^{3}He(^{4}He,γ)^{7}Li at low densities (yield decreasing with η), and through ^{3}H (^{4}He,γ)^{7}Be at high densities (yield of ^{7}Be increasing with η).

The *uncertainties* in BBN yields due to nuclear reaction rates are relatively small: the rates of most of the relevant reactions are known to be better than ~10%, and the impact on the corresponding yields is less than a few percent for D, ^{3}He, and ^{7}Li, and even smaller for ^{4}He.

The results of standard BBN are usually plotted as function of the baryon-to-photon ratio $\eta_B = n_B/n_\gamma$ or the baryonic density Ω_B, and they appear in Figure 2.7. They are compared to the lowest known abundances of the light isotopes, as they are observed (or observationally inferred) in different environments: through emission lines in the most metal-poor extragalactic ionized HII regions for He-4 (which is coproduced with metals in stars); through absorption lines in high-redshift, metal-poor gas clouds for D; and in old (~12 Gy) stars of the Milky Way's halo for Li. In contrast, there is no reliable observation of the primordial abundance of He-3: it is observed in galactic HII regions, but our Milky Way is a metal-rich system and He-3 is also produced by stars along with metals.

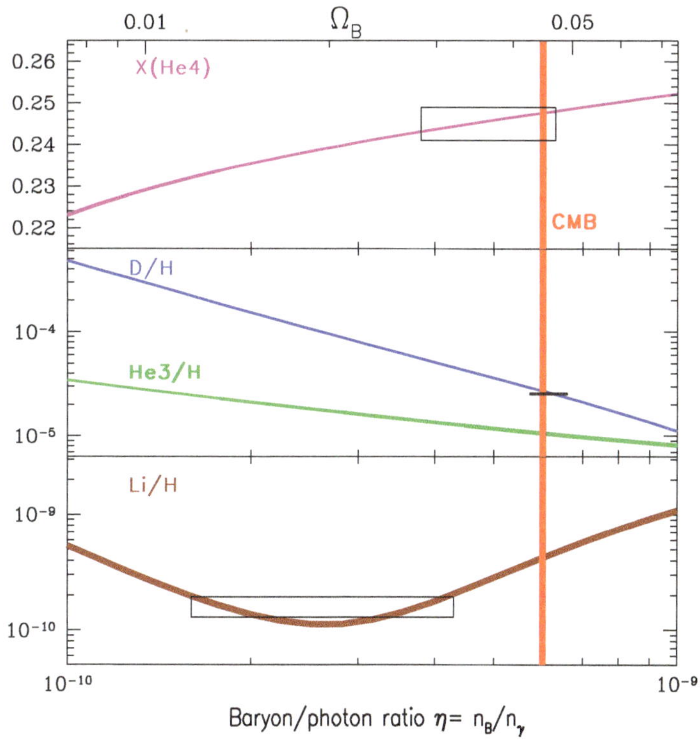

Figure 2.7. Abundances of light nuclei in standard Big Bang nucleosynthesis calculations, plotted as a function of the baryon/photon ratio η (scale on the bottom), or equivalently, the baryonic density parameter (scale on the top) $\Omega_{B,\,0}$. Current observationally inferred abundances are indicated within boxes for each nucleus, while the vertical line shows the value of η determined by observations of the cosmic microwave background. The abundance of He-4 is by mass fraction, all the others are given as number ratios to hydrogen.

The aforementioned observational results are in satisfactory agreement with the results of BBN for a common value of $\eta_B \sim 6 \times 10^{-9}$, which is also suggested by the analysis of the CMB fluctuations. Li is an exception, its abundance of BBN being a factor of ~3 higher than observed in halo stars. The reason for that discrepancy is not yet elucidated. In all probability, it is of stellar origin: Li may be depleted in the convective envelopes of such stars after 12 Gy of evolution, given that it burns at temperatures of $\sim 2 \times 10^6$ K. It is not clear, however, how that depletion may occur in such a uniform way, the observed abundance of Li being fairly constant with the amount of metals in those stars (e.g., Cyburt et al. 2016 and references therein).

2.4 Stars: from the Main Sequence to Red Giants

2.4.1 Basic Stellar Properties

The theory of stellar structure and evolution is arguably the most successful theory in the whole of astrophysics. It relies heavily on the interpretation of the famous Hertzsprung–Russell diagram (H–R diagram), established by E. Hertzsprung and H. N. Russell in the 1910s. This diagram (Figure 2.8) concerns two fundamental

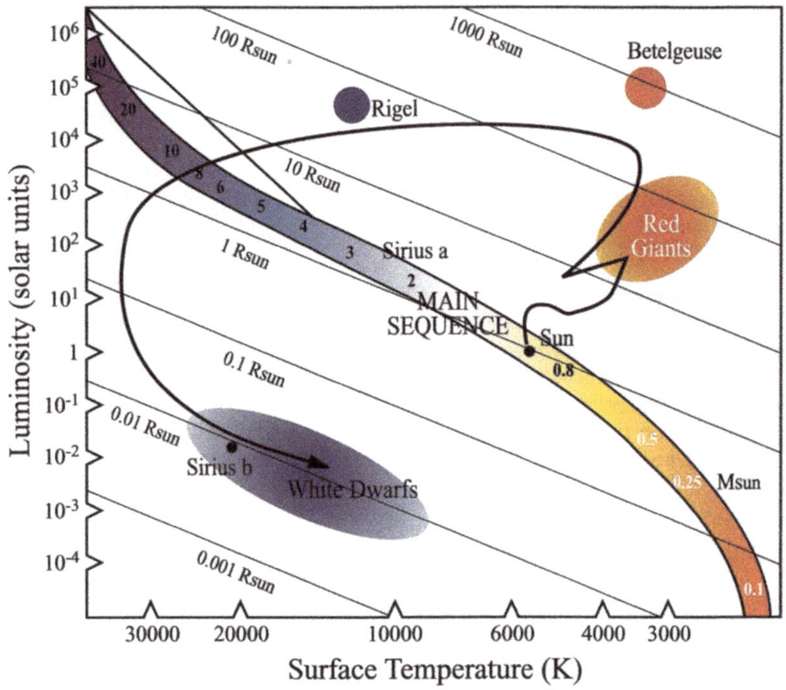

Figure 2.8. Stellar luminosity versus effective temperature. On the main sequence (the quasi-diagonal shaded band, including ~90% of the stars in the solar neighborhood), numbers indicate the corresponding stellar mass (in solar units). Diagonal lines indicate stellar radii (in solar units $R_{\odot}$) at a given position of the diagram. The curve schematically shows the future evolution of the Sun, through the red giant and down to the white dwarf final stage.

stellar properties: the *absolute luminosity L* (derived from the apparent luminosity, once the distance is known) and the surface temperature (measured through the color of the star and called *effective temperature* T_E when the stellar surface is assumed to radiate like a blackbody).

In the solar neighborhood, 90% of the stars lie on the "main sequence," a quasi-diagonal band running from high to low values of L and T_E, with the more massive stars being hotter and more luminous.[7] The remaining 10% are either cold (red) and luminous objects (*red giants*) or hot and subluminous ones (*white dwarfs*).

The interpretation of the H–R diagram became possible once the nature of the energy source of stars was elucidated. The works of H. Bethe in the late 1930s (Bethe & Critchfield 1938; Bethe 1939) established the series of nuclear reactions that produce the energy radiated by the Sun and the other main sequence stars (see Section 2.4.2).[8] The structure of those stars was understood rather well at that time, after the work of A. Eddington in the 1920s.

To a first (and actually, quite good) approximation, a star is a gaseous sphere in hydrostatic equilibrium between

[7] Stellar masses can be accurately determined only in binary systems, by application of Kepler's laws.

[8] Hans Bethe received the 1967 Nobel Prize in Physics for his work on energy production in the Sun and stars.

- the attractive force of its own gravity (depending on its mass),
- the internal pressure, which depends on the physical state of the stellar gas; for a perfect gas (dominating the interiors of main sequence stars), it is proportional to the product of temperature and density.

It turns out that gaseous masses of solar composition between 0.08 and 100 solar masses ($M_\odot$) find equilibrium for central temperatures in the range of 2–30 million K. This has two important consequences:

- Because of the temperature gradient between the center and the surface, the star radiates its internal heat.
- The central temperatures are high enough to induce thermonuclear fusion reactions between the abundant and light hydrogen nuclei, which liberate huge amounts of energy and help keep the stellar interior hot for long durations (see next section).

Note that, thanks to its gravity, the star "controls" the rate of the nuclear reactions in its interior and does not explode; indeed, should the nuclear reaction rate increase, producing more energy, the star would heat up, expand,[9] and cool, thus reducing the energy production and coming back to its previous fuel consumption rate. Thus, a star is a *gravitationally bound thermonuclear fusion reactor*; it shines simply because it is massive,[10] and it shines for long durations because it uses an efficient energy source and it has large amounts of fuel available.

2.4.2 H burning on the Main Sequence

Depending on internal temperature, H burning may take place through different modes inside stars. The overall result, however, is always the same: four protons disappear and give rise to a ^{4}He nucleus, while two positrons and two neutrinos are released, as well as γ-ray photons.

$$4\,p \longrightarrow {}^4\mathrm{He} + 2e^{+} + 2\,\nu_e + Q_\gamma.$$

The energy released, corresponding to the mass difference Δm between the four protons and the ^{4}He nucleus is $E \sim \Delta mc^2 \sim 26$ MeV, i.e., ~6.6 MeV/nucleon or 5×10^{18} ergs g $^{-1}$. Most of the energy is deposited locally and heats the stellar gas, while a minor fraction escapes the star, carried away by neutrinos (which interact only weakly with matter). Neutrinos from the Sun have been detected since the 1960s by several experiments. The detected fluxes are in excellent agreement with predictions of solar models, once neutrino oscillations are taken into account (as suggested by the Sudbury neutrino experiment in 2000); this agreement is a strong and clear piece of evidence that energy production in the Sun is indeed well-understood.

---

[9] For a perfect gas, pressure depends on both density *and* temperature, i.e., $P \propto \rho T$; this is not the case for a degenerate gas, which has explosive consequences for SNIa (see Section 2.6.3).

[10] Even without nuclear reactions, the Sun could shine with its current luminosity $L_\odot$ for ~30 Myr, by slowly contracting and releasing gravitational energy $d(GM^2/R)/dt \sim L_\odot$; the required contraction rate is $dr/dt \sim 7$ m yr^{-1}, too small to be detectable.

In stars of mass <1.2 $M_\odot$ and central temperatures below 20×10^6 K, most of the energy is produced by the p–p chains (Figure 2.9). The first of these reactions involves the conversion of a proton to a neutron, through a weak and very slow interaction, which explains the very long lifetimes of stars powered by this H-burning mode (~10 Gyr for the Sun).

In stars of higher masses and temperatures, H burning occurs through the CNO-cycle (Figure 2.10), where the C, N, and O isotopes (produced from previous stellar

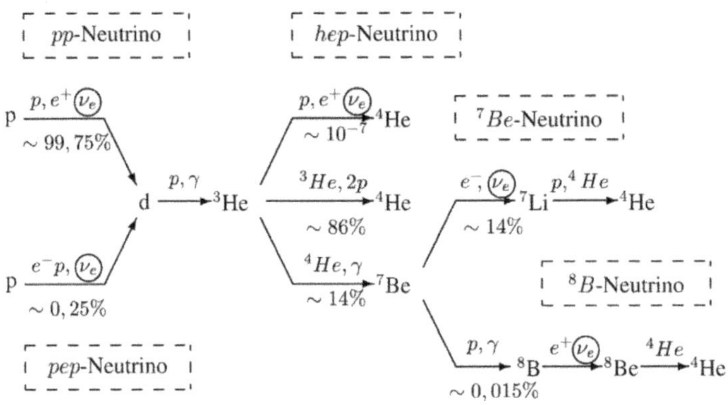

Figure 2.9. Nuclear reactions of the three proton–proton chains, producing one ^{4}He nucleus from four protons and providing most of the energy of low-mass stars on the main sequence. The percentages of occurrence (86% for the first chain, etc.) apply to conditions in the present-day solar interior (central temperature $T_c = 15.7$ MK and density $\rho_c = 160$ g cm^{-3}.

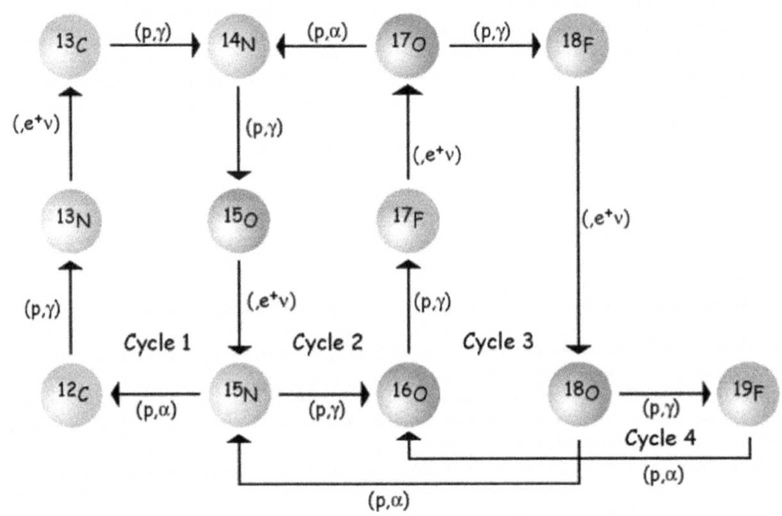

Figure 2.10. The full CNO tri-cycle, using the C, N, and O isotopes as catalysts during H burning. It provides energy in main sequence stars more massive than 1.2 $M_\odot$ (of solar metallicity) and produces ^{14}N from the initial ^{12}C and ^{16}O. (Note: Reaction A + p $\longrightarrow$ B + x can also be written as A(p,x)B; the latter notation is generally adopted in the text, but the former only occasionally.)

generations) act as catalysts. The sum of the abundances of CNO nuclei remains constant throughout H burning, but there is an internal re-arrangement: ^{12}C and ^{16}O turn into ^{14}N, and to smaller extent, into ^{13}C and ^{17}O; these are the main nuclei produced by the CNO cycle.

An important difference between the pp-chain and the CNO cycle concerns the dependence of the respective energy production rates on temperature. In the case of the pp-chains, energy production rate scales as $\epsilon_{pp} \propto T^4$, while in the case of the CNO cycle, it scales as $\epsilon_{\rm CNO} \propto T^{18}$. This difference is due to the effect of Coulomb barriers between reactants, which are higher in the latter case. The strong temperature dependence of $\epsilon_{\rm CNO}$ has an important implication: the energy produced locally by the CNO cycle can only be evacuated by convection, which implies that the stellar core becomes chemically well-mixed (i.e., nuclide abundances are uniform inside the central convective region). On the contrary, in the case of the Sun and low-mass stars, energy is evacuated by radiation; the abundances of reacting nuclei and of their products vary smoothly with radius in the interiors of such stars.

On the main sequence, the luminosity of a star (i.e., its fuel consumption rate) is proportional to some power of its mass ($L \propto M^K$, with $K \sim 3$ in the upper and $K \sim 4$ in the lower main sequence), while the available energy E is only proportional to the mass (a fraction $f \sim$ 10–50% of the mass is "burned," with larger fractions for larger values of M). Thus, the lifetime on the main sequence is $\tau \propto fE/L \propto M/M^K \propto M^{-2}$. The most massive stars (several tens of $M_\odot$) shine for only a few Myr—a short lifetime compared to the Sun (10 Gyr) or to our current closest neighbor, Proxima Centauri (a 0.12 $M_\odot$ star, bound to live for more than 1000 Gyr).

Note that, at central H-exhaustion, the star has "used" ~6.6 MeV/nucleon out of the ~8.8 MeV/nucleon available, i.e., before reaching the ultimate nuclear stability in the Fe peak (see Figure 2.3 bottom). In other terms, it has already spent the largest part of its nuclear fuel. This explains why the H-burning phase is the longest period in a star's life and why most of the stars are found on the main sequence.

2.4.3 He burning in Red Giants

After central H exhaustion, the stellar core contracts and releases gravitational energy, which brings the surrounding H-rich layers to temperatures high enough for H-burning reactions in a thin shell surrounding the He-core. At the same time, the envelope expands and cools and the star turns into a red giant (or supergiant, for the most massive of them, i.e., above 10 $M_\odot$). The envelope of such a star is convective and brings products of the previous central H-burning phase to the surface. Enhanced abundances of ^4He and ^{14}N, as well as modified isotopic ratios of, e.g., $^{13}C/^{12}C$ or $^{17}O/^{16}O$, are the marks of this "first dredge-up" phase, which allows us to compare nucleosynthesis theory to observations.

When the H-exhausted core reaches temperatures of $100–200 \times 10^6$ K (with higher temperatures corresponding to higher stellar masses), He fusion begins. It proceeds

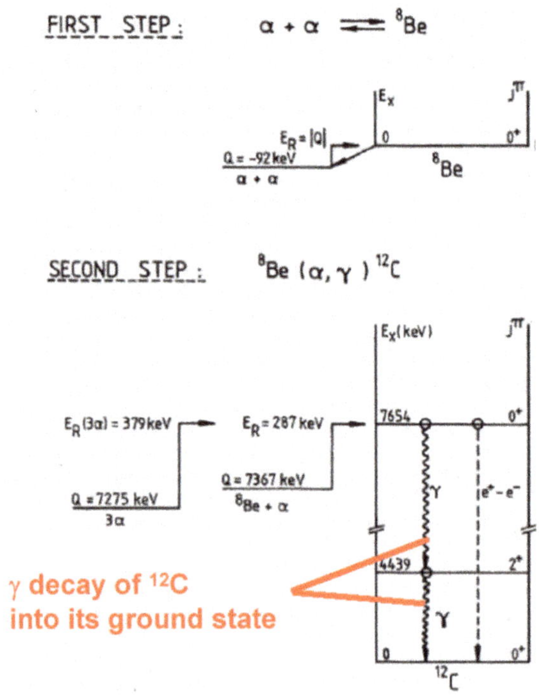

Figure 2.11. The triple-alpha (3α) reaction occurs in two steps: formation of ^{8}Be from two alpha particles (top) and capture of a third alpha by ^{8}Be during the 10^{-16} s of the lifetime of this unstable nucleus (bottom); at characteristic red giant temperatures, the second step becomes rapid enough only because of the existence of the level at 7654 keV of the ^{12}C nucleus, which makes the reaction resonant. (Credit: Rolfs & Rodney 1988.)

in two steps (see Figure 2.11), with the second one involving a *resonant* reaction.[11] The final outcome is the formation of a ^{12}C nucleus from three alpha particles (3α reaction).

The fusion of ^{4}He to ^{12}C is much less energetically efficient than the fusion of H to He. It releases an energy of $[3m(\alpha) - m(^{12}\text{C})]c^2/ \sim 7.3$ MeV or $\sim$0.6 MeV/nucleon, i.e., about ten times less energy per unit mass than H burning. This explains why the number of red giants is so much smaller than the number of main sequence stars, and constitutes another important test of the theory of stellar evolution.

During He burning, ^{12}C nuclei capture α particles to form ^{16}O nuclei; the ^{12}C $(\alpha, \gamma)^{16}$O reaction is not resonant, however, so a considerable amount of ^{12}C is left over in the stellar core at the end of He burning. ^{16}O is usually dominant, but the exact ^{16}O/^{12}C ratio depends on the rate of the ^{12}C$(\alpha, \gamma)^{16}$O reaction, which is still uncertain, as well as on the initial mass of the star (more massive stars produce higher ratios).

[11] The existence of the 7654 keV level in the nucleus of ^{12}C, which renders the 3α reaction resonant, was *predicted* by F. Hoyle (and confirmed by experimenters) in 1953; Hoyle argued that ^{12}C in the universe *should be made* by the 3α reaction in red giant conditions.

Note also that, in the very beginning of He burning, at temperatures $\sim 100 \times 10^6$ K, ^{14}N turns into ^{18}O and later into ^{22}Ne, through successive α captures. Thus, He burning constitutes the production mode of ^{18}O in the universe, because some amount of it survives in the He-shell. Note also that, toward the end of He burning in massive stars, at temperatures $T \sim 250 \times 10^6$ K, neutrons are released through $^{22}Ne(\alpha,n)^{25}Mg$. Due to the lack of Coulomb barriers, those neutrons are easily captured by all nuclei in the star (in proportion to the corresponding neutron capture cross-sections). Part of those nuclei are destroyed in subsequent stages of the evolution of massive stars, but those that survive are ejected by the final supernova explosion. This "weak s-process" leads to the production of the light s-nuclei, with mass number A between 60 and 90, in nature (see also Section 2.7.2).

The most massive stars (with initial masses above 30 $M_\odot$) develop strong stellar winds, due to the high radiation pressure on their envelopes. Losing more than 10^{-5} $M_\odot$ per year, they finally reveal layers that have been affected by nucleosynthesis. The products of core H burning (^4He and ^{14}N) and then of core He burning (^{12}C, ^{16}O, and ^{22}Ne) appear thus with enhanced abundances on the stellar surface. Those stars are called *Wolf–Rayet stars* and their spectroscopy offers an invaluable test of our nucleosynthesis theories.

After core He-exhaustion, the star proceeds in a way similar to the "after H-burning" case: the carbon–oxygen core contracts and ^4He ignites in the surrounding layers, while H may also burn in even more external layers. In the case of stars with masses $M < 8$ $M_\odot$, material from the He-layer is mixed first in the H-layer and eventually in the stellar envelope, which becomes enriched in He-burning products after this "3D dredge-up" phase. The existence of "carbon stars," i.e., of low-mass red giants with high carbon abundances, offers another important test of the theory of stellar evolution and nucleosynthesis.

In the range of intermediate- and low-mass stars ($M < 8$ $M_\odot$) the double shell burning phase is highly unstable; energy is released in "thermal pulses," and in the end of that phase, the the stellar envelope is expelled into space. The star becomes a *planetary nebula* for a few tens of thousands of years. After that, the naked C–O core slowly cools down. Its temperature never rises to the point of carbon ignition because the pressure of its degenerate electron gas can resist its gravity forever. The star becomes less and less luminous, until it ends its life in the "graveyard" of white dwarfs (see Figure 2.8). The fate of stars more massive than 10 $M_\odot$ is quite different (and much more spectacular).

2.5 Advanced Evolution of Massive Stars

Despite their small number (less than 1% of a stellar generation) massive stars constitute the most important agents of galactic chemical evolution. Indeed, most of the "heavy" elements (metals) in the universe are synthesized in the hot interiors of massive stars, particularly during the final *supernova* explosion.

Contrary to their lower-mass counterparts, which stop evolving after burning He in a shell surrounding an inert carbon–oxygen core, stars with mass $M > 8$–11 $M_\odot$ successively burn all the available nuclear fuels in their core, until its composition is dominated by nuclei of the iron peak. However, the duration of all stages subsequent to core He burning is so short that no direct observational tests of the evolutionary status of the core (which is hidden inside a red supergiant envelope) are possible. Only "post-mortem" observations offer the possibility of (indirectly) validating the results of our models.

2.5.1 Neutrino Losses Accelerate Stellar Evolution

One of the most important features of the advanced evolutionary stages of massive stars is the copious production of neutrinos, due to the high temperatures and densities reached in the stellar core after He-exhaustion ($T > 0.8 \times 10^9$ K, $\rho > 10^5$ g cm^{-3}). Neutrino-antineutrino pairs are produced by

- (1) *electron–positron annihilation*, the e$^-$s and e$^+$s being created by the hot thermal plasma (note that only a small fraction of the annihilations leads to $\nu - \bar{\nu}$ pair production);
- (2) *photo-neutrino process*, analogous to Compton scattering, with the outgoing photon replaced by a ν–$\bar{\nu}$ pair;
- (3) *plasma process*, where a *plasmon* (an excitation of the plasma with an energy $\hbar\omega_p$, where ω_p is the plasma frequency) decays into a ν–$\bar{\nu}$ pair.

Neutrinos interact only weakly with matter, with a cross-section $\sigma_\nu \sim 10^{44}$ cm^{-2}. Their mean free path in the stellar core, at densities $\rho \sim 10^5$ g cm^{-3} at carbon ignition (see Section 2.5.2) is $l_\nu \sim (\rho \, N_A \, \sigma_\nu)^{-1} \sim 10^5 \, R_\odot$ ($N_A = 6.023 \times 10^{23}$ being Avogadro's number). Neutrinos escape then from the stellar core, taking away most of the available thermal energy (produced by nuclear reactions and/or gravitational contraction). In order to compensate for that neutrino "hemorrhage," the core has to increase its nuclear energy production by contracting and increasing its temperature. As a result, the evolution of the star is greatly accelerated: the time between C-ignition and the final supernova explosion is less than 0.1% of the corresponding main sequence lifetime.

The extreme sensitivity of nuclear energy production and neutrino losses to temperature allows one to obtain a (relatively) good estimate of the burning temperature of each nuclear burning stage: the so-called "balanced power approximation" (Woosley et al. 1973) states that the local power production $\dot{\epsilon}_{\rm nuc}$ just equals the neutrino loss rate $\dot{\epsilon}_\nu$ (a dot over a symbol indicates a time derivative). Taking into account that both rates depend on temperature and density, one can find the burning temperature by assuming some appropriate density.

By applying the "balanced power approximation" the burning temperatures of C, Ne, O, and Si are derived in Woosley (1986) and are presented in graphical form in Figure 2.12. One sees that C burns at $\widehat{9} \sim 0.85$, Ne burns at $\widehat{9} \sim 1.4$, O at $\widehat{9} \sim 1.9$, and Si at $\widehat{9} \sim 3.4$ (where $\widehat{9}$ is the temperature in 10^9 K). The corresponding densities are $\sim 10^5$ g cm^{-3} for C burning, $\sim 10^6$ g cm^{-3} for Ne burning, a few 10^6 g cm^{-3} for O burning, and a few 10^7 g cm^{-3} for Si burning.

The evolution of the central temperature and density for stars in the 13–25 $M_\odot$ range (with no mass loss) is shown in Figure 2.13. Assuming that the stellar core (of

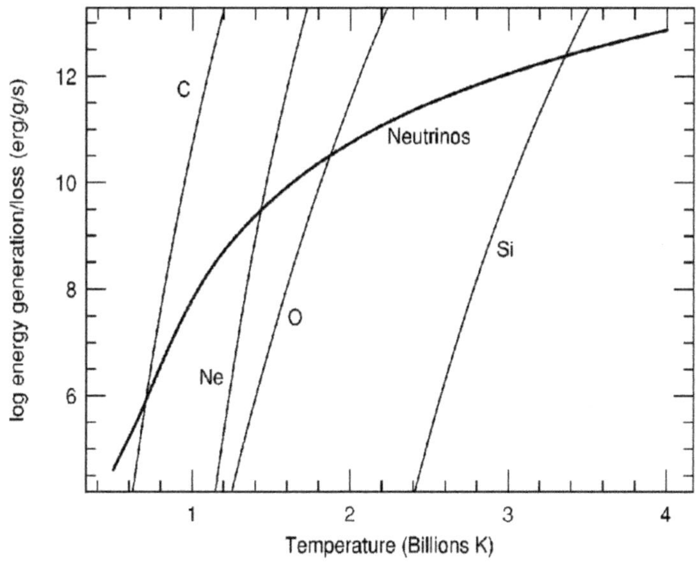

Figure 2.12. Energy production rate (per unit mass) during the advanced evolutionary stages of massive stars (C, Ne, O, and Si burning) and neutrino loss rates as a function of temperature. Actual burning temperatures for each stage are found on the intersections of the neutrino loss curve with the corresponding energy production one. (Credit: Woosley et al. 1986.)

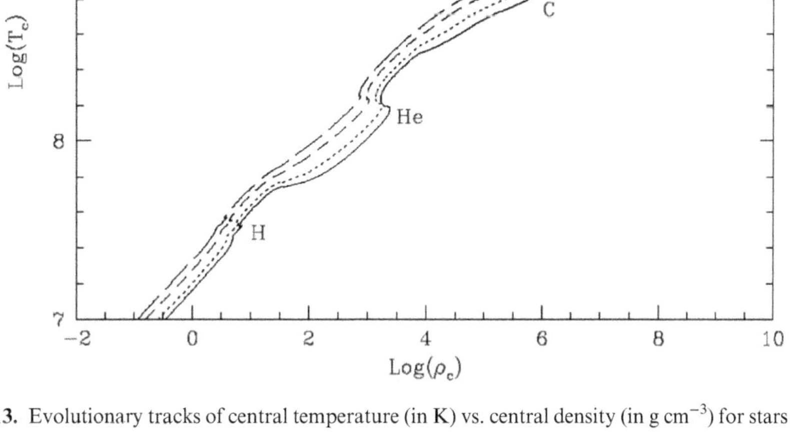

Figure 2.13. Evolutionary tracks of central temperature (in K) vs. central density (in g cm^{-3}) for stars of 13, 15, 20, and 25 $M_\odot$. (Credit: Limongi et al. 2000.)

mass M and radius R) contracts in quasi-hydrostatic equilibrium between two burning phases, one has for the internal pressure $P \propto M^2/R^4$; combined with the density ($\rho \propto M/R^3$), this leads to $d \ln P = 4/3\ d \ln \rho$. Assuming that the core material is in the perfect gas regime ($d \ln P = d \ln \rho + d \ln T$), one finally obtains $T \propto \rho^{1/3}$. The numerical results displayed in Figure 2.13 follow that relation up to the onset of copious neutrino emission (slightly preceding C ignition in the core). In fact, after Ne burning, the very center of the star is in the partially degenerate regime, but in the largest part of the burning, core conditions are still those of the perfect gas. Note that, *during* periods of nuclear energy production, the relation $T \propto \rho^{1/3}$ is not satisfied anymore (the curves turn slightly to the left).

2.5.2 C, Ne, and O burning

Carbon burning in massive stars occurs at $\widehat{9} \sim 0.8$ and $\rho \sim 10^5$ g cm^{-3}. The core composition at C-ignition is dominated by the ashes of He burning, ^{12}C, and ^{16}O (more than 90% of the total) in a proportion that decreases with the stellar mass. The exact proportion of ^{12}C/^{16}O and the exact initial fraction of ^{12}C (which is crucial for the energetics of C burning) depend sensitively on the—still uncertain—value of the ^{12}C(α,γ)^{16}O rate and on the adopted criterion of convection during He burning. The reason for the latter dependence is that every α particle brought in the convective core while He is almost exhausted (and the rate of the 3α reaction too low to produce new ^{12}C nuclei) converts a ^{12}C nucleus into ^{16}O.

The fusion of two ^{12}C nuclei produces a compound nuclear state of ^{24}Mg, which decays by emitting a proton or an α particle (see Figure 2.14). Note that the neutron emission channel, i.e., ^{12}C+^{12}C $\longrightarrow$ n+^{23}Mg, corresponds to an endothermic reaction with a small probability (0.1% at $\widehat{9} = 1$), but it is nevertheless important because the decay of ^{23}Mg to ^{23}Na changes the *neutron excess*

$$\eta = \sum_i (N_i - Z_i)\ Y_i \tag{2.10}$$

Basic:

$$^{12}\mathrm{C}\left(^{12}\mathrm{C}, {\alpha \atop p}\right){^{20}\mathrm{Ne} \atop ^{23}\mathrm{Na}}$$

^{23}Na(p,α)^{20}Ne $\qquad$ ^{23}Na(p,γ)^{24}Mg

Down to 10^{-2} of above:

^{20}Ne(α,γ)^{24}Mg $\qquad$ ^{23}Na(α,p)^{26}Mg(p,γ)^{27}Al

^{20}Ne(n,γ)^{21}Ne(p,γ)^{22}Na(β^+)^{22}Ne(α,n)^{25}Mg(n,γ)^{26}Mg

^{21}Ne(α,n)^{24}Mg $\qquad$ ^{22}Ne(p,γ)^{23}Na

^{25}Mg(p,γ)^{26}Al(β^+)^{26}Mg

Figure 2.14. Main nuclear reactions taking place during C burning. The first four (top) ones produce the quasi-totality of the nuclear energy. The others are less important energetically, because their "reaction fluxes" (the product of the abundances of the reactants times the corresponding reaction rate) are smaller than (down to 10^{-2} of) the flux of the first four reactions; however, they contribute to nucleosynthesis during C burning (from Thielemann & Arnett 1985).

where N_i, Z_i, and $Y_i = X_i/A_i$ are the neutron number, the charge, and the number fraction of nucleus i, respectively (see Section 2.5.3). Moreover, neutron captures on heavy nuclei produce heavier than Fe-nuclei, thus modifying the s-process composition resulting from the previous He-burning phase. Also, protons and alphas released by the $^{12}C+^{12}C$ fusion are captured on the ambient nuclei through dozens of (energetically unimportant) reactions; in particular, α captures on ^{20}Ne produce ^{24}Mg. Thus, the main products of C burning are ^{20}Ne, ^{23}Na, and ^{24}Mg.

Carbon burning releases an energy $q_{nuc} \sim 0.4$ MeV/nucleon or $\sim 4 \times 10^{17}$ erg g^{-1}. The energy production rate (expressed in erg g^{-1} s^{-1}) is

$$\dot{\epsilon}_{nuc} \propto Y_{12}^2 \, \rho \, \lambda_{12,12} \tag{2.11}$$

where $\lambda_{12,12}$ is the $^{12}C+^{12}C$ fusion reaction rate and Y_{12} the number fraction of ^{12}C. For initial number fractions of ^{12}C <2 % (or mass fractions $X = A\,Y_{12} < 0.2$, with $A = 12$), the energy production rate is small enough that ^{12}C burns radiatively. According to detailed numerical models, radiative burning happens for stars more massive than $\sim$19 $M_\odot$; in less massive stars, a convective core is formed. Note that the convection criterion plays also a role in the determination of that critical mass.

After C exhaustion, the composition of the stellar core is dominated by ^{16}O and ^{20}Ne (more than 90% of the total by mass). ^{23}Na and ^{24}Mg also exist at the few percent level (in mass fraction). Despite its smaller Coulomb barrier, ^{16}O is not the next fuel to burn, because it is exceptionally stable (being a doubly magic nucleus with $Z = N = 8$). The photodisintegration of ^{20}Ne nucleus $^{20}Ne(\gamma,\alpha)^{16}O$ becomes energetically feasible at $\widehat{9} \sim 1.5$, i.e., before the fusion temperature of ^{16}O nuclei ($\widehat{9} \sim 2$) is reached. The released α particles are captured on both ^{16}O (to restore ^{20}Ne) and ^{20}Ne (to form ^{24}Mg). The net result of the operation is that this ^{20}Ne "melting" can be described by

$$2\ ^{20}Ne \longrightarrow\ ^{16}O +\ ^{24}Mg.$$

The photodisintegration of ^{20}Ne is endoergic, but the exoergic α captures on ^{16}O and ^{20}Ne more than compensate for the energy lost. ^{20}Ne burning produces $\sim$0.1 MeV/nucleon or 1.1 10^{17} erg g^{-1}, i.e., about 1/4 the specific energy released by C burning. Several other energetically unimportant reactions, induced by p, n, and α particles occur during Ne burning; in particular, ^{23}Na (a product of C burning) disappears through $^{23}Na(p,\alpha)^{20}Ne$ and $^{23}Na(\alpha,p)^{26}Mg$ (see also Figure 2.15).

After Ne burning, the stellar core consists mainly of ^{16}O, ^{24}Mg, and ^{28}Si (the latter being produced mainly through $^{24}Mg(n,\gamma)^{25}Mg(\alpha,n)^{28}Si$). Also, ^{29}Si, ^{30}Si, and ^{32}S are present at the 10^{-2} level (in mass fraction).

The fusion of two ^{16}O nuclei produces a compound nucleus of ^{32}S, which decays through the p, α, and n channels (see Figure 2.16); the corresponding branching ratios and energy released are 58% (7.68 MeV), 36% (9.58 MeV), and 6% (1.45 MeV), respectively. Note that, at high temperatures, the endoergic decay through the deuteron channel is also effective.

Basic reactions:

$^{20}\mathrm{Ne}(\gamma,\alpha)^{16}\mathrm{O}$ $\qquad$ $^{20}\mathrm{Ne}(\alpha,\gamma)^{24}\mathrm{Mg}(\alpha,\gamma)^{28}\mathrm{Si}$

Flows $> 10^{-2}$ *times the above:*

$^{23}\mathrm{Na}(p,\alpha)^{20}\mathrm{Ne}$ $\qquad$ $^{23}\mathrm{Na}(\alpha,p)^{26}\mathrm{Mg}(\alpha,n)^{29}\mathrm{Si}$

$^{20}\mathrm{Ne}(n,\gamma)^{21}\mathrm{Ne}(\alpha,n)^{24}\mathrm{Mg}(n,\gamma)^{25}\mathrm{Mg}(\alpha,n)^{28}\mathrm{Si}$

$^{28}\mathrm{Si}(n,\gamma)^{29}\mathrm{Si}(n,\gamma)^{30}\mathrm{Si}$

$^{24}\mathrm{Mg}(\alpha,p)^{27}\mathrm{Al}(\alpha,p)^{30}\mathrm{Si}$

$^{26}\mathrm{Mg}(p,\gamma)^{27}\mathrm{Al}(n,\gamma)^{28}\mathrm{Al}(\beta^-)^{28}\mathrm{Si}$

Figure 2.15. Main nuclear reactions taking place during Ne burning (from Thielemann & Arnett 1985).

Basic reactions:

$$^{16}\mathrm{O}\begin{cases}(^{16}\mathrm{O},p)^{31}\mathrm{P}\\(^{16}\mathrm{O},\alpha)^{28}\mathrm{Si}\\(^{16}\mathrm{O},n)^{31}\mathrm{S}(\beta^+)^{31}\mathrm{P}\end{cases}$$

$^{31}\mathrm{P}(p,\alpha)^{28}\mathrm{Si}(\alpha,\gamma)^{32}\mathrm{S}$

$^{28}\mathrm{Si}(\gamma,\alpha)^{24}\mathrm{Mg}(\alpha,p)^{27}\mathrm{Al}(\alpha,p)^{30}\mathrm{Si}$

$^{32}\mathrm{S}(n,\gamma)^{33}\mathrm{S}(n,\alpha)^{30}\mathrm{Si}(\alpha,\gamma)^{34}\mathrm{S}$

$$^{28}\mathrm{Si}(n,\gamma)^{29}\mathrm{Si}\begin{cases}(\alpha,n)^{32}\mathrm{S}(\alpha,p)^{35}\mathrm{Cl}\\(p,\gamma)^{30}\mathrm{P}(\beta^+)^{30}\mathrm{Si}\end{cases}$$

Electron captures:

$^{33}\mathrm{S}(e^-,\nu)^{33}\mathrm{P}(p,n)^{33}\mathrm{S}$

$^{35}\mathrm{Cl}(e^-,\nu)^{35}\mathrm{S}(p,n)^{35}\mathrm{Cl}$

Figure 2.16. Main nuclear reactions taking place during O burning (from Thielemann & Arnett 1985).

The energy released by oxygen burning is 5×10^{17} erg g^{-1} or 0.5 MeV/nucleon. As in the previous stages, dozens of reactions induced by n, p, and α occur. However, the increased temperature and density introduce two novel features:

- Electron captures occur, mainly on ^{31}S, ^{30}P, ^{33}S, ^{33}Cl, and ^{37}Ar. These weak interactions substantially modify the neutron excess η, up to $\eta \sim 0.01$ (especially in the lower-mass and denser stars); the final neutron-rich composition is clearly non-solar and should be rarely ejected by the star.
- Photodisintegration reactions also become important and destroy most of the heavier than Fe nuclei that have been built through n captures in the previous burning phases (mostly by s-process during He burning). However, this happens only in the innermost and hottest regions of the stellar core; outside them, products of previous burning stages survive.

2.5.3 Si melting and Nuclear Statistical Equilibrium (NSE)

At O-exhaustion, the composition of the stellar core is dominated by ^{28}Si (30–40% by mass fraction), as well as either ^{32}S and ^{38}Ar (in the more massive cores, where low densities do not favor electron captures) or ^{30}Si and ^{34}S (in the lower-mass stars, below $\sim$15 M$_\odot$).

^{28}Si "burns" at temperature $\widehat{9} \sim 3.2$ (see Section 2.5.1) and in a way that is reminiscent of the ^{20}Ne "burning." The fusion of two ^{28}Si nuclei requires such high temperatures that photodisintegration of all nuclei would result. Instead, part of the ^{28}Si nuclei photodisintegrates, mainly through a sequence of reactions involving α particles: $^{28}\text{Si}(\gamma,\alpha)^{24}\text{Mg}(\gamma,\alpha)^{20}\text{Ne}(\gamma,\alpha)^{16}\text{O}(\gamma,\alpha)^{12}\text{C}(\gamma,2\alpha)\alpha$.

Other photodisintegration reactions releasing p and n also occur, especially in material with neutron excess η substantially different from 0 (due to previous electron captures, i.e., in high-density stellar cores). The released α particles and nucleons are further captured by ^{28}Si and heavier nuclei and an equilibrium is established between direct and inverse reactions, e.g., $^{28}\text{Si}(\alpha,\gamma)^{32}\text{S}(\gamma,\text{p})^{31}\text{P}(\gamma,\text{p})^{30}\text{Si}(\gamma,\text{n})^{29}\text{Si}(\gamma,\text{n})^{28}\text{Si}$.

The mean atomic weight of the mixture progressively increase because free nucleons and α particles are tight to heavier and more strongly bound nuclei (see Figure 2.17); that is why the overall process is better described as "Si melting." In fact, local equilibrium between a few nuclear species is established already by the end of O burning, but, as the temperature increases, the various quasi-equilibrium clusters merge. In early Si melting, there are already two major quasi-equilibrium clusters established: one with nuclei with A = 24–46 and another with Fe-peak nuclei. Toward the end of Si melting, the two clusters merge and the quasi-equilibrium group includes all nuclei above ^{16}O. Note that such reaction sequences also *indirectly* (i.e., *not* through photodisintegration of α particles in free nucleons) bring the abundances of α particles and free nucleons into equilibrium. Assuming

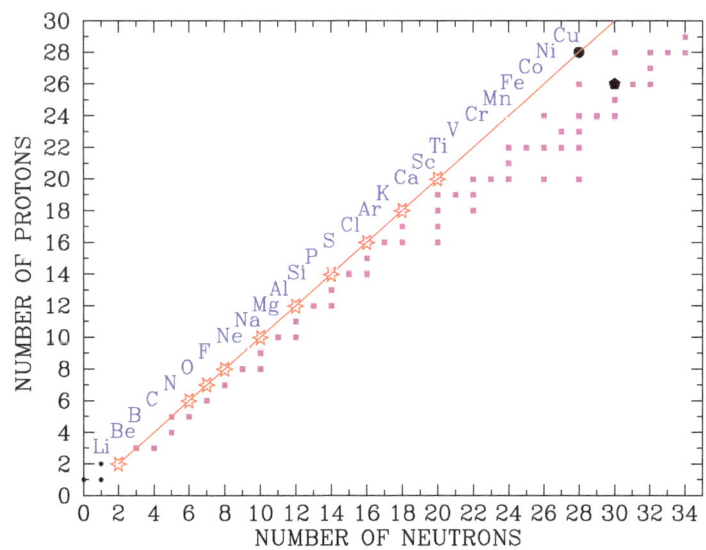

Figure 2.17. Position of stable nuclei on the proton–neutron diagram. Stellar nucleosynthesis proceeds at $Z = N$ up to ^{40}Ca, favoring the strongly bound α-nuclei (*asterisks;* see also Figure 2.3). Weak interactions (electron captures) during late O burning and Si melting modify the neutron excess and progressively shift the composition toward ^{56}Fe (Z = 26, N = 30, *filled pentagon*). In explosive burning conditions, weak interactions have no time to operate and the nuclear flow proceeds at constant neutron excess and $Z = N$ (*solid line*), all the way up to ^{56}Ni ($Z = N$ = 28, *filled dot*).

that two ^{28}Si nuclei turn into one nucleus of the Fe peak, one finds that the energy released by Si melting is ~0.2 MeV/nucleon or $\sim 2 \times 10^{17}$ erg g^{-1}.

Toward the end of Si melting, all electromagnetic and strong nuclear interactions are in equilibrium with their inverses. Because neutrinos still escape freely from the stellar core, neutrino-producing weak interactions never come into equilibrium with their inverses and total equilibrium is not established. The last reactions to reach equilibrium are those linking ^{24}Mg to ^{20}Ne, ^{16}O to ^{12}C, and finally, the reaction 3 $\alpha \longleftrightarrow {}^{12}C$. When this happens, and assuming that all nuclides obey the statistics of an ideal Maxwell–Boltzmann gas, their abundances can be expressed by application of the Saha equation:

$$Y_i(A_i, Z_i) = (\rho\ N_A)^{A_i-1} \frac{G_i}{2^{A_i}} A_i^{3/2} \left(\frac{2\pi\hbar^2}{m_H kT} \right)^{3(A_i-1)/2} e^{B(A_i,\ Z_i)/kT}\, Y_p^Z\, Y_n^{(A-Z)} \tag{2.12}$$

where $G_i = \sum_j (2J_j^i + 1)\exp(-E_j^i/kT)$ is the *partition function* of nucleus i (a summation over all nuclear levels of spin J and energy E_j) and $B(A_i, Z_i)$ its binding energy. Two more equations are required to eliminate the neutron and proton abundances Y_n and Y_p, and these are taken to be the mass conservation equation

$$\sum Y_i\ A_i = 1 \tag{2.13}$$

and the charge conservation

$$Y_e = \frac{n_e}{\rho\ N_A} = \sum_i Z_i Y_i = \frac{1-\eta}{2} \tag{2.14}$$

where Y_e is the *electron mole number* and n_e the electron density. Thus, at nuclear statistical equilibrium (NSE), the composition is described as a function of three parameters: temperature T, density ρ, and electron fraction Y_e or neutron excess η. In fact, Y_e and η are slowly modified under the action of weak interactions (electron captures), and that variation must also be taken into account, e.g., through

$$\frac{dY_e}{dt} = \sum_i -\lambda_i\ Y_i \tag{2.15}$$

where λ_i is the weak interaction rate of nucleus i.

In the conditions of NSE, the Saha equation implies that for temperatures that are not too high ($\widehat{9} < 10$), the most abundant nuclei are those with the largest binding energy for a given value of η. Thus, for $\eta \sim 0$, the most tightly bound nucleus is ^{56}Ni (Figure 2.18). Indeed, in explosive nucleosynthesis, ^{56}Ni is the dominant product of NSE; its radioactive decay has been (indirectly) observed in the case of the supernova SN1987A (see Section 2.6.2), brilliantly confirming the theory. However, in the quiescent burning conditions of late stellar evolution, weak interactions have time enough to modify η to values ~0.06–0.08, corresponding to $Y_e \sim 0.46$–0.47. In those conditions, the most tightly bound nucleus is the stable ^{56}Fe, while ^{52}Cr is also produced in substantial amounts (see Figure 2.18).

When the composition of the stellar core becomes dominated by the strongly bound nuclei of the Fe peak, the star has arrived at the end of its quiescent life.

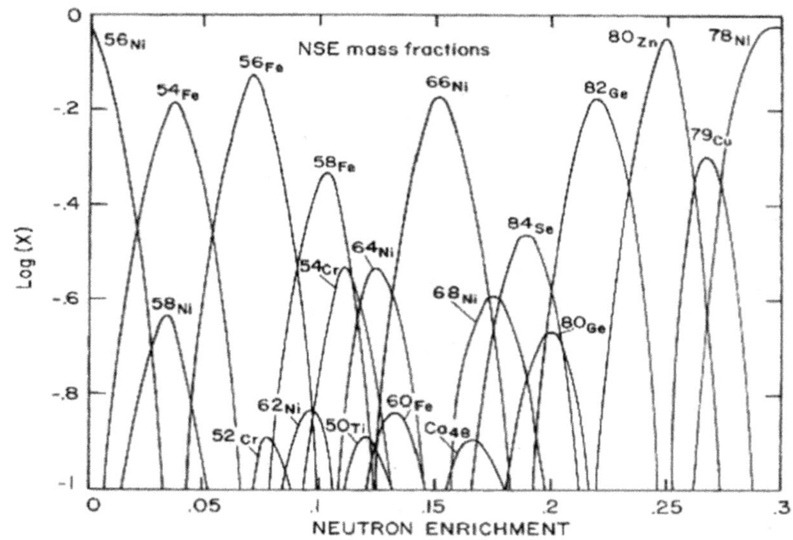

Figure 2.18. Composition at nuclear statistical equilibrium as a function of the neutron enrichment η; composition corresponding to $\eta > 0.1$ is not encountered in the cosmic abundance curve. (Credit: Hartmann et al. 1985.)

Further contraction of the core and increase of its temperature cannot release nuclear energy to sustain internal pressure: reactions involving stable nuclei at the peak of the binding energy curve are endothermic and constitute sinks of thermal energy. Moreover, electron captures proceed at even higher rates and continue removing electrons from the stellar plasma, weakening the main source of resistance against gravity, namely electron pressure. The iron core collapses. The rate and the details of the collapse depend sensitively on the electron fraction Y_e, because the Chandrasekhar mass (i.e., the maximum mass supported against gravity by a gas of degenerate electrons) is given by

$$M_{\text{Ch}} = 3(8\,\pi^2)^{1/4}\left(\frac{\hbar c}{2G}\right)^{3/2}(N_A\ Y_e)^2 \sim 1.45\,(2\ Y_e)^2 M_\odot. \tag{2.16}$$

Obviously, the rates of electron captures in heavy nuclei play an important role in determining the final value of Y_e before collapse and the fate of the stellar core. Modern calculations produce values around $Y_e \sim 0.445$.

2.5.4 Overview of the Advanced Evolutionary Phases

A schematic view of the interior evolution of a massive star is presented in Figure 2.19. The life of a massive star is essentially a series of relatively long quiescent central burning stages interrupted by much shorter periods of core contraction and heating, which lead to the ignition of the next fuel. Shell burning of a given nuclear fuel also takes place at the border of the former convective core, where that fuel has been exhausted; the burning shell progressively migrates outward. However, shell

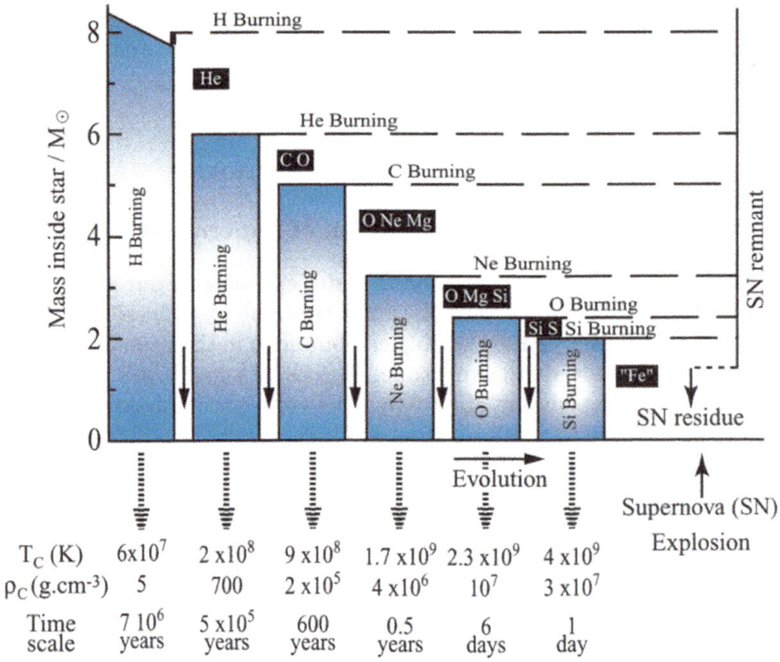

Figure 2.19. Schematic view of the interior evolution of a 25 $M_{\odot}$ star. *Blue regions* indicate convective core burning and *dashed lines* shell burning, while the corresponding main products are inside *boxes*. *Vertical arrows* indicate core contraction between two burning stages. Corresponding central temperatures, densities, and burning timescales (approximate values) are given in the bottom. (Credit: Arnould & Takahashi 1999.)

burning occurs at temperatures higher than the corresponding central burning stage, and the fuel never burns to exhaustion (*incomplete* burning).

In practice, the situation is much more complicated than the schematic view of Figure 2.19. Core burning is not always convective, because it depends on the amount of available fuel, which determines the nuclear energy production (e.g., central C burning and Ne burning may occur radiatively). Further, the various burning shells do not burn steadily, but intermittently. Only full-scale numerical calculations, including detailed nuclear reaction networks and appropriate physical ingredients, can (one hopes) "realistically" describe the advanced stages of massive star evolution (e.g., Woosley et al. 2002). Note that several key ingredients, like the various mixing processes and the role of rotation and mass loss, remain uncertain and may hold important clues as to the fate of stars of all masses (Maeder & Meynet 2012; Chieffi & Limongi 2013, and references therein). Note also that the numerical treatment of oxygen and silicon burning, where convective and nuclear timescales are comparable, is still problematic.

Despite those uncertainties, it is believed that, at the end of its "quiescent" life, a massive star develops an "onion-skin" structure, with heavier nuclei dominating the composition as one moves from the surface to the center. The star retains the "memory" of its previous burning stages in its various layers (i.e., in a way analogous —in fact, opposite—to the various geological strata that keep the "archives" of the earlier periods in the Earth's history). The reason for that layered structure of the star is the increasing sensitivity of nuclear reaction rates to temperature, as the Coulomb barriers of the successive nuclear fuels increase: the nuclear energy is produced in a region more and more confined to the center of the star, so the convective core is smaller at each new burning stage (shell burning does not modify the picture, because it slightly increases the sizes of all the former convective cores).

With no more nuclear fuel available, the Fe core collapses and the star dies in a spectacular supernova explosion.

2.6 Explosive Nucleosynthesis in Supernovae

2.6.1 Main Properties and Classification of Supernovae

Several supernovae have been observed in historical times in our Galaxy by naked eye (e.g., in 1054 in the constellation of the Crab by Chinese astronomers, in 1572 by Tycho, and in 1604 by Kepler), but none in the past four centuries, i.e., after the invention of the telescope. Dozens of supernovae are routinely observed each year outside the Milky Way, most of them in distances of tens of millions of light years.

The peak absolute luminosity of a supernova (SN) is 10^{42}–10^{43} erg s^{-1}, i.e., comparable to that of a whole galaxy, while the time-integrated energy that is radiated away is $\sim 10^{49}$ erg, comparable to that radiated by the Sun in 10^8 yr. A hundred times more energy is released in the form of kinetic energy of the ejecta, which are expelled at several 10^3 km s^{-1} (and up to a few 10^4 km s^{-1} in some case s).

Supernovae are classified on the basis of their spectra (Figure 2.20) as belonging to type I (absence of characteristic H lines) or type II (presence of H). Despite its interest (few stars show no H in their spectra), this classification concerns only the

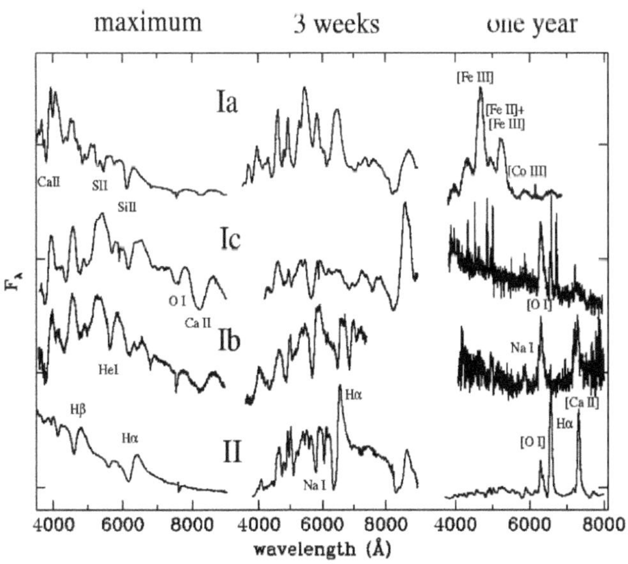

Figure 2.20. Spectra of SN of various types (from top to bottom), taken at three different epochs (from left to right).

"epidermic" properties of SN and tells us very little about the mechanism of the explosion.

SNIa are a subtype of SN distinguished by the presence of Si in their early spectra and quasi-identical peak luminosities, which decline in a characteristic exponential way: first with a period of ~6 days, and then more slowly after a few weeks, with a period of 77 days. This exponential decline of SNIa lightcurves is attributed to the radioactive decay chain $^{56}Ni \longrightarrow ^{56}Co \longrightarrow ^{56}Fe$; the half-lives of ^{56}Ni and ^{56}Co are 6.1 days and 77 days, respectively.

SNIa are encountered in all types of galaxies, including ellipticals, which ceased star formation billions of years ago and contain only old populations of small mass stars. All other types of supernovae are encountered in galaxies with currently active star formation, i.e., spirals and irregulars. Combined with the lack of H, that property of SNIa suggests that they originate from a population of old objects that have lost their H envelope. White dwarfs are the obvious candidates. Indeed, the thermonuclear burning of 1 $M_\odot$ of C–O to ^{56}Ni releases about 10^{51} ergs, sufficient to account for the observed energetics of SNIa.

A different mechanism is at the origin of all other SN types: the gravitational collapse of the Fe core of a massive star, which turns into an explosion in a way that is poorly understood up to now. Assuming that the collapse proceeds to the formation of a compact object (a neutron star with radius $R \sim$ 10 km), the available gravitational energy of a 1 $M_\odot$ Fe core ($E \sim GM^2/R \sim 10^{53}$ erg) is more than sufficient to explain the SN energetics. The considerable differences in the properties of those SN (peak luminosities, lightcurves, spectra) are due to the details of the collapse/explosion and, in particular, to the amount of the H-rich envelope left to the star at the moment of its collapse.

From statistics of SN in external galaxies, it is estimated that their frequency is ~0.25–0.35 SNIa and 1.5–2 core collapse SN per century in a galaxy similar in size and type to the Milky Way. The fact that no SN has been observed in the past 400 years in our Galaxy is attributed to the fact that massive stars are expected to be born and to die mostly in the inner galactic disk (much richer in gas than the outer one), which is hidden from our optical observations by a large column density of gas and dust.

2.6.2 Explosive Nucleosynthesis in Core-collapse Supernovae

The idea that supernova explosions are related to the death of massive stars goes back to the Swiss astronomer F. Zwicky in the 1930s. The first numerical models of such explosions were constructed in the early 1960s and the pioneering work of Colgate & White (1966) showed how the gravitational energy of the collapsing star turns into an explosion. However, the mechanism of core-collapse supernova explosions remains uncertain today (e.g., Janka 2017, and references therein). A schematic view of our current understanding is presented in Figure 2.21.

Iron core collapse proceeds on a timescale of milliseconds. Due to increasingly high temperatures, photodisintegrations tear down Fe nuclei to nucleons and alpha particles, while higher densities favor electron captures and conversion of protons to

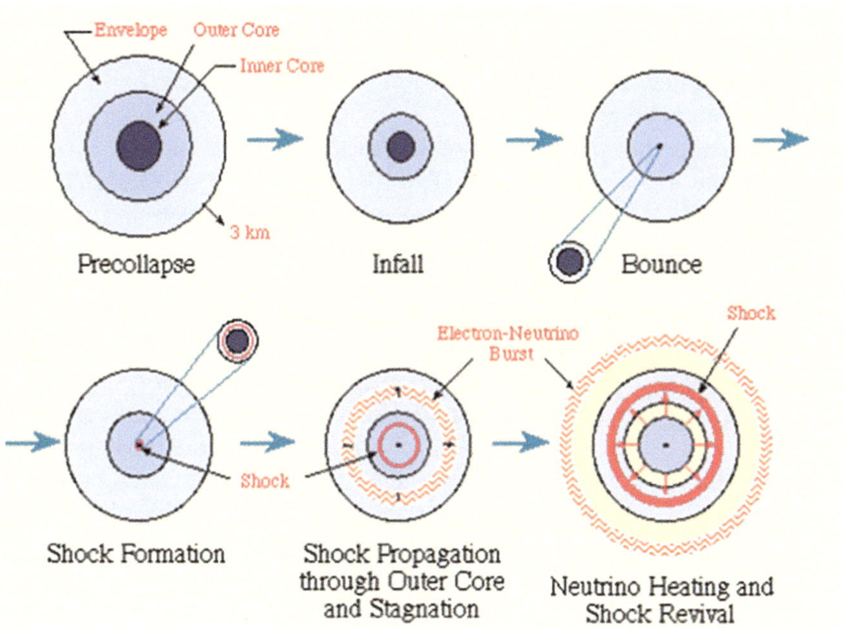

Figure 2.21. Illustration of the various successive stages of the core collapse of a massive star; see Section 2.6.2 for details.

neutrons. When the density of nuclear matter is reached ($\rho \sim 10^{14}$ g cm^{-3}), the repulsive component of the strong nuclear force brings the collapse of the inner core abruptly to a halt. At the boundary of the inner core, a shock wave is formed with an energy approximately equal to the kinetic energy of the core at the moment of the bounce ($\sim 5 \times 10^{51}$ erg, according to results of numerical simulations). As the shock wave propagates outward, into the still infalling *outer core*, it photodisintegrates its Fe nuclei (tearing them down to nucleons) and consequently loses ~8.8 MeV/ nucleon or 1.5×10^{51} erg per 0.1 $M_{\odot}$.

If the mass of the infalling outer core is $M_{OC} = M_{Fe} - M_{IC} < 0.4\ M_{\odot}$, the shock can reach the base of the envelope with enough energy to launch a successful explosion. Only in the smallest massive stars (below 13 $M_{\odot}$, with Fe core mass $M_{Fe} \sim 1.1\ M_{\odot}$) does the mechanism of *prompt explosion* have some chance to succeed. In more massive stars, the prompt explosion fails. One is left with a dense and hot protoneutron star, accreting matter at a rate 1–10 $M_{\odot}\ s^{-1}$. If no other energy source is involved, the star collapses to a black hole (at least in most numerical simulations).

It is currently believed that the supplementary energy source is provided by neutrinos, which are "trapped" in the core during the hydrodynamical collapse for a timescale of a few ms[12] and start diffusing outward after the shock is launched; the

[12] The mean free path of neutrinos becomes temporarily smaller than the dimensions of the collapsing core; however, after 0.1 s, neutrinos diffuse and escape from the "neutrinosphere."

proto-neutron star radiates its thermal content in the form of neutrinos, on a timescale of a few seconds. These neutrinos carry the largest fraction of the gravitational binding energy of the neutron star ($\sim 3 \times 10^{53}$ erg or $\sim$10% of its rest mass); if they manage to communicate a few percent of that energy in the stalled shock wave, they may induce a *delayed explosion*. The detection of about 20 neutrinos in the Kamiokande and IBM detectors on 1987 February 23 from supernova SN1987A in the Large Magellanic Cloud demonstrated that neutrinos are indeed copiously produced during the deaths of massive stars.

If the shock wave manages to reach the base of the stellar envelope with enough energy (i.e., a few 10^{51} erg, to account for the gravitational binding energy of the envelope plus the typical kinetic energy of observed SNII ejecta), the explosion is successful. As the shock wave propagates through the stellar envelope, it heats the various layers to temperatures higher than in the corresponding central burning stages. The peak temperature $T_P(r)$ at radius r can be obtained to a good accuracy by assuming that the kinetic energy of the shock KE is equal to the thermal energy in the radiation field behind the shock $E_{\mathrm{TH}} = \frac{4}{3}\pi r^3 a T_P^4(r)$ (where $a = 7.56 \times 10^{-15}$ erg cm^{-3} K^{-4} is the blackbody constant); this leads to

$$T_P(r) \sim 1.33 \times 10^{10} \left(\frac{KE}{10^{51}\ \mathrm{erg}}\right)^{1/4} \left(\frac{r}{10^8\ \mathrm{cm}}\right)^{-3/4} K. \tag{2.17}$$

In Figure 2.22, it can be seen that this formula gives a very good approximation to the "realistic" peak temperature profile (calculated numerically, right panel of Figure 2.22). Stellar material in the layers hit by the shock wave is heated to those temperatures for a time on the order of the local hydrodynamical timescale[13]

$$\tau_{\mathrm{HD}} \sim 0.446 \rho_6^{-1/2} s$$

where ρ_6 is the mean density interior to radius r in 10^6 g cm^{-3}.

If the nuclear burning timescale at radius r ($\tau_{\mathrm{nuc}} = q_{\mathrm{nuc}}/\dot{\epsilon}_{\mathrm{nuc}}$, see Section 2.5.2) is smaller than the hydrodynamical timescale $\tau_{\mathrm{HD}}(r)$, then *explosive nucleosynthesis* will take place, modifying the composition left from the previous quiescent burning stage. Taking into account the relatively steep density profiles of pre-supernova stars, it turns out that peak temperatures are sufficient for vigorous explosive Si and O burning; the Ne and C layers are less affected (see Figure 2.23) while the He and H layers are not affected at all by the explosion (envelope densities are too low for explosive burning to take place in those layers).

The major reactions in explosive burning for a given fuel are the same as those of the corresponding quiescent burning (i.e., Figures 2.14, 2.15, and 2.16). However, the beta decay lifetimes of unstable nuclei are often larger than the timescales of explosive nucleosynthesis τ_{nuc}. This implies that the cross sections of nuclear reactions on those unstable nuclei are also required, a situation that is not usually encountered in quiescent burning.

---

[13] τ_{HD} is essentially the freefall timescale, i.e., it is assumed that the explosion is "symmetric" in time to the collapse.

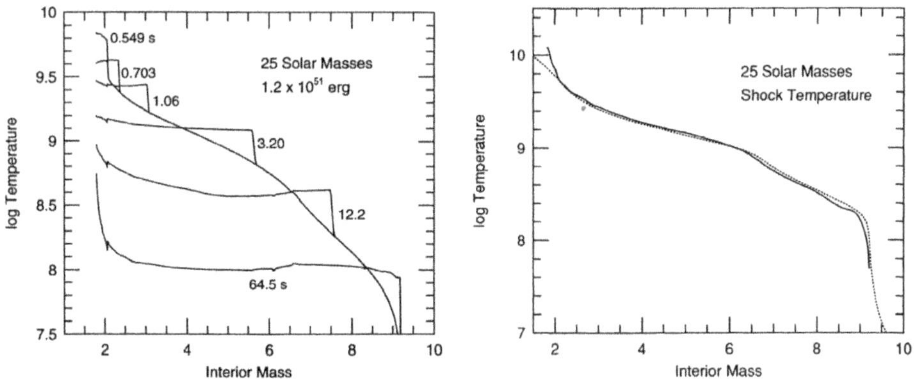

Figure 2.22. Left: Initial temperature profile (*continuous curve*) and snapshots of the temperature profile at different times after the passage of the shock wave in the mantle of a 25 $M_\odot$ star. Right: Peak temperature profile after the passage of the shock wave in the mantle of a 25 $M_\odot$ star, obtained numerically (*solid curve*) and analytically (*dotted curve*), i.e., by assuming adiabatic expansion of the mantle. (Credit: Woosley & Weaver 1995.)

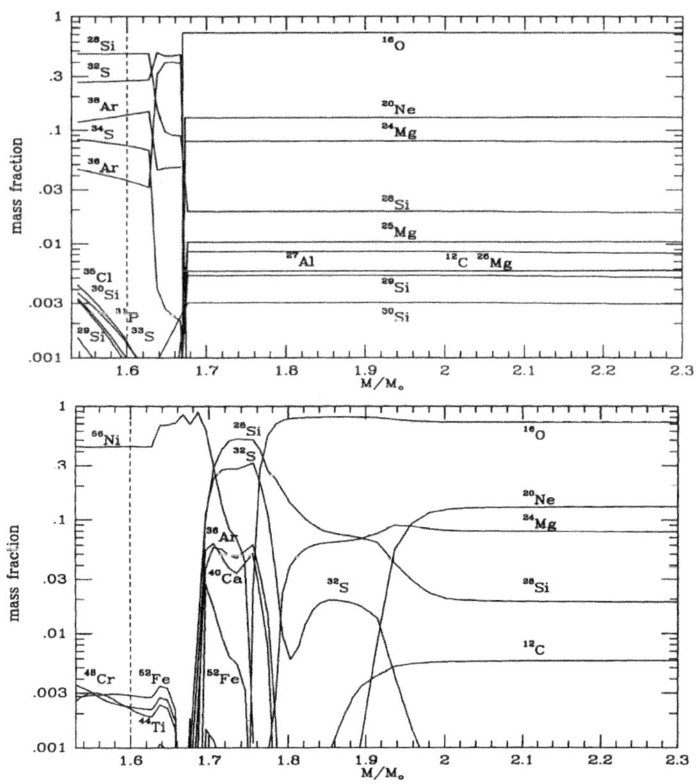

Figure 2.23. Top: Chemical composition profile of the innermost 2.3 $M_\odot$ of a 20 $M_\odot$ star prior to core collapse, with the major products of *hydrostatic* nucleosynthesis; uniform composition (flat abundance profiles) characterize former convective zones. Bottom: Chemical composition profile of the inner 2.3 $M_\odot$ of a 20 $M_\odot$ star after the passage of the shock wave with the major products of *explosive* nucleosynthesis dominating the innermost 2 $M_\odot$. (Credit: Thielemann et al. 1996.)

The products of explosive nucleosynthesis in massive stars have been studied in detail (Woosley & Weaver 1995; Thielemann et al. 1996; Chieffi & Limongi 2013; Nomoto et al. 2013). In general, one may distinguish three classes of elements/isotopes coming out of the explosion (see Prantzos 2000):

(1) The first set (N, C, O, Ne, Mg) results mainly from hydrostatic He and C/Ne burning (layers not affected by explosive nucleosynthesis). The *yields* (ejected mass) of those elements increase with stellar mass, as do the He- and C-exhausted cores. A comparison of those yields to observations (see below) allows us to test essentially the physics of the pre-supernova models.

(2) The second group (Al, Si, S, Ar, Ca) includes elements produced in both hydrostatic and explosive burning. Their yields vary less with progenitor mass than the ones of the previous group. Comparison to observations tests both pre-supernova models and explosion energy.

(3) The third set involves elements of the Fe peak, produced essentially by explosive Si burning. Their yields are highly uncertain at present, because they depend on the mechanism of the explosion, i.e., the shock wave energy, the various mixing processes and the "mass-cut" (the dividing line between the mass of the star finally ejected and the one that falls back to the compact object).

How can the validity of the theoretical stellar yields be checked? Ideally, individual yields should be compared to abundances measured in supernova remnants of stars with known initial mass and metallicity! However, such opportunities are extremely rare. In the case of SN1987A, theoretical predictions for a 20 $M_\odot$ progenitor are in rather good agreement with observations of C, O, Si, Cl, and Ar (Thielemann et al. 1996). SN1987A also allowed the Fe yield (~0.07 $M_\odot$) to be calibrated using the optical light curve (powered at late times by the decay of ^{56}Co, the progeny of ^{56}Ni), extrapolated to the moment of the explosion (e.g., Arnett et al. 1989). More recently, observations of extragalactic supernovae have allowed a rather good correlation between the peak luminosity (i.e., the ^{56}Ni mass) and the kinetic energy of the explosion to be established.

2.6.3 Explosive Nucleosynthesis in Thermonuclear SN

It is widely accepted now that SNIa are thermonuclear explosions of white dwarfs (WD). Because a single WD is eternally stable, a companion star is required. The currently leading model invokes a carbon–oxygen WD close to the Chandrasekhar mass-limit that accretes mass from a companion star, evolved or not (Figure 2.24). Another, less popular, scenario involves merging of two WD, after orbital decay (due to energy loss by gravitational wave emission) in a binary system. Although no self-consistent model currently exists, the former scenario seems to be in much better agreement with observations.

The accretion scenario requires a relatively well-adjusted accretion rate, $\sim 10^{-7}$ $M_\odot\ yr^{-1}$ for several Myr. As the mass of the WD increases, compressional heating rises the internal temperature to several 10^8 K. The high density of the WD ($\sim 10^9$ g cm^{-3})

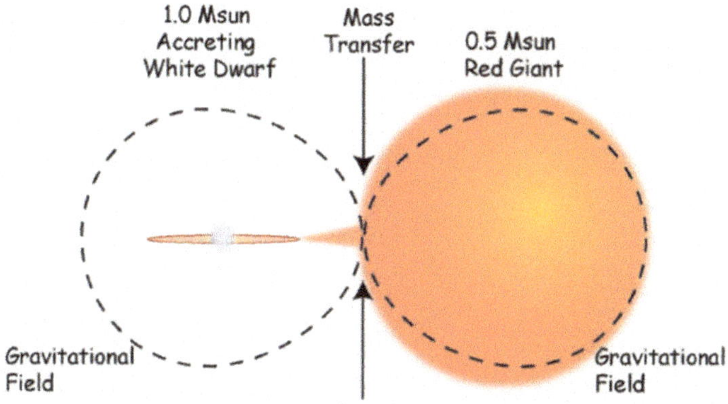

Figure 2.24. Mass transfer in a binary system, progenitor of an SNIa (courtesy F. Timmes).

leads to a very efficient *screening* of the nuclear reactions[14]; $^{12}C+^{12}C$ fusion then occurs in considerably lower temperatures and at much higher rates than in massive stars. Ignition occurs when the $^{12}C+^{12}C$ reaction releases energy faster than neutrino losses (mostly by plasma process; see Section 2.5.1) can carry it away.

Contrary to the case of normal stars (where perfect gas dominates the internal pressure), WD are composed of degenerate gas. Its pressure depends only on density,[15] not on temperature. Thus, heating of the medium (through the energy released by $^{12}C+^{12}C$) is not accompanied by pressure increase, which would lead to gas expansion and cooling; instead, temperature increases steadily, as does the $^{12}C+^{12}C$ reaction rate. In those conditions, a "carbon deflagration" explosively burns the material of the star, all the way toward NSE. The timescale of the explosion is too short to allow for important electron captures, and thus the final outcome is essentially radioactive ^{56}Ni (see Figure 2.17): about 0.7 $M_{\odot}$ of this nuclide are produced (about half of the WD mass), as required by the observed luminosity of SNIa. Only in the central regions, the high densities favor the production of ^{54}Fe and ^{58}Ni through electron captures; the overproduction of those nuclei is one of the current shortcomings of this scenario. In the outer layers, NSE is not reached because of lower temperatures that lead to the production of intermediate mass nuclei, such as Si, Ca, etc.; again, the simplest scenario currently produces insufficient amounts of those nuclei, compared with observations.

The C-deflagration scenario is the most successful on the nucleosynthesis side, but other scenarios concerning delayed-detonation Chandrasekhar-mass explosions, sub-Chandrasekhar-mass detonations, and double-degenerate mergers are also investigated (e.g., Sasdelli et al. 2017, and references therein). From the perspective of galactic chemical evolution, SNIa play a very important role because they produce between 50% and 65% of the Fe-peak nuclei. Their smaller frequency (they

[14] Because of the high electron density, carbon nuclei "feel" a much smaller Coulomb repulsion; as a result, their fusion rate is considerably enhanced.

[15] In non-relativistic degenerate gases, the pressure scales with density as $P \propto \rho^{5/3}$, while in relativistic degenerate gases, $P \propto \rho^{4/3}$.

are ~5 times less frequent than core collapse SN) is compensated by the larger amounts of ^{56}Ni and Fe-peak nuclei produced by their explosion (0.7 $M_{\odot}$ of ^{56}Ni compared to 0.1 $M_{\odot}$ on average for core collapse SN).

2.6.4 Production of Intermediate-mass Nuclei (From C to Fe peak)

In general, it is (and will continue to be) very difficult to test models of stellar nucleosynthesis on a star-by-star basis. However, we already know that, in a statistical sense, current models are not far from reality. Indeed, adopting current stellar yields (of intermediate and massive stars, as well as of SNIa), an appropriate stellar initial mass function (IMF), and running galactic chemical evolution models for the solar neighborhood, one finds that the abundances of elements between C and Zn at solar system formation are satisfactorily reproduced (see Figure 2.25). This is, at present, the most convincing global test of our theory of stellar nucleosynthesis. Still, observations of abundance ratios in old stars of the Milky Way's halo or in remote gaseous systems (formed early in the history of the universe and enriched predominantly by short-lived massive stars) present new challenges to our understanding of those objects.

2.7 The Heavier-than-Fe Nuclei

Nuclei heavier than those of the Fe peak (henceforth called "heavy nuclei") cannot be produced by charged particle reactions: the temperatures needed to overcome the high Coulomb barriers are such that photodisintegrations would tear down all nuclei

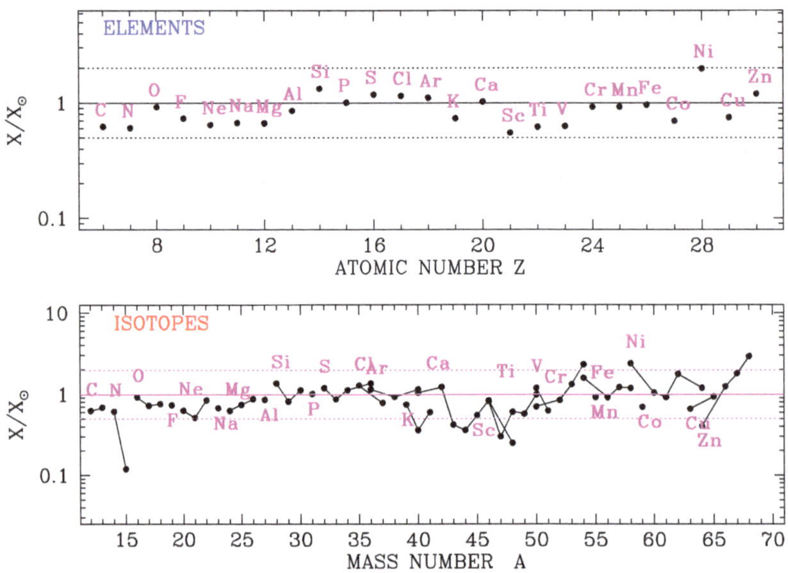

Figure 2.25. Abundances at solar system formation (4.5 Gyr ago) obtained with a galactic chemical evolution model appropriate for the solar neighborhood and compared to the solar ones. Yields from intermediate and massive stars, as well as SNIa, are used. It can be seen that all elements and isotopes up to the Fe peak are nicely co-produced (within a factor of 2); another source (novae?) is needed for ^{15}N. (Credit: Goswami & Prantzos 2000.)

to their constituent particles. Thus, nature has found another mechanism: neutron captures on pre-existing "seed" nuclei.

2.7.1 Production Mechanisms and Classification of Isotopes

In the early works of nucleosynthesis, it was recognized that two different types of neutron capture processes have contributed to the formation of heavy nuclei: the so-called s-process (for *slow*), responsible for the formation of nuclei in the valley of nuclear stability; and the r-process (for *rapid*), responsible for the synthesis of the most neutron-rich isotopes of heavy elements (e.g., Rolfs & Rodney 1988, and references therein).

In the s-process, low neutron densities ($\sim 10^8$ cm^{-3}) and rather long timescales (>1yr) are involved, such that when an unstable nucleus (A, Z) is encountered along the s-process path (Figure 2.26), it has the time to decay into a stable isobar ($A, Z+1$). Indeed, the lifetimes of most of the unstable nuclei close to the valley of stability range from a few seconds to less than a year and are lower than their lifetimes with respect to neutron captures in the above conditions ($\tau_n = (n_n \sigma_n v)^{-1}$, where n_n is the neutron density, σ_n the neutron capture cross section, and v the average neutron velocity). The neutron captures are *slow* with respect to beta decay, and thus the s-process path closely follows the bottom of the nuclear stability valley. Note that the s-process, by its very nature, cannot reach the heaviest nuclei (the long-lived Th and U isotopes), which are separated from the heaviest stable nucleus (^{209}Bi) by more than 20 mass units.

On the contrary, the r-process involves huge neutron densities (>10^{20} cm^{-3}) for timescales of the order of $\sim$1 s (obviously corresponding to an explosive site):

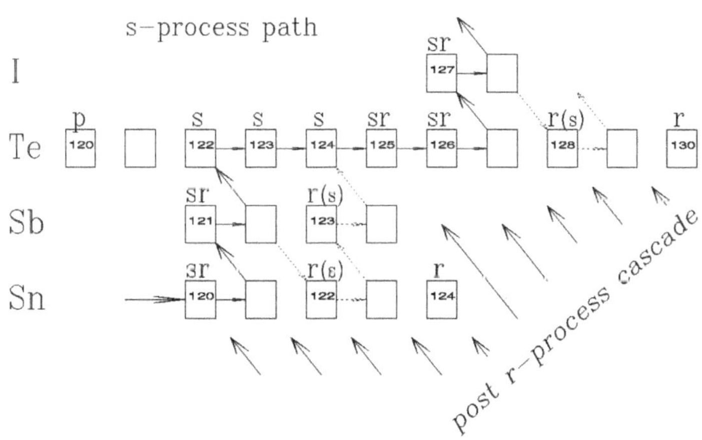

Figure 2.26. Paths in the plane of neutron number N versus proton number Z, leading to the formation of nuclei heavier than those of the Fe peak. Horizontal arrows indicate neutron captures and diagonal ones beta decays (β^-). For pure (or mostly) s-nuclei, the path lies close to the valley of nuclear stability. The most neutron-rich stable nuclei (r-nuclei) are produced by the beta decay of very neutron-rich unstable progenitors, formed by a rapid exposure to high neutron fluxes. The proton-rich p-nuclei cannot be reached by neutron captures. (Credit: Arnould & Takahashi 1999.)

neutron captures are so *rapid* with respect to beta decays that the r-process path drives the stellar material far to the neutron-rich side of the stability valley. When the neutron burst ceases, the unstable nuclei that have been formed decay back to the stability valley. The most neutron-rich nuclei of the heavy elements (pure r-nuclei) are thus formed, as well as the heaviest elements (Th and U, which cannot be reached by the s-process). At the same time, many nuclei that can also be formed by the s-process are synthesized, so the abundances of the majority of the heavy nuclei have contributions from both the s- and the r-processes, i.e.,

$$N(A, Z) = N_s(A, Z) + N_r(A, Z).$$

Obviously, for pure s-nuclei, $N_r(A, Z) = 0$. The abundances of the heavy nuclides, classified as s-, r-, or p- are displayed in Figure 2.27. The three peaks in the s-abundance curve correspond to nuclei with "magic" neutron numbers ($N = 50$, 82, 126; see Figure 2.4). They are accompanied by less sharp peaks in the r-abundance curve, occurring at smaller mass number A. The total amount of each of the two classes (s- and r-) of heavy nuclide abundances is $\sim 10^{-6}$ by mass; taking into account the very different production mechanisms, this is a rather strange coincidence.

The p-nuclei are much less abundant, by two or three orders of magnitude, than either the corresponding s- or r-isotopes (an exception being ^{92}Mo, which constitutes 14% of Mo). These isotopes are not reached by neutron captures on stable nuclei (Figure 2.26) and are thought to be produced mainly by photodisintegration of s-nuclei. The most favorable site appears to be the O–Ne layers of core-collapse SN explosions, reaching peak temperatures of 2–3 $\times 10^9$ K.

2.7.2 The s-process

An analysis of the solar abundances (Figure 2.27) and neutron capture cross sections (Figure 2.28) of nuclei along the s-process path allows us to determine the relevant

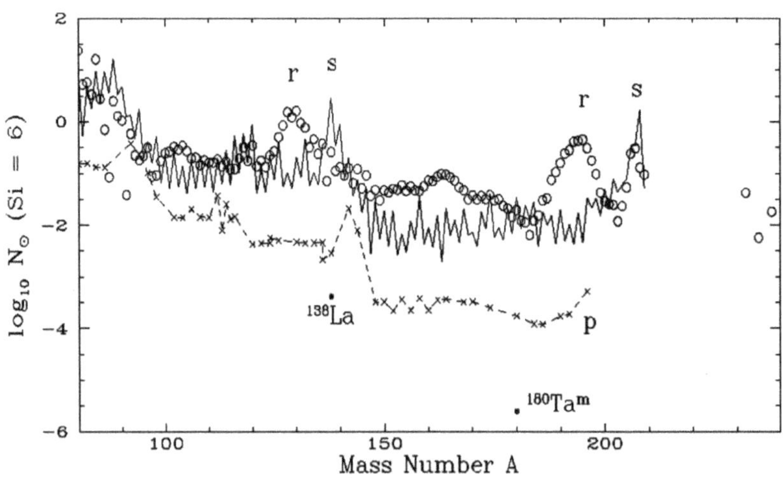

Figure 2.27. Solar system abundances of heavy nuclei classified as s-, r-, and p- (see Figure 2.26); ^{180}Ta (probably produced by the p-process) is the rarest isotope in nature. (Credit: Arnould & Takahashi 1999.)

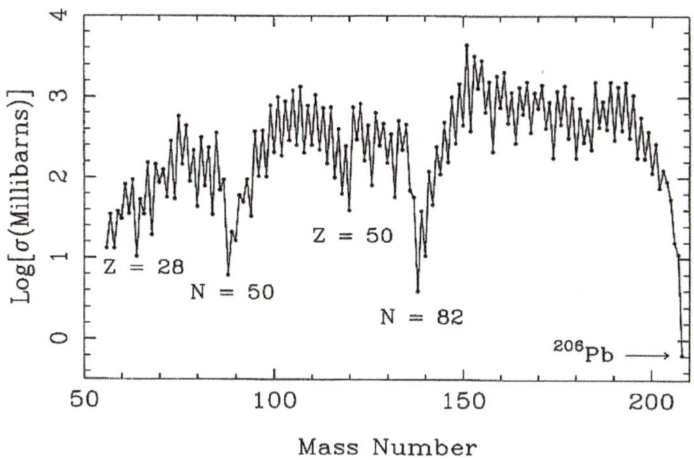

Figure 2.28. Neutron capture cross-sections of s-nuclei as a function of mass number A. Nuclei with "magic" neutron numbers (50, 82, 126) have extremely low cross sections and high abundances with respect to their neighbors (see Figure 2.27). ^{208}Pb is a representative case, with $Z = 82$ and $N = 126$.

physical conditions: neutron densities $n_n \sim 10^8$ cm^{-3}, temperatures $T \sim 1.5–3 \times 10^8$ K, densities $\rho \sim 2–12 \times 10^3$ g cm^{-3}, and timescales $\tau_S \sim 10–10^3$ yr. These conditions correspond to the phase of quiescent He burning. Two relevant sites have been identified:

(1) Central He burning in massive stars, where the neutron source is ^{22}Ne(α,n); this reaction operates toward the end of He burning, at $T \sim 2.5 \times 10^8$ K. The number of neutrons released allows to synthesize only nuclei in the mass range $60 < A < 90$ (the so-called "weak" component of the s-process). Part of those nuclei survive the advanced evolutionary phases of massive stars and are ejected in the interstellar medium by the final supernova explosion. From a theoretical perspective, the He-core s-process is quite robust (despite the uncertainties in the ^{22}Ne(α,n) rate), but it lacks direct observational support.

(2) Double shell (H- and He-) burning in low-mass stars, evolving in the asymptotic giant branch (AGB) and running very close and parallel to the red giant branch. During the thermal pulse phase (Figure 2.29), protons from the H-shell are mixed in regions rich in α and ^{12}C (produced from the $3\alpha \longrightarrow {}^{12}$C in the He-shell). The coexistence of p,α and ^{12}C allows the production of abundant neutrons through the reaction series p+^{12}C$\longrightarrow$^{13}N, ^{13}N(β^+)^{13}C, ^{13}C(α,n)^{16}O. The recurring pulses allow the production of a large number of neutrons per seed Fe-nucleus and the synthesis of the "main component" of the s-process, i.e., nuclei with $90 < A < 210$. Between thermal pulses, the convective stellar envelope penetrates deep in the star and "dredges up" material from the s-processed region to the surface, from where the strong AGB wind expels s-nuclei to the interstellar medium. This rather complex scenario is supported by observations of enhanced s-abundances in the surfaces of low-mass AGB stars. However, the mixing

of protons with material from the He-zones, as well as the 3D dredge up, are poorly understood at present (see Käppeler et al. 2011, and references therein).

2.7.3 The r-process

For sufficiently high neutron densities, successive neutron captures convert seed nuclei of a given element to very neutron-rich ones. This continues until nuclei with very low neutron separation energies S_n are encountered (Figure 2.30). At this point,

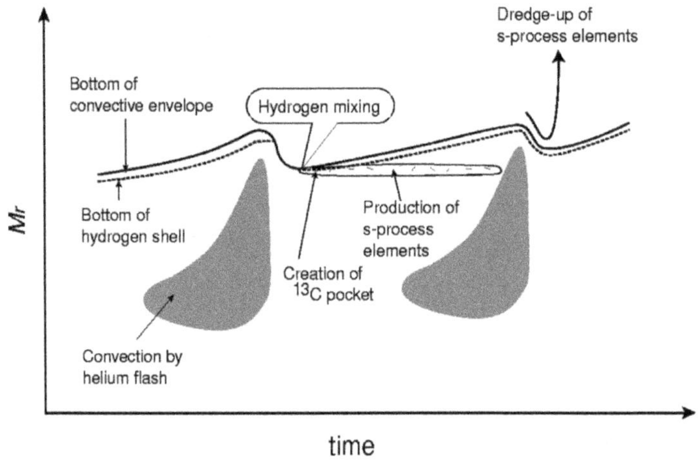

Figure 2.29. Structure of a thermally pulsing AGB star. After a thermal pulse, protons from the H-shell are mixed with He-shell material (rich in ^{4}He and ^{12}C). This leads to neutron production through ^{12}C(p,γ)^{13}N (β^+)^{13} C(α,n) and s-process nucleosynthesis. After the next thermal pulse, s-nuclei are brought to the stellar surface by the convective envelope and expelled in the interstellar medium.

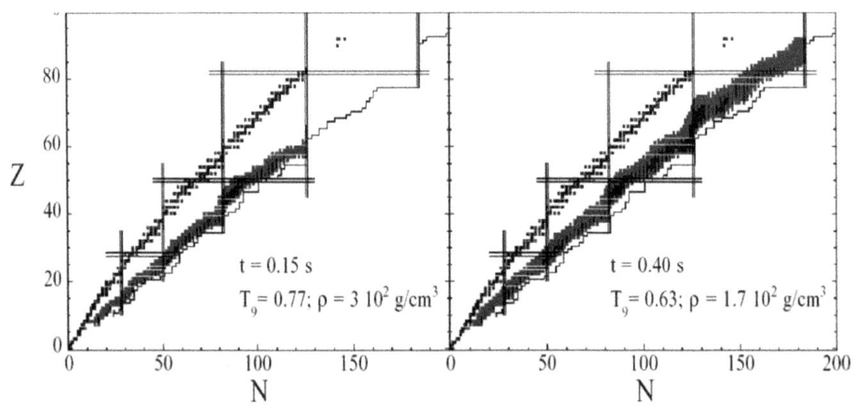

Figure 2.30. Nuclear flow in a calculation of the "dynamical" r-process, at two different instants. Neutron captures drive material away from the nuclear stability valley, toward the neutron-drip line (where neutron separation energy S_n = 0; represented by the curve on the right of the diagrams). Straight lines indicate "magic" nuclei. (Credit: Arnould et al. 2007.)

the inverse reactions (γ,n) balance (n,γ) reactions. In the simplest picture of the r-process, a (n,γ)$\leftrightarrow$(γ,n) equilibrium is reached in each isotopic chain. Analogously to the NSE (see Section 2.5.3), the isotopic abundances of a chain of given Z depend then on S_n, temperature T, and neutron density n, as well as on nuclear partition functions. In those conditions, a large amount of material is accumulated in unstable neutron-rich nuclides with "magic" neutron numbers. Material slowly shifts to higher Z values through beta decays, which govern the speed of the nuclear flow along the r-process path. At the end of the neutron irradiation, all unstable nuclei beta decay toward the nuclear stability valley. The peaks in the r-abundance curve (shifted by a few mass units with respect to the corresponding s-abundance peaks) are obtained from the beta decay of very neutron-rich material "stored" in nuclei with "magic" neutron numbers during the r-process.

Even for the simplest r-process model, a large body of nuclear data is required (masses and beta decay rates of very unstable neutron-rich nuclei). More sophisticated models drop the assumption of (n,γ)$\leftrightarrow$(γ,n) equilibrium and also require neutron capture and photodisintegration rates, along with very large nuclear reaction networks (involving several thousands of nuclei).

The site of the r-process is not yet identified, despite more than thirty years of intense investigation. Required neutron densities and timescales suggest an explosive event, while observations of the composition of the oldest stars in the Milky Way suggest that the r-nuclei were produced early on, i.e., that their source was short-lived. Core-collapse SN satisfy both those requirements, but neither the neutron source nor the relevant stellar layers have yet been identified in that site. In recent years, the case of neutron star mergers appears to be more promising in that respect, because the solar distribution of r-nuclei results "naturally" in that site. However, the rate of that source is not well-understood at present, and more work is needed before its role is confirmed. The r-process remains the last poorly understood nucleosynthetic process in nature (see Arnould et al. 2007, and references therein).

2.8 Summary

The basic picture of stellar nucleosynthesis is well-established nowadays, at least qualitatively. It can be summarized as follows:

Massive stars (M > 10 $M_\odot$) synthesize in their various burning layers elements from C to Ca and eject them into the interstellar medium in the final SN explosion. During the explosion, a large fraction of Fe-peak isotopes (from Ti to Zn) is also produced.

Intermediate-mass stars (1.5–7 $M_\odot$) produce important amounts of He and CNO isotopes (except ^{16}O, ^{18}O, and ^{15}N), as well as nuclei heavier than Fe peak that lie on the nuclear stability valley (s-nuclei, produced by neutron captures). All those nuclei are convected to the surface of the star and ejected into the interstellar medium through the stellar winds, mostly during the AGB and planetary nebula phases.

The nuclei heavier than Fe peak that lie relatively far from the nuclear stability valley are thought to be produced during the explosions of massive stars, either in stellar layers submitted to large neutron fluxes (neutron-rich or r-nuclei) or in layers where photodisintegrations of already synthesized s-nuclei play an important role (proton-rich or p-nuclei); the former process is also responsible for the formation of the heaviest stable nuclei, the Th and U isotopes.

Finally, a large fraction of the Fe-peak nuclei (>50%) is synthesized in thermonuclear SN (SNIa), white dwarfs in binary systems, that explode upon reaching the Chandrasekhar limit M_{Ch} (through accretion from a companion star).

Taking into account the lifetimes (a few 10^6–10^7 yr for massive stars, a few 10^8–10^9 yr for intermediate-mass stars, and $\sim 10^9$ yr for the average SNIa), as well as the relative numbers of the various nucleosynthesis sites (the so-called "stellar initial mass function," which favors the less massive stars), one may construct models of "galactic chemical evolution"[16] and compare their results to various observables.

The success of this scenario lies mainly in its ability to reproduce (within reasonable assumptions) the solar distribution of intermediate-mass elements (C to Fe peak; see Figure 2.25) and s-elements. It also explains reasonably well most of the abundance ratios of such elements that are observed on the surfaces of stars of various ages in the Milky Way.

However, several open questions remain, e.g., the proton mixing at the origin of the $^{13}C(\alpha,n)$ neutron source for the s-process in AGB stars, the stellar layers of the core-collapse SN where the r-process takes place, etc. Furthermore, uncertainties in key nuclear data (e.g., the $^{12}C(\alpha, \gamma)$ rate for He burning, the nuclear masses for the r-process) or in important stellar physics processes (e.g., mixing processes in the stellar core and envelope, the explosion mechanism in core collapse SN, the role of rotation, mass loss, and even magnetic fields in overall stellar evolution, etc.) prevent us from a satisfactory quantitative picture at present. Progress is expected from continuous improvements in the fields of numerical simulations, nuclear data, and spectroscopic observations in all wavelengths.

The theory of stellar nucleosynthesis is certainly important to exobiology studies, to the extent that life (at least as we currently understand it) requires the presence of various heavy elements, e.g., oxygen, carbon, etc. However, a quantitative assessment of that requirement is difficult at present: the absolute amount of heavy elements (i.e., their abundance with respect to H) that is required for the formation of telluric planets is very poorly constrained, either by theory or by observations.

References

Arnett, W. D., Bahcall, J. N., Kirshner, R. P., & Woosley, S. E. 1989, ARA&A, 27, 629
Arnould, M., Goriely, S., & Takahashi, K. 2007, PhR, 450, 97
Arnould, M., & Takahashi, K. 1999, RPPh, 62, 395
Asplund, M., Grevesse, N., Sauval, A. J., & Scott, P. 2009, ARA&A, 47, 481
Bethe, H. A. 1939, PhRv, 55, 434
Bethe, H. A., & Critchfield, C. L. 1938, PhRv, 54, 248
Burbidge, E. M., Burbidge, G. R., Fowler, W. A., & Hoyle, F. 1957, RvMP, 29, 547
Cameron, A. G. W. 1957, PASP, 69, 201

[16] "Chemical evolution," in this context, means "the evolution of the abundances of the various chemical elements."

Chieffi, A., & Limongi, M. 2013, ApJ, 764, 21
Coc, A. 2016, JPhCS, 665, 012001
Colgate, S. A., & White, R. H. 1966, ApJ, 143, 626
Cyburt, R. H., Fields, B. D., Olive, K. A., & Yeh, T.-H. 2016, RvMP, 88, 015004
Goswami, A., & Prantzos, N. 2000, A&A, 359, 191
Hartmann, D., Woosley, S. E., & El Eid, M. F. 1985, ApJ, 297, 837
Janka, H.-T. 2017, in Handbook of Supernovae, ed. A. Alsabti, & P. Murdin (Cham: Springer)
Käppeler, F., Gallino, R., Bisterzo, S., & Aoki, W. 2011, RvMP, 83, 157
Limongi, M., Straniero, O., & Chieffi, A. 2000, ApJS, 129, 625
Lodders, K., Palme, H., & Gail, H.-P. 2009, in Landolt Börnstein (Berlin: Springer)
Maeder, A., & Meynet, G. 2012, RvMP, 84, 25
Nomoto, K., Kobayashi, C., & Tominaga, N. 2013, ARA&A, 51, 457
Prantzos, N. 2000, NewAR, 44, 303
Rolfs, C. E., & Rodney, W. S. 1988, Cauldrons in the Cosmos: Nuclear Astrophysics (Chicago, IL: University of Chicago Press)
Sasdelli, M., Hillebrandt, W., Kromer, M., et al. 2017, MNRAS, 466, 3784
Thielemann, F. K., & Arnett, W. D. 1985, ApJ, 295, 604
Thielemann, F.-K., Nomoto, K., & Hashimoto, M.-A. 1996, ApJ, 460, 408
Weizsäcker, C. F. V. 1935, ZPhy, 96, 431
Woosley, S. E. 1986, Saas-Fee Advanced Course 16: Nucleosynthesis and Chemical Evolution, ed. J. Audouze, C. Chiosi, & S. E. Woosley, 1
Woosley, S. E., Arnett, W. D., & Clayton, D. D. 1973, ApJS, 26, 231
Woosley, S. E., Heger, A., & Weaver, T. A. 2002, RvMP, 74, 1015
Woosley, S. E., & Weaver, T. A. 1995, ApJS, 101, 181

Chapter 3

Gas-phase Chemistry: Reactive Bimolecular Collisions

Sébastien Le Picard, Ludovic Biennier, Maurice Monnerville and Hua Guo

3.1 Introduction

Context. The interstellar medium (ISM) is an extremely diluted environment. In molecule-bearing clouds, the density essentially ranges from 10^2 to 10^6 molecules per cubic centimeter. Consequently, the collisional frequency, i.e., the average rate at which two reactants collide, is low and the vast majority of gas-phase reactions occur as the result of binary collisions, also called bimolecular reactions. Three-body collisions are therefore extremely unlikely in the ISM except at the surface of grains.

When the two colliding molecules react to form one or more new molecules, the collision is said to be "reactive." Current networks for astrochemical models contain a few thousands of gas-phase reactions involving a few hundreds of atomic and molecular species, which comprise neutral, positively, and negatively charged species. Ion and electron chemistry dominates the chemical evolution in many areas; however, in environments of very low ionization fraction, the neutral–neutral reactions are of major importance, especially when they involve radicals, because they tend to have larger rates when temperature decreases.

Even when considering only the gas phase, the chemistry taking place in the interstellar medium is highly complex, as many individual reactive and non-reactive processes are at play. In order to provide a quantitative understanding of the gas-phase chemistry, a detailed knowledge of elementary reactions is required. Basically, what we wish to know is how the two colliding species of a bimolecular reaction interact with each other and evolve into products forming new chemical species. The key quantitative data needed in astrochemical models therefore are the intrinsic rates and the proportion of each accessible product channel of the chemical reactions involved, at the low temperatures (down to less than 10 K) prevailing in the ISM.

doi:10.1088/2514-3433/aae1b5ch3

Kinetic parameters. Let us define the corresponding kinetic parameters, which are, for each reaction, the **total rate coefficient** and its **branching ratio**. A "forward" bimolecular reaction leading from R_1 and R_2 reactants to P_1 and P_2 products can be represented by the equation

$$R_1 + R_2 \rightarrow P_1 + P_2. \tag{3.1}$$

The rate of this reaction is proportional to the reactant concentrations, $[R_1]$ and $[R_2]$, the proportionality constant being the bimolecular rate coefficient, $k(T)$, where (T) is included to emphasize that rate coefficients generally depend on temperature:

$$-\frac{d[R_1]}{dt} = -\frac{d[R_2]}{dt} = +\frac{d[P_1]}{dt} = +\frac{d[P_2]}{dt} = k(T)[R_1][R_2]. \tag{3.2}$$

Here, the square brackets denote concentrations in units of molecule cm^{-3}, or simply cm^{-3}, so that $k(T)$, the rate constant or rate coefficient for the reaction between R_1 and R_2, has units of cm^3 $molecule^{-1}$ s^{-1}, or simply cm^3 s^{-1}.

A bimolecular reaction can lead to the formation of several product channels. Let us consider the simple case where two exit channels are accessible for the reaction:

$$A + B \xrightarrow{k_{C+D}} C + D, \tag{3.3}$$

$$\xrightarrow{k_{E+F}} E + F. \tag{3.4}$$

The total rate coefficient of the reaction is $k(T) = k_{C+D}(T) + k_{E+F}(T)$ and the branching ratio for the product channels leading to C + D and E + F are defined respectively by the following equations

$$BR(C + D) = \frac{k_{C+D}(T)}{k(T)}, \tag{3.5}$$

$$BR(E + F) = \frac{k_{E+F}(T)}{k(T)}. \tag{3.6}$$

For most reactions, the total rate coefficient and their branching ratio, as well as their temperature dependence, are not readily accessible with accuracy, either experimentally or theoretically. Therefore, only a fraction of the total rate coefficients of the reactions involved in astrochemical models, as well as a very small proportion of their branching ratios, have been studied either experimentally or theoretically in such extreme conditions of temperature.

Challenges. From an experimental point of view, the first difficulty lies in the generation of a gas at very low temperature, without significant condensation or clustering of the molecular species present in the reactor. In addition, absolute concentration determination of molecular or radical species, often formed in small amounts, in some cases with short lifetimes, are needed to derive branching ratios.

Such absolute concentrations are not readily accessible experimentally. Furthermore, many of the reactions of astrophysical interest involve the collision of two radicals, which is particularly challenging because their study requires both the production of one of the radicals in excess and the measurement of its concentration, in order for absolute rate coefficients of the reaction to be extracted.

From a theoretical point of view, because the reactions of astrochemical interest frequently involve either ions or free radicals as reactants and often form reaction intermediates before decomposing to products, an accurate calculation of the corresponding low-temperature rate coefficients is quite challenging, particularly because certain quantum effects are magnified in cryogenic conditions. In particular, tunneling underneath a classical barrier is enabled in quantum mechanics by the wave-particle duality of the atom, particularly one with a small mass. In addition, molecular vibrational modes have intrinsic energies even at the lowest energy level (zero-point energy or ZPE), due to the uncertainty principle, and they can strongly affect near-threshold reactivity. Finally, resonances are metastable quantum states embedded in the continuum, which could greatly enhance the reaction rate in a narrow energy range. The proper inclusion of these quantum effects, which have no classical analogs in theoretical treatments remains a challenge.

Content of this chapter. The purpose of this chapter is to give an overview of the current experimental and theoretical methods used to determine or estimate rate coefficients and branching ratios of elementary chemical reactions at the low temperatures of the interstellar medium ($\leqslant$ 100 K). This chapter is organized as follows. Section 3.2 introduces briefly some basic concepts of bimolecular reaction kinetics. Section 3.3 describes various experimental approaches dedicated to low-temperature reaction kinetic studies, for ion–molecule and neutral–neutral reactions. Section 3.4 describes state-of-the-art theoretical methods, including quantum and approximate quantum approaches.

3.2 Basics in Bimolecular Reaction Kinetics

In order to describe a chemical reaction between two atomic or molecular species, one needs to know how their nuclei and electrons interact during the collision, ideally at all distances and for all geometries. The interactions between nuclei during the collision are described by the potential energy surface (PES) of the reaction. Once known, it is possible to infer the probability for the chemical reaction to happen when the two species collide. This is described by the collision cross section of the reaction. The thermal rate coefficient of the reaction is then determined considering the number of atoms or molecules impinging each other per unit time per unit area as well as their relative velocity.

3.2.1 Interaction Potential—Potential Energy Surface

The rates of elementary reactions are driven by the intermolecular forces at play during the collision. Intermolecular forces are attractive at long distance and become strongly repulsive at shorter internuclear separations. As for a chemical bond, these interactions can be schematically depicted by a potential energy curve. The depth of the minimum, however, is typically two orders of magnitude smaller than for a chemical bond, and this minimum is found at much longer internuclear separation, typically a few Bohr radii ($a_0 \simeq 53$ pm).

To accurately describe a reactive process, the whole range of geometries, from separated reactants to the strong interaction region (at least up to a point from which back dissociation to the reactants is negligible), has to be covered. The values of the potential energy of the system are therefore represented by a multidimensional surface. When the fragments approach, their electronic clouds interact until a common electronic cloud is formed. In addition, to characterize the products of the reaction, the PES must also cover the range of geometries associated with the separated products.

Generally, it is possible to separate the motion of the electrons from the motion of the nuclei, because the nuclei move more slowly due to their higher masses. This consideration is known as the Born–Oppenheimer or electronically adiabatic approximation. If one assumes that the nuclear motion proceeds without change in the quantum state of the electron cloud, the potential energy is only a function of the nuclear coordinates. The motion of nuclei and electrons is therefore considered to be separable. In this framework, the quantum dynamics study of a molecular process requires knowledge of the PES associated with the electronic state of interest for all geometries likely to be probed by the nuclei. The electronic energy can be computed ab initio by solving the electronic Schrödinger equation at fixed nuclear geometries. Nowadays, electronic structure theory is sufficiently reliable to generate the PES for small molecular systems within the chemical accuracy ($\leqslant 1$ kcal mol^{-1}, 1 kcal = 4.18 kJ). Two high-level ab initio methods are often used. One is the coupled cluster method with single, double, and perturbative triple excitations (CCSD(T)), while the other is the multireference configuration interaction (MRCI) method. For systems with more than three or four atoms, recently developed explicitly correlated methods, such as (R)CCSD(T)-F12x (x = a, b), are much less expensive in computational time and remain an accurate technique for mapping the short- and long-range multidimensional PESs of single configurational electronic states. The quality of the results is also determined by the choice of the basis set to build the atomic orbitals—it must be sufficiently large, but not too extended, in order to keep computation time reasonable: aug-cc-pVTZ, aug-cc-pVQZ, aug-cc-pV5Z (for augmented correlation-consistent valence triple, quadruple, or quintuple zeta). The choice of the method often depends on the problem at hand. Multireference methods, for instance, are usually required when forming/breaking bonds, or even at particular minima when the electronic structure is not well-described by a single configuration.

It is worth noting that, for many bimolecular reactions, particularly reactions involving atoms, radicals, ions, or electronically excited reactants or products, the Born–Oppenheimer (adiabatic) approximation is no longer valid. Consequently, the dynamics of such reactions does not proceed on a single potential energy surface. This means that more than one PES drives the reaction dynamics, and although quantum mechanical methods have been developed, they are very costly and only a few atom–diatom reactions were computed using multiple-PES quantum scattering calculations.

Once an ab initio PES has been obtained, an analytical representation of the PES has to be built that will be used to derive the kinetics parameters of the reaction. Fitting procedures therefore must be developed and care has to be taken to keep the benefits of the accuracy of the PES calculation. In particular, it is important to carefully extend the analytical PES values to the long-range part for deriving dynamic parameters at low temperatures. Significant progress has been made in analytical representation of the PES from a large number of ab initio points (Braams & Bowman 2009; Manzhos et al. 2015; Jiang et al. 2016). One of the important issues in representing PESs is the permutation symmetry, which guarantees the invariance of the PES under exchange of identical atoms. Permutation-invariant PESs can be constructed by either interpolation or quasi-interpolation methods. The latter is typically more efficient for large polyatomic systems.

3.2.2 Cross Sections and Thermal Rate Coefficients

Once the PES has been determined, the collisional cross sections and rate coefficients are derived from the solution of the nuclear Schrödinger equations.

The concept of collision cross section can be easily apprehended when considering that the two particles colliding are hard spheres with no further interactions. Consider a spherical molecule A of diameter d_{A} traveling through a stationary gas of spherical molecules B of diameter d_{B}. A circular area can be defined in which every trajectory of molecule A will lead to a collision with molecule B, provided it lies within this area. This circular collision cross section has a diameter $d_{\mathrm{AB}} = \frac{1}{2}(d_{\mathrm{A}} + d_{\mathrm{B}})$ and an area

$$\sigma_{\mathrm{AB}} = \pi d_{\mathrm{AB}}^2. \tag{3.7}$$

In this very simple model, it is stipulated that, as long as the distance between the two particles is larger than d_{AB}, there is no interaction and therefore no chemical reaction. At all shorter distances, the reaction occurs.

The thermal rate coefficient of the reaction is defined as the product of the magnitude of the relative velocity v_r of the two partners by the cross section:

$$k(T) = v_r \sigma_{\mathrm{AB}}. \tag{3.8}$$

In this expression, the magnitude of the relative velocity between the two molecules can be expressed using the kinetic theory of gases via the following expression:

$$v_r = \left(\frac{8k_B T}{\pi \mu_{\mathrm{AB}}}\right)^{1/2}, \tag{3.9}$$

where k_B is the Boltzmann constant and μ_{AB} is the reduced mass between A and B.

In reality, the reaction cross sections will depend on both the quantum states of the colliding partners and their relative velocities. The state-to-state rate coefficient is related to the state-to-state cross section by the following equation:

$$k_{i,f}(v_r) = v_r \sigma_{i,f}(v_r), \tag{3.10}$$

where i and f subscripts refer to initial and final states.

Finally, the thermal rate coefficient, $k(T)$, is obtained by summing over the initial states weighted by the population of the reactant species in those initial states $p(i)$:

$$k(T) = \sum_i p(i)k_i(T), \tag{3.11}$$

where

$$k_i(T) = \sum_f k_{i,f}(T). \tag{3.12}$$

It is worth noting that the term *thermal* implies that the reactant species are in thermal equilibrium at a given temperature. Traditional astrochemical networks use thermal rate coefficients, and therefore assume that, on average, collisions involve species with a Boltzmann distribution of kinetic, rotational, and vibrational energies corresponding to a specific temperature.

As for the PES determination, the standard scattering theory takes place within the Born–Oppenheimer approximation for the separation of electronic and nuclear motions, considering the large ratio of the nuclear mass to the electronic mass. Scattering cross sections are obtained by solving for the motion of the nuclei on PES, which is independent of the masses and spins of the nuclei. First, the electronic Schrödinger equation is solved using a quantum chemical method. The nuclear motion is solved separately in a second step, using a quantum or (semi)classical scattering method. The most accurate one is quantum dynamics; the details of such an approach are discussed in Section 3.4.1 and in a recent review (Zhang & Guo 2016). Briefly, the reaction cross sections are first computed in an energy grid, and the thermal rate coefficient can then be obtained as follows:

$$k(T) = \frac{1}{k_B T}\sqrt{\frac{8}{\pi \mu_R k_B T}} \int e^{-E_c/k_B T}\sigma(E_c)E_c dE_c, \tag{3.13}$$

where E_c is the collision energy, μ_R the reduced mass between reactants, and k_B the Boltzmann constant. A grid has to be chosen with a given number of points, in a given collision energy range (fraction of meV to a few hundred of eV) with increasing steps. The convergence of the cross sections has to be checked. The deeper the potential wells on the PES, the more difficult is to get converged cross sections.

3.2.3 Reaction Mechanisms

It is worth noting that chemical reactions taking place in the interstellar medium are often classified in several types. A direct reaction involves the simultaneous breaking of an old bond and forming of a new one. Such reactions are often associated with a barrier. A direct reaction is unlikely to occur at low temperatures. However, tunneling is a possible mechanism through which a light atom (e.g. H) could tunnel under the barrier, even at low temperatures. On the other hand, an insertion reaction forms two new bonds while breaking an old one. The net gain in bonding results in a potential well, and the final products are formed by breaking one bond. These reactions are often barrierless and also called complex-forming reactions. They are indirect in the sense that the reactions typically go through a long-lived intermediate phase.

3.2.4 Thermodynamic Considerations

In the low-temperature conditions of the ISM, a chemical reaction has to be exothermic in order to proceed, i.e., its enthalpy change (heat of reaction at constant pressure) has to be negative, meaning that the reaction releases heat. On the other hand, an endothermic reaction would need to absorb energy from its environment, and cannot proceed at the very low temperatures of the cold interstellar medium. The enthalpy of a reaction can be obtained at a given temperature from the enthalpies of formation of the reactants and products. For most atom or molecules with up to about a few tens of electrons, enthalpies of formation can be computed with an accuracy of $\sim$1 kcal mol^{-1}.

As mentioned above, very little is known about reaction branching at low temperature. In astrochemical models, as in combustion, in the absence of experimental or theoretical data, the total rate of reaction is often partitioned in reaction channels considering the most exothermic channels to have the largest branching ratios. In other cases, analogies are drawn with similar systems for which some information are known. However, recent results regarding branching ratio determination have shown that both criteria are not adequate because it is the reaction dynamics along the underlying PESs that controls whether or not a specific reaction pathway is accessible by the system, and to what extent competing pathways occur.

3.2.5 Temperature Dependence of the Thermal Rate Coefficients

Knowledge of the rate coefficients of bimolecular reactions at the temperature where molecular species are observed is of crucial importance for understanding the past and current physics and chemistry at play in interstellar objects. Thanks to major progress in the last decades in both experimental measurements and theoretical calculations, our understanding of the temperature dependence of the rate coefficients of bimolecular reactions has dramatically improved. Indeed, when considering a large range of temperatures, from 10 to 1000 K, the values of $k(T)$ rarely adhere to the simple Arrhenius law: $k(T) = A\exp(-E_{act}/k_B T)$, where A is the Arrhenius pre-exponential factor, k_B is the Boltzmann constant, T the temperature, and E_{act} is the activation energy of the reaction (Smith 2008). This is not that surprising, as this law was established at the end of the 19th century when the range of temperatures over which measurements could be made was very narrow.

The temperature dependence of the kinetics of a reaction, especially at low temperatures, i.e., low collisions energies, depends on the detail of the potential energy surface(s) involved. The temperature dependence of a bimolecular collision depends on the potential energy variation along the minimum energy path leading from reactants to products. Only fast reactions ($k \geqslant 10^{-11}$ cm^3 s^{-1}) are generally of importance in the interstellar medium, unless they involve very abundant species such as H_2 for instance. At very low temperatures like those prevailing in the cold cores of dense interstellar clouds where the chemistry is very rich, there must be no (or at least, no significant) barrier along the minimum energy path leading from reactant to products for the reaction to be rapid. At such temperatures, and therefore low collisional energies, long-range molecular interactions are often considered dominant and the rates of elementary reactions are supposed to be controlled by the ability of the long-range intermolecular forces to bring the reactants into a close collision: a process generally referred to as "capture." If this assertion turns out to be adequate for reactions involving ions, the recent progress in measuring reaction rate coefficients over wide ranges of temperature have shown that the temperature dependence can be much more complex, especially for reactions involving neutral species only. Indeed, experimental results have shown a large variety of temperature dependences for fast reactions at low temperatures (see Section 3.3). This can be explained by the roles played by both short-range repulsion and chemical bonding, which are more important than expected. These effects can only be revealed by ab initio electronic structure calculations.

Long-range interactions: the capture rate. Classical statistical theories (see Section 3.4.2) are widely used to derive capture rates because they require much less computational resources than dynamical theories. When the reaction occurs on a single electronic state or when non-adiabatic effects can be neglected, classical trajectory simulations provide accurate predictions of the capture rate as long as the temperature is not too low and the electronic PES is of good quality. At low temperatures, quantum mechanical approaches described in Section 3.4.1 are required.

Short-range interactions. Short-range interaction treatment requires quantum mechanic calculations. A picture of the short-range interplay between the various interactions involved has been interpreted as follows by Georgievskii & Klippenstein (2007): at short range, when a chemical bond forms, in order to maximize the overlap of the electronic orbitals, a specific orientation of the reactants involved is required. The restriction in the range of reactive orientations, and corresponding decrease in the entropy, increases as the distance between the reactant decreases. Meanwhile, the strength of the chemical bond increases with decreasing separation.

A comprehensive interpretation has been given in the last two decades by development of the transition state theory (TST) along with related kinetic theoretical tools, with great success in predicting rate coefficients over a wide range of temperatures. In the TST formalism, different optimal orientations of the reactants are required at short- and long-range separations, giving rise to two transition states (Georgievskii & Klippenstein 2005, 2007). The breaking of an old bond and the forming of a new one give rise to a transition state, called an "inner transition state." On the other hand, the optimal orientation of the reactant for the long-range interaction gives rise to a van der Waals potential minimum, called an "outer transition state." The relative importance of the short-range interactions, for which a detailed and high level ab initio treatment is required, depends on the nature of both reactants: ionic, radical, or molecular.

In the case of many bimolecular ion–molecule reactions, a large variety of formulations for predicting the capture rate coefficients have been established, the simplest analytical one being the Langevin formula for the reaction of an ion with a non-polar molecule (ion-induced dipole potential):

$$k_{\mathrm{L}} = 2\pi e(\alpha/\mu)^{1/2}, \tag{3.14}$$

where α is the polarizability of the molecule assumed independent of its orientation, e is the charge of the ion, and μ is the reduced mass of the two species. A typical value for a Langevin rate coefficient is 10^{-9} cm^3 s^{-1}.

For charge–dipole capture, various approaches, including statistical methods such as the statistical adiabatic channel model (SACM) or classical trajectory (CT) calculations briefly described in Section 3.4.2, have been developed. Analytical relationships have been deduced that yield essentially the same results in the temperature range relevant to ISM, in good agreement with available measurements. It is worth noting that, even for reactions involving ions, if the long-range attractions are weak enough, the inner transition state can be important at low temperatures, and at the lowest temperatures (1–10 K), deviations from the capture analytical formula are expected.

To summarize, the most accurate treatments to infer the temperature dependence of the rate coefficient of a bimolecular reaction are based on quantum mechanics and electronic structure calculations using ab initio quantum chemistry for determining the PES. Alternative approaches can be used when quantum methods are not applicable, for instance because there are too many atoms or a too large number of states involved. It is worth noting that capture calculations and the TST approach do not require a global PES. The question of the accuracy needed for astrophysical modeling is crucial for choosing the appropriate theoretical approach. The temperature range also must be considered. At very low temperatures ($\leqslant$ 50 K), quantum effects can not be neglected. For quantum methods, recent comparisons between theory and experiment show that an agreement of 10–20% is reached for the best possible calculations. However, such an agreement is obtained for only a limited number of systems that have been the subject of intensive studies. The number of atoms involved is also important for the choice of the method. An overview of the current theoretical approaches for methods current will be the object of Section 3.4.

In kinetic databases, such as KIDA (KInetic Database for Astrochemistry; http://kida.obs.u-bordeaux1.fr/), a modified version of the Arrhenius equation, also referred as the Kooij equation (Kooij 1893), is often used to express the temperature dependence of bimolecular rate coefficients:

$$k(T) = \alpha(T/300)^{\beta}\exp(-\gamma/k_B T), \tag{3.15}$$

where α, β and γ are simply parameters that define the temperature dependance of a particular rate coefficient.

3.3 Experimental Methods

As mentioned in Section 3.2.2, astrochemical models need both rate coefficients and product branching ratios at the relevant temperatures of the astrophysical objects studied. A brief overview of the experimental methods available for low-temperature bimolecular investigation is presented below.

3.3.1 Total Thermal Rate Coefficient Determination

Chemical networks use mostly thermal rate coefficients, therefore assuming local thermal equilibrium (LTE) for treating the chemistry. From an experimental point of view, this entails the use of reactors with gas densities high enough to ensure sufficient collisions for maintaining local thermal equilibrium. Two general approaches allow one to cool gases to a low temperature: cryogenic cooling and expansion techniques. In addition, when charged species are involved, electromagnetic fields can be used to confine them far from the walls of the reactor or to drive beams of molecules.

Pseudo-first-order conditions. Let us first review how to define the rate coefficient of a reaction between two reactants R_1 and R_2 forming two new species P_1 and P_2:

$$-\frac{d[\mathrm{R_1}]}{dt} = -\frac{d[\mathrm{R_2}]}{dt} = +\frac{d[\mathrm{P_1}]}{dt} = +\frac{d[\mathrm{P_2}]}{dt} = k(T)[\mathrm{R_1}][\mathrm{R_2}]. \tag{3.16}$$

From an experimental point of view, most of the time, the concentration of only one species is monitored directly by various spectroscopic or mass spectrometric techniques. Therefore, care must be taken to ensure that the time evolution of the monitored reactant or product is actually due to the chemical reaction of interest. In other words, side or secondary reactions have to be eliminated or reduced as much as possible, and in any case, they have to be taken into account in the data analysis if they cannot be neglected. Usually, conditions are chosen so that the initial concentration of one of the reactants, let us say $[R_1]$, is much less than the other, $[R_2]$. In these conditions, $[R_1]$ decays exponentially with time while $[R_2]$ can be considered constant. Hence,

$$-\frac{d[\mathrm{R_1}]}{dt} = -k_{1\mathrm{st}}, \tag{3.17}$$

and

$$k = \frac{k_{1\mathrm{st}}}{[\mathrm{R_2}]}, \tag{3.18}$$

where k is the (second-order) rate coefficient for the reaction.

The two main advantages of being able to perform a kinetic experiment in these pseudo-first-order conditions are the following: first, there is no need to know the absolute concentration of the R_1 species, which is usually an experimental challenge. second, interference by other reactions is significantly reduced if not eliminated. In order to extract the rate coefficient from this equation, the absolute concentration of the species in excess R_2 must be accurately known.

In most kinetic studies, R_1 is an atom, a molecular radical, or an ion (negative or positive), usually produced by various methods such as thermal decomposition, microwave/radio-frequency discharge, or flash or pulsed photolysis. Care has to be taken to ensure that the quantum state of the generated radicals or ions is known. When excited species are formed, which is often the case in microwave/radio-frequency discharges for instance, a relevant quencher can be added into the reactor in order to relax them. On the other hand, R_2 is usually a stable molecule because it is not straightforward to produce a radical or an ion in such a large amount as to be in excess compared to R_1, and it is even more challenging to accurately determine its concentration. This explains why radical–radical or ion–radical reactions are difficult to study. Once these pseudo-first-order conditions are established, the relative concentration of the ion or radical produced has to be monitored. The most sensitive techniques, widely used for this reason, are laser-induced fluorescence and mass spectrometry.

In the next section, we present the main experimental methods currently used to perform kinetic experiments for ion–molecule and neutral–neutral reactions.

Ion–molecule Reactions

Specific methods have been developed to explore the reactivity of ions, firstly to ensure their efficient production and secondly to allow their detection (Geppert & Larsson 2013). We can distinguish three broad families of methods: flow reactors, ion traps and crossed molecular beams. A selection of the techniques the most widely employed is briefly detailed below.

Flowing afterglows. Introduced in the late 1960s by Ferguson et al. (1969), flowing afterglows have been successfully employed to measure the rate of ion–molecule reactions at room temperature. In this conceptually simple technique, the ions first generated upstream in the flow by some excitation method (microwave discharge, electron gun) are rapidly thermalized through collisions with bath gas atoms (He or Ar). Ions are then offered the possibility to react with some neutral gas introduced further away. The initial and product ions are then sampled downstream by mass spectrometry. The rate coefficient is derived by examining the reactant ion population with varying initial neutral co-reactant density. The strength of the method is that it decouples the ion production, ion reaction, and ion detection zones. Many developments followed, including variable temperature (up to 1800 K and down to 80 K), variable kinetic energy, and movable detection to name a few (Graul & Squires 1988). The addition of a Langmuir probe to measure the electron number density extended the application of the method to dissociative recombination (Adams 1993) and electron attachment (Smith & Spanel 1994; Figure 3.1).

However, the technique of flowing afterglow suffers from the presence of the precursor gas, which may react with the ions, and sometimes from the occurrence of multiple reactant ions, both of which impede the identification of the reaction products.

Selected ion flow tube (SIFT). In the 1970s, to circumvent these limits and extend the investigation of the reactivity to a greater variety of ions, Adams and Smith developed the selected ion flow tube method (Adams & Smith 1976). The SIFT represented a major breakthrough in the study of ion–molecule reactions. The desired ions are

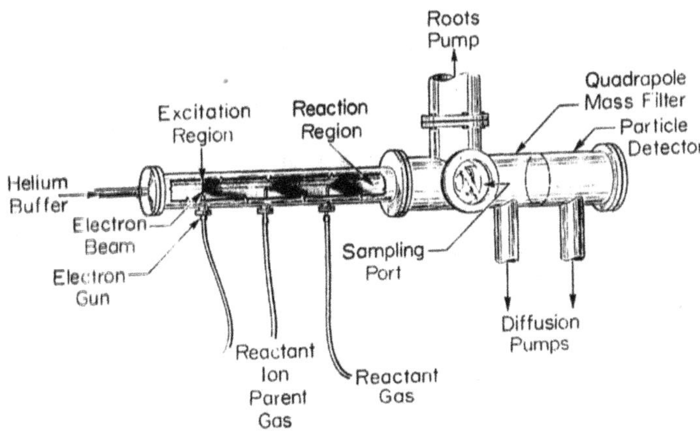

Figure 3.1. Drawing of the original flowing afterglow (Credit: Ferguson et al. 1969).

produced in a remote source, selected according to their mass to charge ratio by a quadrupole mass filter, and injected into a fast-flowing carrier gas by the means of a Venturi inlet. The SIFT can also be coupled to a variable-temperature flow tube.

One of the important applications of this method to the chemistry of the interstellar medium is the study of the reactivity of ions with neutral atoms. Because of the high abundance of atomic H, C, N, and O, their reactions with ions play an essential role in chemical networks. In the laboratory, hydrogen atoms are, e.g., produced by thermal dissociation, which prevents the formation of metastable species. Nitrogen atoms $N(^4S)$ are generated by introducing pure nitrogen through a microwave discharge. Oxygen atoms $O(^3P)$ can be formed by adding NO downstream of the nitrogen plasma through the reaction $N + NO \rightarrow O + N_2$ (Snow & Bierbaum 2008). The continuous production of carbon atoms in their ground state $C(^3P)$ in sufficient amounts to guarantee pseudo-first-order conditions (i.e., the concentration of carbon atoms is at least two orders of magnitude higher than those of the ions) is extremely difficult and hampers such measurements. This is mainly due to the high reactivity of the carbon atoms, which are consumed on their way from the source to the flow tube. An additional complication of these kinetic experiments involving atoms comes from the necessary determination of the absolute concentration of the concentration of the neutral atoms.

The potential of the method can be illustrated by studies of the reactivity of negative ions with atomic species in relation with the detection of carbon containing anions in the interstellar medium. In particular, the SIFT method has been employed to explore the reaction of C_x^- ions (x= 2–7) with atomic nitrogen at room temperature. These reactions proceed at 20–40% of the Langevin rate (Eichelberger et al. 2007). The dominant exit channel for the reaction of N atoms with bare carbon chains is the production of CN^-. Above $x \geqslant 4$, a channel toward $C_{x-1}^- + CN$ opens. For C_x^- anions with $x \geqslant 5$, C_3N^- and C_5N^- are also produced. Associative detachment leading to $C_xN + e^-$, which is an energetically accessible pathway, is also observed for all anions but remained undetectable under the configuration adopted.

$$C_x^- + N \rightarrow CN^- + C_{x-1} \quad (3.19)$$

$$C_{x-1}^- + CN \text{ for } x \geqslant 4 \quad (3.20)$$

$$C_3N^- + C_{x-3} \text{ for } x \geqslant 5 \quad (3.21)$$

$$C_5N^- + C_{x-5} \text{ for } x \geqslant 5 \quad (3.22)$$

$$C_xN + e^-. \quad (3.23)$$

This study is just a small sample of the collection of ion–molecule reactions of interest for the chemistry of ionospheres and of the interstellar medium, which were studied with the help of flowing afterglows and selected ion flow tubes (Anicich & Huntress 1986; Wakelam et al. 2015), demonstrating the capacity of the approach. However, in practice, measurements are limited to cryogenic temperatures of ~80 K and light neutral reactants. In cryogenic cooling techniques, any gas present at concentrations above its saturated vapor pressure condenses onto the walls of the reactor.

Supersonic uniform expansions: the CRESU technique. Low temperatures can advantageously be achieved by replacing the fast flow tubes, in which the velocity is subsonic, with supersonic uniform expansions. This original method, dubbed CRESU (French acronym for Kinetics of Reactions in Supersonic Uniform Flows), was developed by Marquette and Rowe to explore ion–neutral reactions down to 8 K (Rowe et al. 1985). This technique is based on the supersonic expansion of a non-condensable buffer gas (usually He, Ar or N_2) through a de Laval nozzle (see Figure 3.2). Cooling arises from conservation of energy, which implies that, for a gas flow under adiabatic conditions, the sum of specific enthalpy and kinetic energy remains constant during the expansion. For a perfect gas with both a constant specific heat capacity c_p and a well-defined temperature T, the energy conversion is driven by the following equation:

$$c_pT_0 = c_pT + \frac{u^2}{2}, \quad (3.24)$$

where T_0 and T are the temperatures of the gas before and after expansion, respectively and u is the supersonic flow velocity.

Introducing the Mach number M (ratio of flow velocity to the speed of sound) of the supersonic flow this expression can also be written as follows

$$\frac{T_0}{T} = 1 + \frac{\gamma - 1}{2}M^2, \quad (3.25)$$

where γ is the ratio of specific heats of the buffer gas at constant pressure and volume.

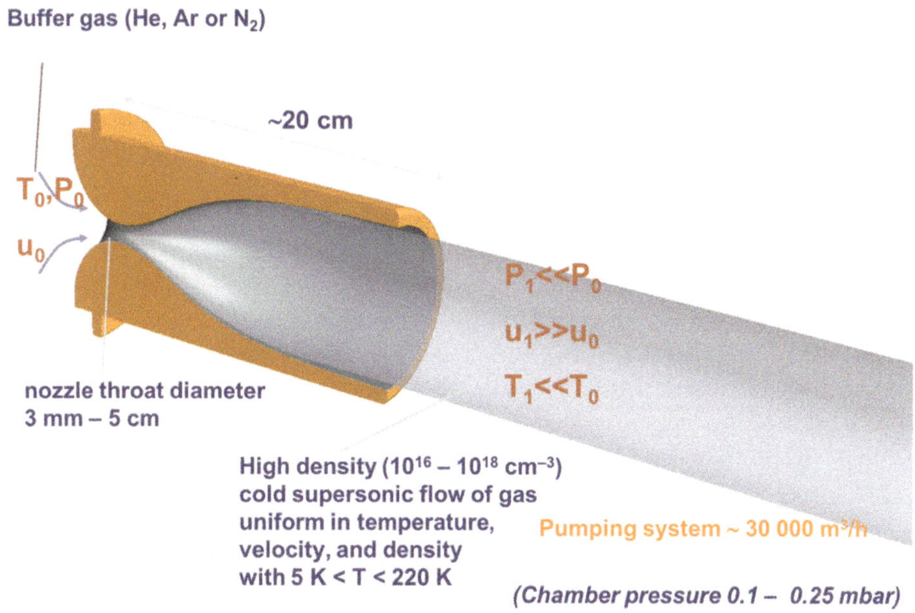

Figure 3.2. Schematic diagram showing a de Laval nozzle generating a supersonic flow of buffer gas of relatively high density, which ensures that frequent collisions take place during the expansion and subsequent flow, maintaining thermal equilibrium. In its continuous configuration (here at Rennes), the required mass flow rates of a few moles per minute, at the relatively low pressure needed to avoid excessive clustering, necessitate large pumping capacities.

The strength of the technique lies in the properties of the supersonic flow generated: its density is high enough to ensure thermal equilibrium is maintained due to frequent collisions; additionally, the temperature, density, and celerity of the flow are uniform over several tens of centimeters, corresponding to a few hundreds of microseconds downstream of the nozzle exit. Due to strongly supersaturated conditions in the subsequent flow and its rapid expansion through the nozzle, condensation can be essentially avoided. This technique therefore provides an excellent environment to study fast collisional processes ($k \gtrsim 10^{-13}$ cm^3 s^{-1}) at extremely low temperatures: down to 15 K when the buffer gas is at room temperature before expansion, and down to 5 K when it is precooled at 77 K. To reach lower temperatures ($\lesssim$ 50 K), however, the CRESU technique requires large pumping capacities. For this reason, various pulsed versions of the CRESU technique have been developed, using either one or two pulsed valves supplying gas to a reservoir on which a Laval nozzle is mounted (Atkinson & Smith 1995; Spangenberg et al. 2004; Lee et al. 2000; Taylor et al. 2008; Oldham et al. 2014) or pulsing the flow just after the throat of the nozzle by means of a spinning disc (Morales 2009; Ely et al. 2013; Jiménez et al. 2015).

Until recently, under its initial configuration, the CRESU apparatus dedicated to ion–molecule reaction studies was equipped with an electron gun that generates a high-energy, high-current electron beam (12 kV, 200 μA) crossing the flow transversely (Figure 3.3). Through electron impact, carrier gas cations are produced downstream the nozzle exit by the electron beam. They are then quickly converted into reactant ions by charge transfer by adding their neutral parents in the reservoir. The ions, generated in large amounts with densities in the 10^8–10^9 molecule cm^{-3} range, thermalize quickly through collisions and can react with a neutral co-reactant introduced in the reservoir at various concentrations. The population of reactant and product ions was monitored with a movable quadrupole mass spectrometer. The ion sampling into the spectrometer was achieved by skimming the flow with a cone with a 80 μm aperture and guiding the ions with electrostatic lenses. The kinetic rate coefficient was simply derived from the analysis of the reactant ion spatial profile in the flow for different initial concentration of neutral reagents.

The CRESU method is not confined to cations and can be employed to investigate the reactivity of anions. In that case, the anions are produced by dissociative attachment of cold secondary electrons ($\leqslant$ a few tens of meV) onto neutral precursors following the mechanism $AP + e^- \rightarrow A^- + P$. The difficulty resides in the selection of the precursor, which should fulfill a number of criteria, including high vapor pressure, chemical stability, high electron attachment rate at low electron energy, exoergicity of the exit channel leading to the desired ion, and favorable branching.

Following the recent discovery of molecular anions in the interstellar medium, the kinetics of proton transfer reactions between cyano-polyamide anions $C_{2n+1}N^-$ ($n = 0, 1, 2$) and formic acid HCOOH was investigated with this technique (Joalland et al. 2016). The results, obtained from room temperature down to 36 K, show a surprisingly weak temperature dependence of the CN^- reaction rate, in contrast with longer chain anions. The CN^- + HCOOH reaction was further studied theoretically via a reduced dimensional quantum model that highlighted a tendency of the reaction probability to decrease with temperature, in agreement with

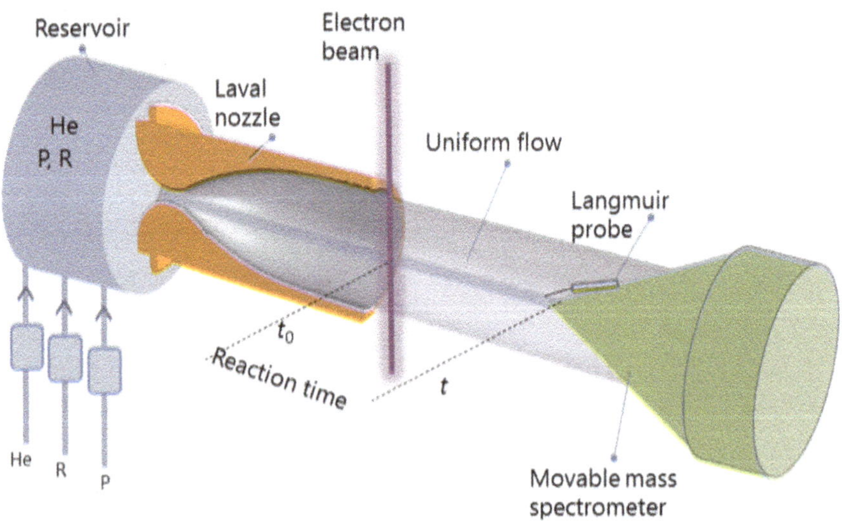

Figure 3.3. Schematic diagram of the CRESU apparatus configured for the investigation of ion-molecule reactions, showing the retractable Laval nozzle, the ionization region, and the movable quadrupole mass spectrometer/Langmuir probe assembly. P: ion precursor, R: reactant.

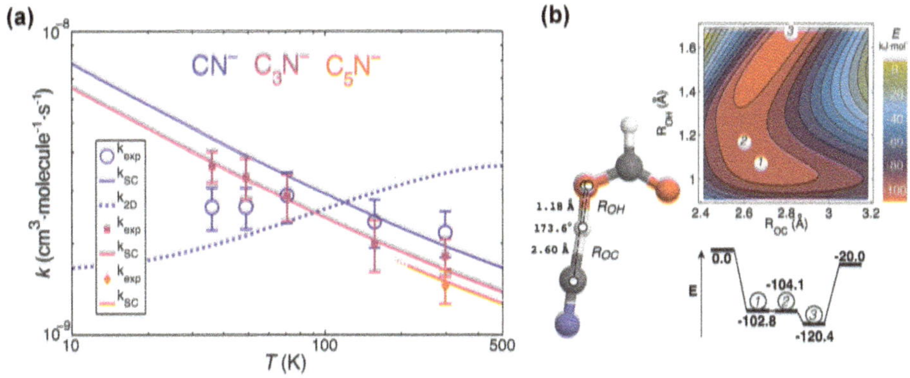

Figure 3.4. (a) Rate coefficients for the $C_{2n+1}N^-$ + HCOOH reactions measured from room temperature down to 36 K (k_{exp}) and estimated from the Su and Chesnavich capture model (k_{SC}, continuous line) (Su & Chesnavich 1982). The result of a two-dimensional quantum modeling of the CN^- + HCOOH reaction is also shown (k_{2D}, dashed line). (b) Contour map of the relaxed PES along the R_{OH} and R_{OC} coordinates for proton transfer between CN^- and HCOOH, calculated at the CCSD/aug-cc-pVDZ level of theory, shows a reef on the reaction coordinate that could explain the slowdown of the reaction at low temperatures. Similar PES landscape is calculated for C_3N^- but stronger dipole–dipole interactions for this polar anion must preempt this effect. (Credit: Joalland et al. 2016. Copyright 2016 American Chemical Society.)

experimental data but in opposition to conventional long-range capture theories. In return, comparing HCOOH to HC_3N as target molecules suggests that dipole–dipole interactions must play an active role in overcoming this limiting effect at low temperatures. Beyond providing rate coefficients for astrochemical models, this work offered new fundamental insights on prototypical reactions between polar anions and polar molecules (Figure 3.4).

The method is however limited by the availability of suitable precursors and the possible presence of reacting ions. In an effort to overcome these restrictions, a mass-selective ion source has been coupled to the CRESU. The idea of injecting mass-selected ions into the core of a uniform supersonic flow was initially suggested by Rowe et al. in the late 1980s (Rowe et al. 1989). However, the drawback of this original experimental setup is that the ion source is incorporated into the Laval nozzle, which has to be changed every time the reaction is investigated at a different temperature, making the operation rather burdensome. In the current configuration, the ions are first produced in a hollow cathode discharge source, transported by an octopole, filtered by a quadrupole, and then guided and injected into the supersonic flow just downstream the exit of the nozzle. Thanks to electrostatic lenses, the ions are drifted into the core of the uniform flow. Their population is then monitored with a movable quadrupole mass spectrometer. The main goal is to achieve the same versatility that has been demonstrated for the SIFT.

In parallel of the early CRESU takeoff, an ion–molecule reactor based on free jet expansion was developed in the 1990s by Smith & Hawley (1992). The main strengths of this alternative method, which operates under a pulsed mode, are its low pumping capacity and gas consumption requirements. Ions are generated by laser multiphoton ionization of a precursor downstream of the nozzle, where the velocity has reached a value close to its hypersonic limit, $\sqrt{\gamma c_p T_0}$. Reactions of the ions with neutrals are monitored by time-of-flight mass spectrometry as a function of the distance from the laser beam. However, the strong density gradients and the low associated collision frequency, both characteristic of free jet expansions, have to be adequately taken into account to extract the kinetic rate coefficients. To complicate matters, because of the rarity of collisions with the gas, the temperature is not well-defined and rotational temperature can diverge quite far. These shortcomings have understandably limited the dissemination of the free jet expansion method for the quantitative study of ion–molecule reactive collisions.

Ion traps. The second type of method to study the kinetics of ion–molecule reactions is based on ion trapping. Many different ion traps have been devised to perform such measurements, including ion cyclotron resonance cells. However, the most remarkable results were obtained with low-temperature traps. The method mostly developed by Gerlich (1992) allows the investigation of ion–molecule reactions at temperatures as low as 10 K. The device in this approach is cryogenically cooled and constituted of 22 pole electrodes on which non-homogeneous radio frequency fields are applied in order to produce an attractive potential to trap the ions (Gerlich 1995).

After being created, usually by electronic impact or by some more exotic methods, the ions are mass selected by a quadrupole filter and then injected into the trap's interior. The reaction zone is cooled through collisions with the buffer gas, which is itself thermalized by collisions with the cryogenic surfaces. Effective potentials produced by the 22 pole trap have steep radio frequency walls well-suited to cool ions (Figure 3.5). During the trapping time, which can vary between milliseconds and a few minutes, the ions react with the neutral gaseous molecules. The density of the neutral reactant can be varied between 10^9 molecule cm^{-3} and 10^{14} molecule cm^{-3}. It is important to mention that, by choosing an appropriate combination of these two parameters, i.e., the trapping time and the density of the neutrals, this technique can be used to measure rate coefficients of both fast and slow bimolecular reactions. At the end of the reaction time, the ions are extracted from the trap and analyzed by a second quadrupole filter before reaching the detector.

Sophisticated setups have been designed to exploit the capabilities of the 22 pole ion trap. In particular, a new instrument has recently been developed that combines a 22 pole ion trap with a cold effusive H-atom beam. The stored ions are relaxed to temperatures of $T \geqslant 12$ K, whereas H atoms, produced in a radio frequency discharge, are slowed down to temperatures as low as 7 K. The effective density of atomic and molecular hydrogen in the trap is determined in situ using chemical probing with CO_2^+. The apparatus has been employed, for instance, to explore the reaction of $CH^+ + H$ (Plasil et al. 2011). The methylidyne cation CH^+ is a key species of carbon chemistry in diffuse interstellar clouds. Its high abundance in these environments remains unexplained. While it is believed to be mainly produced by the reaction,

$$C^+ + H \rightarrow CH^+ + h\nu \tag{3.26}$$

the kinetics of one of the destruction pathways,

Figure 3.5. Photograph of a 22 pole ion trap (Mikosch 2007). Reprinted with permission from the author.

$$CH^{+}+H \rightarrow C^{+}+H_2 \tag{3.27}$$

had not been previously investigated at low temperatures. Ion trap experiments derived a low rate coefficient of 5.4×10^{-11} cm^3 s^{-1} at the coldest temperature ($T_{beam} \sim 7$ K; $T_{trap} \sim 12$ K), suggesting that non-rotating CH^+ is protected against attacks of H atoms. This surprising result is not yet understood and remains a matter of debate because recent quantum scattering calculations have shown the opposite: a reactivity enhancement with rotational excitation (Werfelli et al. 2015).

Similar to the variable temperature flow tube method previously discussed, the ion trapping technique uses cryogenic cooling to obtain low temperatures, which can lead to the condensation of the neutral reactants onto the walls of the chamber when using reactant gases other than H_2 or He. This method will therefore be limited to the study of ion reactions with light neutrals.

Investigations of ion–molecule reactions are not restricted to cold conditions. In planetary ionospheres, cations are generated by ionization through cosmic rays, photons, or magnetospheric electrons, which can lead to the formation of fast and/or excited ions with long lifetimes. At high energies, branching ratios can evolve and new products can emerge. Reactions of excited species are also of interest for photon-dominated regions (PDRs) and shock regions. Reaction rates measured so far overwhelmingly concern the ion ground states. Tightly related to *ion traps*, the method of *guided ion beams* was developed in the 1970s by Gerlich (1992) and can measure ion reactions at high collision energies. The technique is well-suited to investigate the reactivity of state-selected ions produced by multistep laser ionization or VUV single-photon ionization generated by synchrotron radiation.

Crossed and merged beams. Through electron impact, carrier gas cations are produced downstream the nozzle exit by the electron beam. They are then quickly converted into reactant ions by charge transfer by adding their neutral parents in the reservoir. Merged molecular beams represent an attractive technique for reaching very low relative energies in order to study ion–molecule reactions. In the early 1990s, Gerlich (1993) proposed an original setup combining an ion trap, a neutral beam, and a guided ion beam. In this configuration, the collimated neutral beam is produced by dual skimming of a supersonic jet generated by a pulsed valve while ions are prepared in a storage ion source, mass selected, and thermalized in a variable-temperature ion trap. The ions are then transferred and merged with the neutral beam via magnetic deflection. The interaction takes place in the merged beam region where the ion beam is guided by an RF field. Primary and product ions are detected with the help of a quadrupole mass spectrometer. Integral reactive cross sections can be obtained for very low kinetic energies down to 1 meV.

The method was adapted recently with fast beams to specifically investigate anion–neutral reactions (Bruhns et al. 2010). The apparatus was employed to measure the associative detachment reaction

$$H^{-}+H \rightarrow H_2^{-} \rightarrow H_2 + e^{-} \tag{3.28}$$

which happened to be the dominant pathway in the early universe for the formation of molecular hydrogen H_2. Despite its importance and apparent simplicity, the rate for this reaction at low temperature is debated. In their sophisticated setup, Bruhns et al. (2010) implemented an ion source to generate a fast H^- anion beam (typically 10 keV). A laser is then employed to photo-detach and neutralize a portion of the H^- parent beam:

$$H^{-}+h\nu \rightarrow H + e^{-}. \tag{3.29}$$

Center-of-mass energies down to the meV range can be achieved as the beams co-propagate with minimal divergence. The merged beams finally enter a gas cell filled with helium, which converts H_2 into H_2^+ through the process

$$H_2 + He \rightarrow H_2^{+} + He + e^{-} \tag{3.30}$$

and is employed to monitor the reaction. This new approach is well-suited to chemical studies with systems such as H, C, and O, which can all be generated via photo-detachment of a parent atomic anion beam. The associative detachment rate coefficients measured agree well with the most recent theoretically calculated cross sections, which were then used in cosmological simulations to predict masses for stars in the early universe (Kreckel et al. 2010).

Neutral–neutral Reactions

Cryogenic cooling and expansion techniques are currently the two main methods used to cool neutral species in the gas phase to low temperature at sufficiently high density to ensure thermal equilibrium to be reached.

Cryogenically cooled cells. Condensation is a major issue when using cryogenically cooled cells, which cannot be applied at temperatures relevant to interstellar chemistry. This kind of reactor has been used to provide reaction rate coefficients in a few favorable cases down to below 100 K, especially where the co-reagent is not easily condensable. The cryogenic method is more designed for atmospheric chemistry applications with temperatures ranging from 180 to 300 K. The technique is also well-suited to the study the kinetics of slow reactions (from $k = 10^{-15}$–10^{-13} cm^3 s^{-1}) because reaction times of several milliseconds can be reached and relatively high concentrations of reagents can be introduced into the cell. Although these methods are limited in temperature, they provide a valuable complement to supersonic expansion techniques.

Supersonic uniform expansions: the CRESU technique. The CRESU technique, first dedicated to ion–molecule reaction kinetics as described in the previous section, was adapted to investigate neutral–neutral reactions in the early 1990s (Sims et al. 1994) as illustrated in Figure 3.6. The implementation of pulsed-laser-photolysis laser-induced-fluorescence technique enabled to perform kinetic measurements on radical–neutral reactions down to 5 K over the last 25 years. The measured rate coefficient of a large number of radical–radical metathesis, radical–unsaturated molecule, and radical–saturated molecule reactions have had a significant impact on the astrochemical models because they allowed a large variety of temperature dependences of such reactions to be shown, motivating a great number of state-of-the-art theoretical studies.

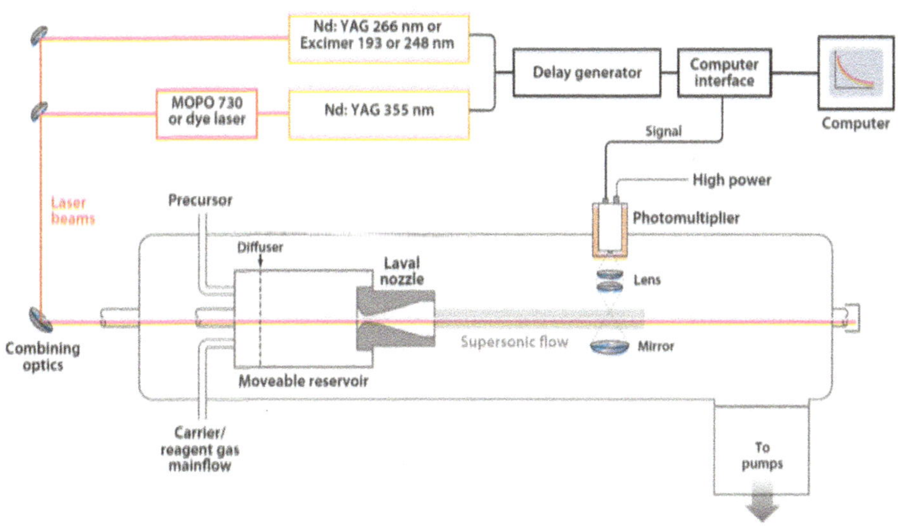

Figure 3.6. Sketch of a CRESU apparatus configured for the study of neutral–neutral reactions. Radicals are here generated by photolysis of a suitable precursor using radiation from a fixed-frequency pulsed laser, and are detected by laser-induced fluorescence excited by tunable radiation from an other pulsed laser.

Let us illustrate the neutral–neutral kinetic studies via three examples where both experiments and theoretical work synergistically improved our understanding of bimolecular reactivity.

CN + C_2H_6. Figure 3.7 displays the temperature dependence of the rate coefficient for the reaction CN + $C_2H_6 \rightarrow C_2H_5$ + HCN (Georgievskii & Klippenstein 2007). Because of the importance of CN radical reactions in astrochemistry as well as in combustion, this reaction has been studied experimentally over a broad range of temperature (25–1140 K). In this figure, the experimental results below room temperature were obtained using the CRESU technique (Sims et al. 1993). At high temperature, the temperature dependence of the reaction rate coefficient adheres to a modified Arrhenius equation. At low temperature, on the other hand, the rate coefficient increases when the temperature is lowered. In between, near 200 K, the rate coefficient of the CN + C_2H_6 shows a strong minimum in the rate coefficient. The calculations based on the transition state theory (TST) led by Georgievskii & Klippenstein (2007) allow to explain this temperature dependence of the rate coefficient and illustrate the changing importance of the "inner" and "outer"

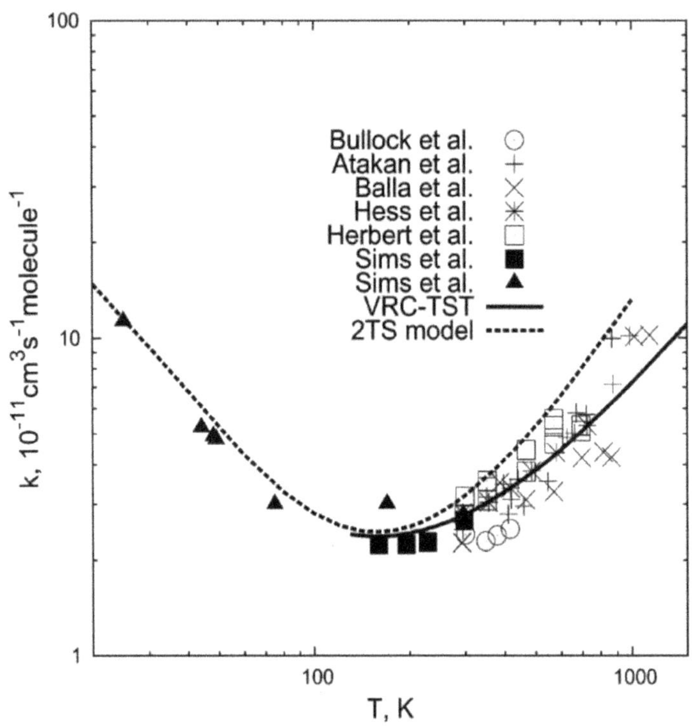

Figure 3.7. Temperature dependence of the rate coefficient for the rate coefficient of the reaction CN + $C_2H_6 \rightarrow C_2H_5$ + HCN studied by various experimental techniques, from 25 to 1140 K. CRESU data (black triangles) were obtained from 25 K to room temperature. A two-transition-state model (dashed line) was proposed by Klippenstein and Georgievskii to explain the unusually strong minimum in the rate coefficient near 200 K. (Credit: Georgievskii & Klippenstein 2007. Copyright 2007 American Chemical Society.)

transition states with increasing temperature, i.e., the subtle interplay of long- and short-range interactions mentioned in 3.2.5. In brief, at low temperatures, the rate is determined by the rate for forming the van der Waals complex ($CN–C_2H_6$), which reaches the collision limit value of a few times 10^{-10} cm^3 s^{-1}. With increasing temperature, the inner transition state plays an increasingly important role, due to its much lower entropy. This leads to a reduction in the rate coefficient. Ultimately, however, at higher temperatures, the effect of the outer transition state becomes negligible. The temperature dependence is then determined by that of the partition function for the transitional modes and the rate coefficient of the reaction increases with the temperature.

$F(^2P_J) + H_2$. Figure 3.8 displays the temperature dependence of the rate coefficient for the reaction $F + H_2 \rightarrow HF + H$ from room temperature down to 11 K (Tizniti et al. 2014). This reaction is of great astrophysical interest because HF is ubiquitous in the interstellar medium and is used as a tracer for H_2, which has no dipole moment and is thus difficult to detect directly. The reaction $F + H_2$ reaction is actually the sole source of HF in the interstellar medium. However, it presents a high barrier to reaction ($\simeq$ 800 K). From room temperature to about 100 K, the rate coefficient shows an expected Arrhenius behavior. At lower temperatures, however, the rate coefficient reaches a plateau that can only be explained by quantum

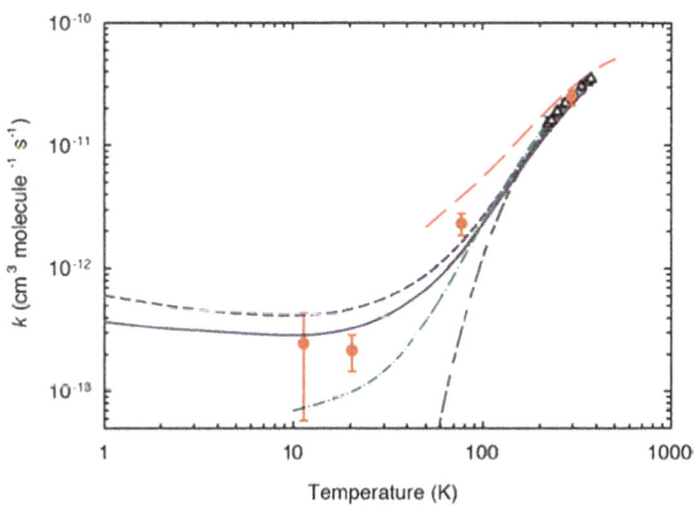

Figure 3.8. Temperature dependence of the rate coefficient of the reaction $F(^2P_J)$ + n-H_2 → HF + H. Comparison of rate coefficients k as a function of temperature. Filled red circles present experimental results obtained using a CRESU apparatus. The solid and dashed blue lines represent the predictions of scattering calculations with, respectively, inclusion and exclusion of non-BO coupling, assuming a ratio of ortho ($j = 1$) to para ($j = 0$) H_2 of 3:1 (Tizniti et al. 2014). Also shown (alternating short–long dashed dark green line) is the Arrhenius temperature dependence proposed by Persky & Kornweitz (1997), the predictions of the single-PES quantum scattering calculations of Zhu and colleagues (long-dashed magenta line) (Zhu et al. 2009), and the quantum reactive scattering calculations of Aquilanti et al. (2005, 2012; light green dashed–dotted line) performed on their PES III, which is a modification of the original Stark–Werner PES.

mechanical tunneling. The CRESU technique was used to study the kinetics of this reaction from 298 to 11 K, and state-of-the-art quantum reactive scattering calculations involving multiple PESs, have been undertaken independently (Tizniti et al. 2014). This non-Born–Oppenheimer treatment includes coupling between the three FH_2 PESs, as well as the presence of non-zero electronic and spin angular momenta. The theoretical calculations shed light on the importance of these non-adiabatic effects at low temperatures on the kinetics of this reaction. These joint results are currently used to improve the prediction of the abundance of HF in various objects, resulting in a better estimation of the abundance of H_2, and therefore of the total mass distribution in various regions of the interstellar medium (Gerin et al. 2016).

$N(^4S)$ + $OH(^2P)$. The barrierless reaction $N(^4S) + OH(^2P) \rightarrow H(^2S) + NO(^2P)$ is thought to play a key role in the interstellar N_2 formation. It is also one of the very few reactions involving two radicals that has been studied both experimentally and theoretically at low temperature (Daranlot et al. 2011). Figure 3.9 displays the temperature dependence of the rate coefficient for this reaction. Experiments were performed in a CRESU apparatus, from room temperature down to 56 K. A microwave-discharge was used to generate atomic nitrogen, and a relative-rate method to follow the reaction kinetics, using the previously determined rate coefficient of the reaction $N(^4S) + NO(^2P) \rightarrow N_2(^1\Sigma^+) + O(^3P)$. Theoretical calculations were performed, first using a time-independent (TI) quantum mechanical method employing hyperspherical coordinates (see Section 3.4.1), and second, an exact time-dependent (TD) quantum mechanical calculations including contributions from all angular momenta J in the N + OH Jacobi coordinates (see Section 3.4.1). An excellent agreement was found between theoretical predictions and experimental rate coefficients determined using the CRESU technique. Disagreements with previous results are discussed in Daranlot et al. (2011). These results show that this gas-phase N_2 formation mechanism involving both $N(^4S)$ + $OH(^2P)$ and $N(^4S)$ + $NO(^2P)$ reactions is probably less efficient than it was previously thought.

3.3.2 Reaction Product Identification and Branching Ratios

Two crucial pieces of information essentially missing in astrochemical networks include the nature of the products and the branching ratio of the bimolecular reactions at play. This lack of data can have dramatic effects, especially when exit channels are missing for some reactions. In addition, knowledge of reaction products and their branching ratio yields crucial information on the mechanism of elementary reactions and is therefore of fundamental interest. Experimental techniques used for this purpose have to be multiplex so that ideally all potential products, or at least most of them, can be detected within the same set of experiments, in the same experimental conditions. They also have to be very sensitive, as some products can be formed in very small amounts. Finally, they have to bring quantitative information in order to determine product branching ratios.

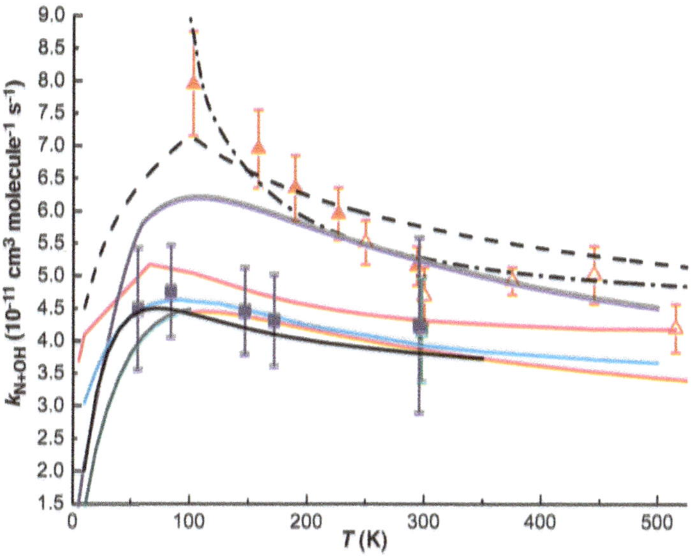

Figure 3.9. Rate coefficients for the $N(^4S) + OH(^2P)$ reaction as a function of temperature. The solid blue squares are kinetic measurements obtained using a CRESU apparatus and the solid black and green lines are theoretical predictions using time-dependent and exact time-independent quantum mechanical methods. Other symbols and lines correspond to previous works and are discussed in Daranlot et al. (2011). (Credit: Daranlot et al. 2011.)

Crossed beam experiments. The crossed-beam scattering technique developed in the 1960s (see Herman & Futrell 2015, and references therein) is one of the methods of choice for exploring reaction products, both for ion molecule and for neutral–neutral reactions under single-collision conditions. The principle is to cross two molecular beams with a given angle of intersection. Most crossed molecular beam machines work with beams at right angles, with corresponding collisional energies of a few to a few tens of kcal/mol, well above those encountered in the molecule-rich regions of the ISM (a collisional energy of 1 kcal mol^{-1} corresponds roughly to 300 K). Only the molecular beam apparatus at the University of Bordeaux and CNRS (France) enables to continuously vary the intersection angle between 90 and 12.5°, facilitating very low relative translational energies with supersonic beams. In order to identify the products, as well as their velocity and angular distributions and possibly their internal energies or energy distributions, crossed-beam experiments have been associated with various detection techniques, such as ionization/photoionization coupled to mass spectrometry, laser-based spectroscopic techniques (laser-induced fluorescence, LIF, or resonance-enhanced multiphoton ionization, REMPI), and velocity map imaging (VMI; Chandler & Houston 1987; Eppink & Parker 1997), which simultaneously and with high resolution provide details on the velocity and angular distribution of charged particles (Chandler & Houston 1987; Eppink & Parker 1997). A recent review (Pan et al. 2017) describes improved universal crossed molecular beam methods that permit to identify primary reaction products, characterize their formation dynamics, and determine the branching ratios for multichannel bimolecular neutral–neutral reactions, including reactions between two radicals. Despite the high collision energies at which these experiments are usually performed, they deepen our understanding of chemical reactivity and offer a bridge between crossed molecular beam dynamics and thermal kinetics, providing important information for the fields of both combustion and astrochemistry.

Flow reactor experiments. Experimental techniques to quantitatively explore the products of reactions and their branching ratio in multiple collision conditions have also greatly improved in the last few years. Among the various techniques that have been used for monitoring products of reactions in such conditions, most of them are optical spectroscopy, mass spectrometry, or microwave/millimeter spectroscopy.

The main drawback of optical techniques is that it is difficult to monitor a large range of species with one single apparatus. In some favorable cases, the branching ratio of reactions can be determined by monitoring the H atom production (via VUV LIF or VUV resonance fluorescence) compared to a reference reaction for which the H atom yield is known. For instance, the branching ratio of reactions involving a CN radical with various hydrocarbons has been determined by Gannon et al. (2007), down to 195 K; using the reactions of CN with C_2H_2 and C_2H_4 as references, a 100% H atom production was found for both reactions at room temperature.

Mass spectrometry has the advantage of being more universal. In particular, Leone and co-workers from the University of California (Berkeley) and Sandia National Laboratories (Livermore) used the Advanced Light Source at Lawrence Berkeley National Laboratory (a continuous-wave synchrotron) as a source for photoionization. This source allows threshold photoionization limiting, if not preventing fragmentation, so that the parent ion radical can be monitored. Its tunability allows discrimination of species with the same mass, potentially including different isomers (Osborn et al. 2008). A large amount of results have been obtained by Taatjes, Osborn, and coworkers at room temperature or above for atmospheric and combustion applications. A few low-temperature experiments were performed at Advanced Light Source (ALS; Berkeley) down to 70 K using a pulsed CRESU apparatus coupled to a quadrupole mass spectrometer (Soorkia et al. 2011), essentially for reactions of interest for Titan' s atmosphere. A new CRESU apparatus, named CRESUSOL, is under development at the Institute of Physics of Rennes and synchrotron SOLEIL Paris (DESIRS beamline) that couples CRESU flows to a photoelectron photoion coincidence (PEPICO) spectrometer in order to detect reaction products and derive branching ratios at very low temperatures, down to 20 K (see Figure 3.10).

Provided they possess a permanent dipole, reaction products can also be detected using rotational spectroscopy, which has the advantage of being highly specific. It has traditionally suffered, however, from a lack of sensitivity compared to laser-based techniques that has prevented its use in reaction kinetics and dynamics. Recently, the Chirped Pulse Fourier-Transform MicroWave (CPFTMW) technique invented by Pate and co-workers at the University of Virginia, has open new perspectives by improving the rate of data acquisition by several orders of

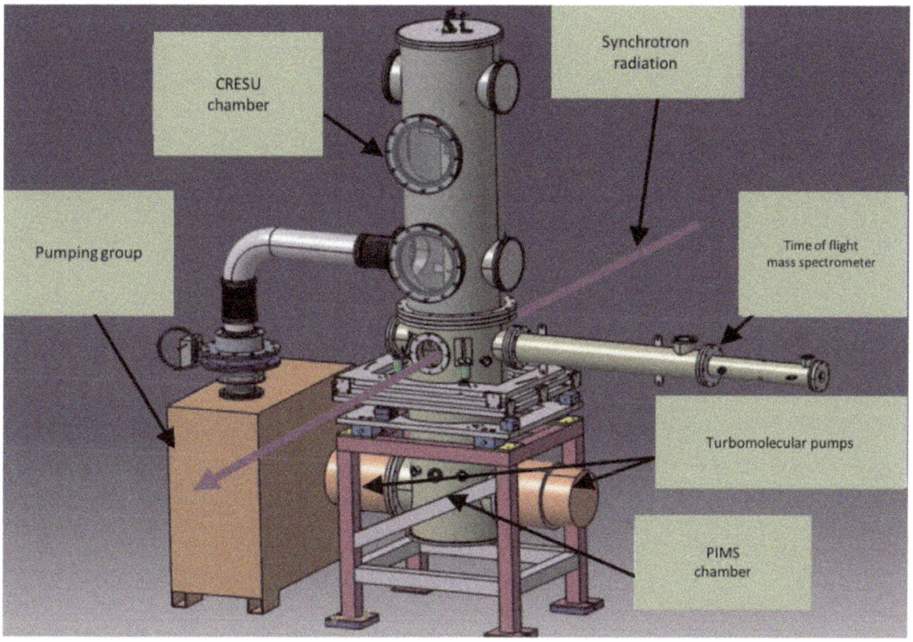

Figure 3.10. Sketch of the CRESUSOL apparatus configured for reaction product identification of neutral-neutral reactions, including clusterization. The center of the supersonic flow generated in the upper (CRESU) chamber is skimmed before entering a lower (PIMS) chamber where the molecular beam generated is photoionized by the synchrotron radiation. Photoions and photoelectrons are both detected in a homemade time-of-flight mass spectrometer and a coincidence technique is used to build up a mass spectrum allowing the identification of various species present into the flow.

magnitude, as well as by covering a wide range of frequencies that enables simultaneous detection of multiple species (Brown et al. 2008; Park & Field 2016). The principle is as follows: a waveform generator is used to produce a short microwave pulse (ca 1 ms or less) with a frequency sweep, the probed species absorbs at all rotational transitions within the frequency range of the chirp and is polarized by the radiation. Free induction decay (FID) of the polarization ensues and the emitted FID radiation is collected, down-converted, and averaged in the time domain by a high-bandwidth oscilloscope. The collected signal is then Fourier-transformed to give the rotational spectrum at MHz resolution. This technique has been recently coupled to the CRESU technique (Oldham et al. 2014), whose flow characteristics present many advantages: the very cold and thermalized environment provided by CRESU flows allows rotational cooling of the reaction products (resulting from exothermic reactions) and the collapse of their rotational state distributions to much lower values of rotational energy. This results, at least for small-to-medium-sized polyatomic molecules, in a shift of the maximum intensity of their rotational spectra to below 100 GHz, into the region accessible by modern CPMW spectrometers. In addition, this cold and multiple-collision environment leads to a large increase in difference in adjacent rotational-level populations. Referred to as the CPUF (chirped pulse in uniform flow) technique (see Figure 3.11),

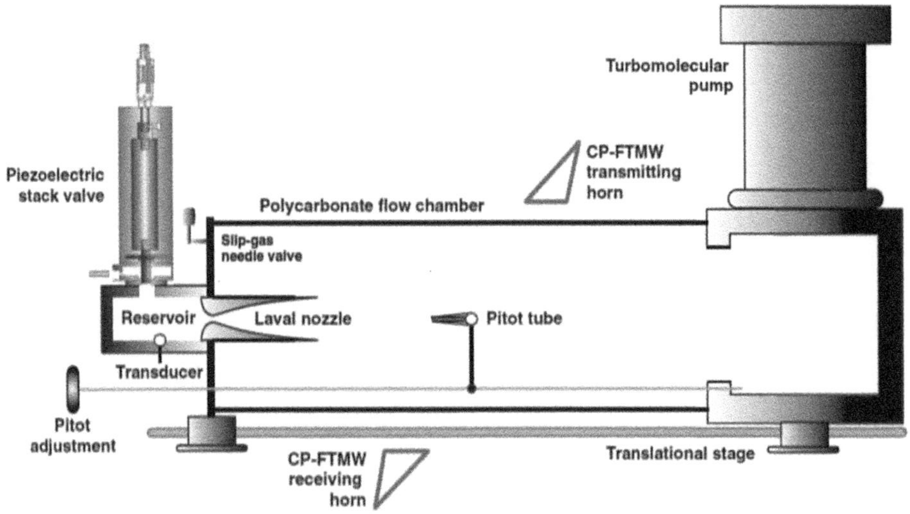

Figure 3.11. The chirped-pulse, uniform flow (CPUF) instrument. A piezoelectric stack actuator is used to generate a pulsed supersonic flow through de Laval nozzles. The cylindrical reaction chamber is made of a polycarbonate tube that is transparent to visible and microwave radiations. Microwave antennas on opposite sides of the chamber transmit and receive microwave radiation. The figure is not to scale. (Credit: Abeysekera et al. 2014.)

a recent development coupling CRESU flows with CP-FTMW spectroscopy has allowed a demonstration of the instrument's capability to study bimolecular reaction products (Abeysekera et al. 2014). More recently (Abeysekera et al. 2015), this apparatus was used to derive the branching ratio of a CN + propyne (C_3H_4) reaction, showing the ability of CPUF to obtain branching among competing reaction exit channels. Quantitative branching ratios indeed, were observed for all products as 12(5)%, 66(4)%, 22(6)%, and 0(8)% into HCN, HCCCN, CH_3CCCN, and CH_2CCHCN, respectively.

3.4 Theoretical Methods

3.4.1 Quantum Calculations

The initial stages to describe quantum mechanically reactive collision processes were dominated by time-independent quantum (TIQ) methods that provide the entire S-matrix at a given energy. Following initial studies on constrained[1] collisions (Truhlar & Kuppermann 1970; Miller & Light 1971; Schatz et al. 1973; Kuppermann et al. 1976), along with the availability of experimental rovibrationally resolved state-to-state integral and differential cross sections, several authors, such as Webster & Light (1989), Kuppermann & Hipes (1986), Pack & Parker (1987), Launay & Lepetit (1988), Launay & Le Dourneuf (1989), Zhang & Miller (1987),

[1] Constrained collisions are for example: (i) collinear collisions where the three atoms A, B and C are constrained to move on a line. The rotation is then not taking into account. (ii) The motion of the three atom can also be constrained to move in a plane.

Haug et al. (1986), Alexander et al. (2000), and Lique et al. (2009, 2011), have developed full-dimensional accurate and efficient methodologies to solve the Schrödinger equation for triatomic reactive systems on a single Born–Oppenheimer PES. One of the pioneering time-dependent descriptions of a reactive scattering process is credit to McCullough & Wyatt (1971a, 1971b), regarding the collinear exchange H + H_2 reaction. Due on the one hand to its pictorial representation of the dynamics and its ability to handle a high density of quantum states or a continuum of coupled scattering channels, and on the other hand to its favorable computational effort scaling and the development of vector and parallel machines, the time-dependent wave packet (TDWP) method has in recent years become the method of choice for the treatment of molecular processes. Therefore, several reactions have been studied by this approach, including restricted $J = 0$ calculations, full converged integral and state-to-state cross-sections, and differential cross sections (Zhu et al. 1997; Gray & Balint-Kurti 1998; Peng et al. 1998; Monnerville et al. 2000; Althorpe 2001; Cvitas & Althorpe 2011; Liu et al. 2011), as well as non-Born–Oppenheimer calculations (Adhikari & Varandas 2013; Saho et al. 2014; Ghosh et al. 2015). From a numerical point of view, if N_i denotes the number of basis functions used to describe the ith internal coordinate, and J the value of the total angular momentum, the numerical effort can be estimated to scale as $[J\Pi_i N_i]^3$ for a TIQ calculation compared to $J\Pi_i N_i \log N_i$ for a TDWP calculation. This scaling makes TDWP methods more efficient than TIQ ones for calculations involving high energies, and high-dimensional systems for which a large number of channels are open. For low-temperature data, for which low-energy calculations are required, TIQ methods are nevertheless more accurate and more efficient than TDWP ones, due to the long de Broglie wavelengths associated with the low collision energy.

For astrophysical applications, global observables like reactive initial state-selected cross sections, associated rates, and thermal rate coefficients of elementary reactions are mostly required. This section will focus on methods commonly used for computing these data. In the following, for the sake of simplicity, the A + BC atom-diatom collision process will be used to illustrate the main features of these theories. The fundamental problem of quantum reactive scattering theory compared to quantum inelastic or quasi-classical trajectory (QCT) ones (see Section 3.4.2) is referred to as the "coordinates problem." Indeed, coordinates that best describe the A + BC reactants channel are not the same as those best describing the B + AC or C + AB products channels. This leads to technical difficulties in the quantum calculations where all regions have to be, in principle, treated simultaneously. This is the main reason why QCT and quantum inelastic scattering programs were available before quantum reactive ones. Two main coordinate systems are used in time-dependent wave packet and TIQ calculations of reactive bimolecular collisions: the well-known Jacobi (Hirschfelder & Dahler 1956; Dahler & Hirschfelder 1959; Smith 1962; Pack & Parker 1987) and hyperspherical (Delves 1958; Smith 1962; Whitten & Smith 1968; Whitten 1969; Kuppermann 1975; Kuppermann & Hipes 1986; Hipes & Kuppermann 1987), Johnson (1980, 1983), Zickendraht (1965, 1967), Pack & Parker (1987) coordinate systems. Jacobi coordinates are well-suited to describe the asymptotic reactants or products channels, and hyperspherical coordinates allow smooth passage from the reactant to product channels. While Jacobi coordinates are intensively used in TDWP calculations, the hyperspherical ones are preferred in TIQ calculations. To the best of our knowledge, only some authors have used hyperspherical coordinates in TDWP calculations (Billing & Marković 1993; Marković & Billing 1994; Echave 1996; Adhikari & Varandas 2013; Crawford & Parker 2013; Saho et al. 2014; Ghosh et al. 2015).

In this section, the TDWP approach and the TIQ method to calculate cross sections and rate coefficients are briefly described.

Jacobi Coordinates

As we said before, the choice of the coordinate system is the first main step in reactive scattering calculations. Two kinds of Jacobi coordinates are used in molecular calculations: the mass-unscaled and mass-scaled Jacobi coordinates, mass-scaled being deduced from mass-unscaled coordinates by a simple multiplicative mass factor. As can be seen in Figure 3.12, each arrangement channel τ, namely A + BC reactants and AB + C or AC + B products, are described by a different set of mass-unscaled Jacobi coordinates $(\vec{r}_\tau, \vec{R}_\tau)$ $(\tau = \mathrm{A, B, C})$, where $\vec{r}_\tau$ is the diatomic internuclear vector and $\vec{R}_\tau$ is the vector joining the diatom center of mass to the atom τ. The complete description of the system requires three internal coordinates $(R_\tau = |\vec{R}_\tau|, r_\tau = |\vec{r}_\tau|, \theta_\tau)$, where θ_τ is the angle between the two vectors $\vec{r}_\tau$ and $\vec{R}_\tau$, and three Euler angles $(\alpha_\tau, \phi_\tau, \gamma_\tau)$. These Euler angles define the orientation of the $(x, y, z)_\tau$ body-fixed (BF) frame with respect to the space-fixed (SF) one.

The major problem of the Jacobi coordinate system is that reactant coordinates $(R_\mathrm{A}, r_\mathrm{A}, \theta_\mathrm{A})$ are well-suited to correctly describing an atom A approaching a diatom BC (channel A + BC) but are not suitable to efficiently describe, for example, the products C + AB where a C atom goes far from an AB molecule, the best coordinates for this specific product being the $(R_\mathrm{C}, r_\mathrm{C}, \theta_\mathrm{C})$ ones. As will be shown below, the "coordinate problem" disappears if only cross sections and rate coefficients are required, because only one reactant or product Jacobi coordinate system can be used. On the other hand, if state-to-state transition probabilities or differential cross sections are required, this "coordinate problem" is a real numerical difficulty and a change of coordinates from reactants to products is necessary. This leads to time-consuming interpolation schemes. Several approaches have been

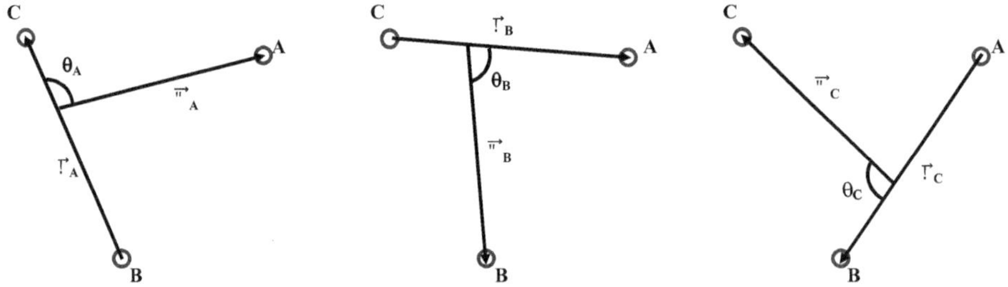

Figure 3.12. Jacobi coordinates for A + BC, B + AC, and C + AB, with $M_\mathrm{C} > M_\mathrm{B} > M_\mathrm{A}$.

developed to solve this reactant–product coordinate transformation problem. The first ones, in which the change of coordinates were performed as the wave packet reaches the strong interaction region, were proposed by several authors, including Judson et al. (1990), Göğtas (1996), Sun et al. (2009a), and Gómez-Carrasco & Roncero (2006). The most efficient method, called reactant–product decoupling (RDP), was proposed by Zhu et al. (1997) and Peng et al. (1998), and has been successfully applied to tetratomic reactive collisions (Cvitas & Althorpe 2011; Liu et al. 2011). Despite this disadvantage, Jacobi coordinates are useful and numerically efficient because of the diagonal form of the Hamiltonian operator enabling the use of the numerically very efficient grid method approaches briefly reviewed in Section 3.4.1.

Hyperspherical Coordinates
For a triatomic system, the hyperspherical coordinates are defined by three Euler angles (α, ϕ, γ) and three internal coordinates, a hyperradius ρ that is a measure of the global size of the system, and two angles θ and χ that specify the shape of the molecule. Two kinds of hyperspherical coordinates have been proposed: the asymmetrical coordinates of Delves (1958), Kuppermann (1975), Kuppermann & Hipes (1986), Hipes & Kuppermann 1987), and the symmetrical ones proposed by Smith (1962), Whitten & Smith (1968), Whitten (1969), Johnson (1980, 1983), Zickendraht (1965, 1967), and Pack & Parker (1987). While the angles and the axes of the asymmetrical hyperspherical coordinates are those of one of the three arrangements A + BC, B + AC, or C + AB, the symmetrical hyperspherical coordinates are based on the principal axis frame and allow smooth switching from the reactant to the product channels. Due to the fact that the Delves hyper angle χ depends on the choice of the arrangement, the different arrangements are not treated equivalently in the asymmetrical hyperspherical coordinates whereas they are in the symmetrical ones. Other mathematical properties describing the reaction region of a general triatomic system are obtained with the symmetrical hyperspherical coordinates, as pointed by Launay and coworkers (Lepetit et al. 1986).

Basis Functions and Action of the Hamiltonian Operator
In TDWP and TIQ calculations, the action of the Hamiltonian operator H on the time-dependent $\Psi(\vec{X}, t)$ or time-independent $\Phi(\vec{X})$ nuclear wave function is computed using local representations of each operator, namely a discrete variable representation (DVR; Light et al. 1985; Lill et al. 1986; Kosloff 1988) for the potential energy operator, and a finite basis representation (FBR; Light et al. 1985; Lill et al. 1986; Kosloff 1988) for the radial and angular kinetic energy operators. The FBR is usually built from eigenfunctions of one-dimensional (1D) zero-order Hamiltonians like harmonic oscillator Hamiltonian, Coulomb-like Hamiltonian, rotation, or second-derivative operators, etc. The DVR representation is then built from this FBR in the manner established by Light et al. (1985), Lill et al. (1986), and Kosloff (1988). In a multidimensional problem, the DVR is built as a direct product of these 1D functions.

Time-dependent Wave Packet Calculations
In this approach, a wave packet initially prepared in the reactant region is propagated using the time-dependent Schrödinger equation:

$$i\hbar \frac{\partial \psi(\vec{X}, t)}{\partial t} = H\psi(\vec{X}, t), \tag{3.31}$$

until it reaches the product region. Three steps can be then identified in this simulation process. The first step is to determine the initial wave packet, the second is devoted to time propagation, and the third is dedicated to physical data calculations. Below, we briefly present the most commonly used approaches for the calculation of reactive cross sections and reactive rate coefficients, which are the most relevant data for astrophysical models. As mentioned earlier only these data are required, the "coordinates problem," disappears because it is possible to compute the total reaction probability using only one set of Jacobi coordinates—either those of the reactants or of the products.

Preparation of the Initial Wave Packet
It is more convenient to construct the initial state representing an atom A far away from the diatom BC in an initial rovibrational quantum state (v_0, j_0, l_0) in the SF rather than in the BF. Then, in the SF, the initial wave packet is given by:

$$\Psi^{JMp}_{v_0 j_0 l_0}(\vec{R}, \vec{r}, t = 0) = G(R)\, \varphi_{v_0 j_0}(r)\, F^{JMp}_{j_0 l_0}(\hat{R}, \hat{r}), \tag{3.32}$$

where $F^{JMp}_{j_0 l_0}(\hat{R}, \hat{r})$ is the total angular momentum eigenstate in the SF representation, M the projection of the total angular momentum J on the SF Z axis, and $p = (-1)^{j_0 + l_0}$ the parity. The function $\varphi_{v_0 j_0}(r)$ is the eigenfunction of the molecule AB in its rovibrational (v_0, j_0) state, and $G(R)$ is a Gaussian wavepacket describing the relative translational motion of A with respect to BC. The time propagation being numerically more efficient in the BF representation, one should express this initial SF wavepacket in the BF using angular momentum transformations.

Wavepacket propagation. Several approaches have been proposed to solve the time-dependent Schrödinger Equation (3.31). One is the multiconfiguration time-dependent Hartree method (MCTDH) developed by Meyer et al. (1990), Manthe et al. (1992), Beck et al. (2000), Meyer & Worth (2003), Meyer et al. (2009), Meyer (2012), and Worth et al. (2000). Another, called the "standard approach," has been developed by several authors.

MCTDH propagation. The MCTDH approach (Meyer et al. 1990; Manthe et al. 1992; Beck et al. 2000; Worth et al. 2000; Meyer & Worth 2003; Meyer et al. 2009; Meyer 2012) is a very efficient method to solve the time-dependent Schrödinger equation for high-dimensional systems (Huarte-Larrañaga & Manthe 2002b; Viel et al. 2004; Bhattacharya et al. 2010; Schiffel & Manthe 2010; Evenhuis & Manthe 2011; Westermann et al. 2011). The MCTDH approach precisely treats triatomic (Manthe et al. 1992) and tetratomic systems (Bhattacharya et al. 2010) and is the only choice to treat systems involving more than five atoms (Huarte-Larrañaga &

Manthe 2000, 2001, 2002a, 2002b; Schiffel & Manthe 2010). In this approach, the wave function is expanded as a linear combination of Hartree products of time-dependent functions called single-particle functions (SPF):

$$\psi(x_1, \ldots x_f, t) = \sum_{j_1=1}^{n_1} \cdots \sum_{j_f=1}^{n_f} A_{j_1 \ldots j_f}(t) \prod_{\kappa=1} \varphi_{j_\kappa}^{(\kappa)}(x_\kappa, t), \tag{3.33}$$

where f is the number of degrees of freedom of the system, $x_1, \ldots, x_f$ denote the nuclear coordinates, $A_{j_1 \ldots j_f}(t)$ the time-dependent MCTDH expansion coefficients, and $\varphi_{j_\kappa}^{(\kappa)}(x_\kappa, t)$ the n_κ time-dependent SPFs associated with the degree of freedom κ. These SPFs are represented on the time-independent DVR basis set defined in Section 3.4.1 with N_κ points for the κth degree of freedom. The multiconfigurational nature of Equation (3.33) ensures a correct description of the correlation between the degrees of freedom if a sufficiently large number of configurations is taken into account, which means that the number of n_κ SPFs is large enough. Each degree of freedom κ is associated with a small number of SPFs with $n_\kappa \ll N_\kappa$, which, through their dependence on time, provides them with enough flexibility to describe the entire process. This latter point plays a major role in the numerical efficiency of the MCTDH approach to treat large systems. From the Dirac–Frenkel variational principle (Dirac 1930; Frenkel 1934), a set of coupled non-linear differential equations of the first order is derived, leading to the equations of motion. In order to make the MCTDH method efficient, the Hamiltonian operator is better written as a sum of products of single-particle operators. The kinetic energy operator when expressed in the appropriate coordinates usually has the required form. For the potential energy operator, which generally does not have this form, several algorithms exist: (i) for small systems, POTFIT (Jäckle & Meyer 1996; Beck et al. 2000; Meyer et al. 2009); and for large systems, Multigrid POTFIT (Peláez & Meyer 2013; 2017) and Monte Carlo POTFIT (Schröder & Meyer 2017), allowing transformation of the potential into a sum-of-products form for subsequent MCTDH calculations. For more details on the MCTDH algorithm, we refer the interested reader to the review papers (Beck et al. 2000; Meyer et al. 2009; Meyer 2012). An alternative approach is provided by the correlation DVR (CDVR) method of Manthe (1996).

The MCTDH method has been used in several studies of large systems, such as the 6D and 12D reactive scattering of H_2 + OH (Bhattacharya et al. 2010), H + CH_4 (Schiffel & Manthe 2010), and O + CH_4 (Huarte-Larrañaga & Manthe 2002b).

Standard Approach. In the standard approach, the time evolution operator is either (i) approximated by a product of terms, each one being diagonal in the given representation momentum or coordinate, or (ii) expanded in terms of orthogonal polynomials.

The most popular approximated time evolution operator extensively used in interstellar reactive TDWP calculations is the second-order split operator (SO) initially proposed by Feit et al. (1982) and Feit & Fleck (1983). In this scheme, the propagation in time is discretized with uniform time steps Δt and the wave packet at time $(n+1)\Delta t$ is obtained by applying the following approximate evolution operator on the wave packet at time Δt.

$$\exp\left[\frac{-iH\Delta t}{\hbar}\right] \approx \exp\left[\frac{-iV\Delta t}{2\hbar}\right].\exp\left[\frac{-iT\Delta t}{\hbar}\right].\exp\left[\frac{-iV\Delta t}{2\hbar}\right]. \tag{3.34}$$

This SO propagator exhibits interesting properties, like its numerical stability, the conservation of the norm of the wave packet, its accuracy up to second order, and its simplicity.

However, the most popular expansion for the time-evolution operator is based on orthogonal Chebyshev polynomials and was proposed by Tal-Ezer & Kosloff (1984). The convergence of this expansion is assured by Bessel functions, which decrease exponentially when the order N is greater than their argument, thus guaranteeing the numerical stability of the method. Despite the fact that this propagator is not intrinsically unitary, propagations can be as accurate as machine precision and then the norm serves as a test for convergence. This scheme can be used with a small or large time step Δt.

Closely related to the Chebychev expansion, some years ago Gray & Balint-Kurti (1998) developed a new, accurate approach in which a real instead of complex wave packet is propagated. Because observables like cross sections $\sigma(E)$ do not explicitly depend on time, it is permissible to substitute a modified time evolution equation for the time-dependent Schrödinger. The basic idea of this approach is to map the Hamiltonian operator as $f(H) = -\frac{\hbar}{\Delta t}\cos^{-1}(H_{\mathrm{norm}})$ and to propagate the wave packet using a modified time-dependent Schrödinger equations. In this latter form, H_{norm} is the normalized Hamiltonian. This specific mapping of the Hamiltonian operator allows the propagation in time to be exactly accomplished by a simple Chebyshev iteration with only a single evaluation of the normalized Hamiltonian on a real wave packet. Similar approaches have been developed by Huang et al. (1994, 1996), Mandelshtam & Taylor (1995a, 1995b), Kroes & Neuhauser (1996), Chen & Guo (1996), and Althorpe (2001). Sun and coworkers have shown that this approach is advantageous, particularly for reactions dominated by long-lived resonances, in terms of both efficiency and accuracy due to the use of real arithmetics and the fact that no interpolation of the propagator is required (Sun et al. 2009b).

Standard Calculations of Quantum Reactive Cross Sections and Rate Coefficients
If only initial state-selected $k_{v_0 j_0}(T)$ or thermal $k(T)$ rate coefficients and no state-to-state cross sections or differential cross section are needed, the final analysis of the wave packet can be performed using a flux-based approach (Miller 1974; Zhang & Zhang 1994; Zhang 1999). For each initial (v_0, j_0) state of the reactant diatom, the total reaction probability $P^{J\Omega p}_{v_0 j_0}(E)$, summed over all final (v', j') states of the product molecule, is calculated as the energy-dependent flux through a dividing surface placed at the product channel:

$$P_{v_0 j_0}^{J\Omega p}(E) = \left\langle \Phi_{v_0 j_0}^{J\Omega p}(E) \mid F(r_d) | \Phi_{v_0 j_0}^{J\Omega p} \right\rangle \\ = \frac{\hbar}{\mu_r} \operatorname{Im}\left[\left\langle \Phi_{v_0 j_0}^{J\Omega p}(E) \,\middle|\, \delta(r - r_d)\frac{\partial}{\partial r} \,\middle|\, \Phi_{v_0 j_0}^{J\Omega p}(E) \right\rangle\right], \tag{3.35}$$

where $F(r_d)$ is the quantum flux operator for the dividing surface defined by $r = r_d$. The TI scattering wave function $\Phi_{v_0 j_0}^{J\Omega p}(E)$ is obtained from the TD wave packet $\Psi_{v_0 j_0}^{J\Omega p}(t)$ by a "time-to-energy" Fourier transform. This method is very efficient when coupled with the use of complex absorbing potentials to avoid unphysical reflections of the wave packet at the edges of the grids in r and R. A polynomial form for the complex absorbing potentials as proposed by Riss & Meyer (1998) is usually used. The initial state-selected integral reaction cross section (ICS) corresponding to an initial rovibrational (v_0, j_0) state of the reactant diatom can be written in term of the total reaction probabilities as in Pack (1974), Schatz & Kuppermann (1976), Zhang & Miller (1989), Alagia et al. (2000), and Aslan et al. (2012):

$$\sigma_{v_0 j_0}(E) = \frac{\pi}{k_{v_0 j_0}^2} \frac{1}{(2j_0 + 1)} \sum_{J=0}^{J_{\max}} (2J + 1)\left[2 \min(J, j_0) + 1\right] P_{v_0 j_0}^{J}(E), \tag{3.36}$$

where $\hbar^2 k_{v_0 j_0}^2 / 2\mu_R = (E - \epsilon_{v_0 j_0})$.

An exact determination of $P_{v_0 j_0}^{J}(E)$ would require calculating the state-selected probabilities $P_{v_0 j_0}^{J\Omega p}(E)$ for all possible helicity states Ω of the reactants. For reactions involving atoms heavier than hydrogen, the number of partial waves $J_{\max}$ necessary to obtain converged cross sections can be quite large (for example, up to 270 for the Si + OH reaction in the energy range $[10^{-3}, 1]$ eV; see Rivero Santamaría et al. 2017). An explicit calculation of the individual reaction probabilities $P_{v_0 j_0}^{J}(E)$ for $J > 0$ would be very computationally demanding in a rigorous treatment including all possible helicity states Ω and Coriolis couplings between adjacent components Ω and $\Omega \pm 1$. Several approximated treatments can be used to reduce the computational effort, such as: the centrifugal sudden (CS) approximation (Pack 1974), where Coriolis couplings are neglected; the standard J-shifting approximation (Sun et al. 1990); a somewhat J-shifting approximation based on the capture model, proposed by Gray et al. (1999); and the uniform J-shifting approximation (Zhang & Zhang 1999).

The ICS values obtained through Equation (3.36) are used to calculate the state-selected rate coefficients $k_{v_0 j_0}(T)$ corresponding to a Maxwell distribution of the reactants relative velocities at the temperature T:

$$k_{v_0 j_0}(T) = f_e(T)\left(\frac{8k_B T}{\pi \mu_R}\right)^{\frac{1}{2}} \times \int_0^{\infty} \sigma_{v_0 j_0}(E_c)\frac{E_c}{k_B T} \exp\left(-\frac{E_c}{k_B T}\right) d\left(\frac{E_c}{k_B T}\right), \tag{3.37}$$

where $E_c = (E - \epsilon_{v_0 j_0})$ is the collision energy, k_B is the Boltzmann constant, and $f_e(T)$ is the electronic partition function, which takes into account the fine-structure states of the reactant. The state-selected rate coefficients obtained in this manner are used to compute the thermal rate coefficients $k(T)$ corresponding to a Boltzmann distribution of the rovibrational states of the reactant molecule BC(v, j) at the temperature T:

$$k(T) = \frac{1}{Q_r(T)} \sum_{v=0}^{\infty} \sum_{j=0}^{\infty} (2j + 1) \exp\left(-\frac{\epsilon_{v,j}}{k_B T}\right) k_{vj}(T), \tag{3.38}$$

where $Q_r(T)$ is the reactant partition function of the rovibrational states.

A series of barrierless reactions proceeding through deep potential energy wells between open-shell atoms (C: Bulut et al. 2009; Jorfi & Honvault 2010a, N: Daranlot et al. 2011, O: Xu et al. 2007; Quéméner et al. 2008; Lique et al. 2009, S: Jorfi & Honvault 2010b, Si: Rivero Santamaría et al. 2017), and the OH radical have been studied using previous formalisms.

Flux Correlation Function Calculations of Quantum Reactive Cross Sections and Rate Coefficients

$k_{v_0 j_0}(T)$ formalism of determining thermal rate coefficient $k(T)$ can no longer be applied to larger systems (more than four atoms), even if the CS and J-shifting approximations are intensively used. In order to solve this problem, approaches allowing a direct calculation of $k(T)$ based on the formalism of flux correlation functions initially introduced by Yamamoto (1960) and reviewed by Miller (1974), Miller et al. (1983), and Manthe & Miller (1993) in the context of reactive scattering have been developed. In concept, this approach can be considered as a rigorously correct, quantum mechanical generalization of the transition-state theory (TST). The thermal rate coefficient can be directly obtained as the Boltzmann average of the cumulative reaction probability (CRP) $N(E)$ at total energy E,

$$k(T) = \frac{1}{2\pi Q_r(T)} \int_{-\infty}^{+\infty} \exp\left(-\frac{E}{k_B T}\right) N(E) dE, \tag{3.39}$$

here with $\hbar = 1$.

Miller and coworkers have shown that the CRP can be rigorously, accurately, and efficiently calculated using flux-flux correlation functions (Miller 1974; Miller et al. 1983; Manthe & Miller 1993). Welsch et al. (2012) have developed an efficient approach using the MCTDH algorithm and this flux–flux correlation function in order to calculate the thermal rate coefficient, initial state-selected rate coefficient, and state-to-state probabilities for triatomic and tetratomic molecules and complex molecules with more than four atoms, such as H + CH_4 (Schiffel & Manthe 2010; Huarte-Larrañaga & Manthe 2000, 2001, 2002a), O + CH_4 (Huarte-Larrañaga & Manthe 2002b), and H_2 /HD/D_2 + CH_3 (Nyman et al. 2007).

Quantum Time-independent Calculations

The time-independent wave function $\psi(\vec{Q})$ is calculated by solving the time-independent Schrödinger equation

$$H\psi(\vec{Q}) = E\psi(\vec{Q}), \tag{3.40}$$

using a close-coupling (CC) scheme obtained by expanding the total wave function on a basis set (Pack & Parker 1987). The majority of TIQ methods are based on hyperspherical coordinates. TIQ calculations provide the entire S-matrix at a given energy while TDWP calculations give only one column of the S-matrix at a time, but for a range of energies defined by the initial wavepacket. Two majors numerical programs have been developed using this TIQ approach: one was proposed by Lepetit et al. (1986) and Honvault & Launay (2005); the other was developed by Schatz (1988) and Skouteris et al. (2000), and implemented in the well-known ABC program. Two differences between these two approaches can be pointed out. The first difference concerns the choice of the hyperspherical coordinates system used, and the second is related to the basis functions used to expand on the total wave function. In the following paragraphs, we briefly present the approach of Launay and will discuss its differences with that of the ABC code.

Launay's approach. In this approach, the symmetrical hyperspherical coordinates, also called democratic coordinates (Section 3.4.1), are used and the Schrödinger equation is solved using the diabatic-by-sector method (Lepetit et al. 1986) in which the hyperradius ρ is discretized in sectors $[\rho_m - \Delta\rho/2, \rho_m + \Delta\rho/2]$. In each sector, the total wave function at a given parity $p = \pm 1$ and given values J of the total angular momentum and M of its projection in the SF Z axis, is expending on basis functions $\Phi^p_{k\Omega}(\rho_m; \theta, \chi)$, which are eigenfunctions associated with eigenvalues $\epsilon_{k\Omega}(\rho_m)$ of the 2D Hamiltonian $H_\Omega(\rho_m)$:

$$H_\Omega(\rho_m) = \frac{\hbar^2}{2\mu\rho_m^2}\left(-\frac{4}{\sin 2\theta}\frac{\partial}{\partial\theta}\sin 2\theta\frac{\partial}{\partial\theta} - \frac{1}{\cos^2\theta}\frac{\partial^2}{\partial\chi_\tau^2} + \frac{4\Omega^2}{\sin^2\theta}\right) + V(\rho_m, \theta, \chi). \tag{3.41}$$

For each symmetry block $\{\Omega, p\}$ and each value ρ_m of ρ, the eigenvalue problem $H_\Omega(\rho_m)\Phi^p_{k\Omega} = \epsilon_{k\Omega}(\rho_m)\Phi^p_{k\Omega}$ is solved by variationally expanding on a primitive basis of pseudo-hyperspherical harmonics (Launay & Le Dourneuf 1989), leading to hyper-radial functions $f^{Jp}_{k\Omega}(\rho_m; \rho)$, solutions of a set of coupled second-order differential equations. The resolution of the radial problem is achieved by propagating the matrix of the logarithmic derivative of $f^{Jp}_{k\Omega}(\rho_m; \rho)$ outward in each sector using the Manolopoulos (1986) algorithm. The asymptotic behavior of the wave function being defined in Jacobi space-fixed coordinates, it is then necessary to transform the wave function $\psi^{JMp}(\rho_{\max}, \theta, \chi, \alpha, \phi, \gamma)$, initially in body-fixed hyperspherical democratic coordinates, into the space-fixed Jacobi coordinates wave function $\Psi^{JMp}(\vec{R_\tau}, \vec{r_\tau})$. By matching this transformed wave function with its asymptotic form for ρ_{max}, all the scattering S-matrix elements $S^{Jp}_{v'l'j' \leftarrow vlj}(E)$ are then obtained and the state-to-state cross sections are given by:

$$\sigma_{v'j' \leftarrow vj}(E) = \frac{\pi}{k_{vj}^2} \sum_{Jpll'} (2J + 1) |S^{Jp}_{v'l'j' \leftarrow vlj}(E)|^2, \tag{3.42}$$

where k_{vj}^2 is the initial wavenumber. The initial state-selected integral cross sections $\sigma_{vj}(E)$ are then obtained by summing Equation (3.42) over all final rovibrational states (v', j'). The main advantage of this approach is the reduced number of channels in the close-coupling expansion due to the "sector-adiabatic" basis functions, which are independent of total angular momentum and allow a contraction scheme. This approach has been successfully applied for both direct (Lepetit et al. 1986; Launay & Lepetit 1988; Launay & Le Dourneuf 1989; Launay & Padkjær 1991) and indirect reactions (Honvault & Launay 2001, 2005; Jorfi et al. 2009; Jorfi & Honvault 2009, 2010a, 2010b).

ABC's approach. In this approach, the Fock Delves hyperspherical coordinates defined in Section 3.4.1 are preferred, and in each sector $[\rho_m - \Delta\rho/2, \rho_m + \Delta\rho/2]$, the total wave function is developed, taking into account all possible arrangements τ, on the basis of functions $\Xi_{\tau jv}(\rho_m; \chi_\tau)$, which are solutions of the equation:

$$\left[\frac{1}{2\mu\rho_m^2}\left(-\frac{\partial^2}{\partial\chi_\tau^2} + \frac{j_\tau^2}{\sin^2 2\chi_\tau}\right) + V(\rho, \chi_\tau) - \epsilon_{\tau jv}(\rho_m)\right]\Xi_{\tau jv}(\rho_m; \chi_\tau) = 0, \tag{3.43}$$

where $V(\rho, \chi_\tau)$ is the diatomic energy potential. Functions $\Xi_{\tau jv}(\rho_m; \chi_\tau)$ can be then interpreted as the vibrational eigenfunctions of the diatom in the hyperspherical coordinates. By inserting this form in the Schrödinger equation, one obtains a set of CC equations for the hyperradial coefficients $f^{Jp}_{\tau' jv\Omega_\tau}(\rho_m; \rho)$, solved using the log derivative (Manolopoulos 1986) algorithm. Additionally, if reactive scattering boundary conditions are used with the helicity representation, the scattering matrix $S^{Jp}(E)$ can then be extracted. State-to-state cross sections $\sigma_{v'j' \leftarrow vj}(E)$ and initial state-selected integral cross sections $\sigma_{vj}(E)$ are obtained according to Equation (3.42). The initial state-selected rate coefficient $k_{vj}(T)$ and thermal rate coefficient $k(T)$ are obtained by Equations (3.37) and (3.38).

Similar to Launay's approach, the use of the helicity representation allows for reduction of the computational effort because convergence can be achieved by reducing the number of ω values in the development of the total wavefunction. This approach has been successfully applied for both the direct (Alexander et al. 2000; Che et al. 2007; Lique et al. 2011) and indirect reactions (Lique et al. 2009; Warmbier & Schneider 2011; Werfelli et al. 2015). The disadvantage of ABC code, as compared to Launay's, comes from the simultaneous expansion of the wave function in Delves hyperspherical coordinates of the three channels. This results in the use of larger basis sets, which can become a severe limitation for simulation of insertion reactions involving deep wells and heavy atoms, like C + OH (Bulut et al.

2009; Jorfi & Honvault 2010a; Rajagopala Rao et al. 2013) or Si + OH (Dayou et al. 2013; Rivero-Santamaría et al. 2014; Rivero Santamaría et al. 2017).

3.4.2 Semi-classical and Statistical Calculations

Potential Energy Surface

For large polyatomic systems, quantum approaches are no longer applicable. For interstellar reactions, the entrance channel is often the most important in determining the rate coefficients, because these reactions are exothermic and typically have no intrinsic barriers on the PES. For ion–molecular reactions, the long-range electrostatic terms dominate (Gioumousis & Stevenson 1958). Even for barrierless neutral reactions, the dipole-induced dipole and/or dipole–quadrupole interactions may also be overwhelming at low temperatures. Significant progress has been made in analytical representation of the PES from a large number of ab initio points (Varandas 1988). Direct fitting of the ab initio points at large intermolecular distances is usually not needed. Instead, analytical forms of the long-range interactions (Stone 2013) are often used, which are switched to the short-range ab initio PES via smooth functions with asymptotic values of 0 and 1.

In a recent study of the astronomically important $H_2O^+ + H_2 \rightarrow H_3O^+ + H$ reaction involved in water formation in interstellar media, a full-dimensional PES was developed based on this approach (Li & Guo 2014). The short-range PES was determined by fitting many CCSD(T) points distributed in a large configuration space, using the recently developed permutation invariant polynomial-neural network (PIP-NN) method (Jiang et al. 2016). In the reactant and product asymptotes, on the other hand, analytical ion–quadrupole and ion-induced dipole terms are used to describe the long-range interaction. It has been demonstrated that the short-range PIP-NN PES and long-range terms agree very well in the intermediate range, where the switching function is used. Dynamical calculations on this PES yield excellent agreement with both experimental thermal rate coefficients and energy-dependent cross sections for several different isotopes (Ard et al. 2014; Song et al. 2016).

Rate Coefficient Determination

Once a PES is available, the most accurate method to determine rate coefficients is quantum dynamics (QD), which solves the nuclear Schrödinger equation, with appropriate boundary conditions. The details of such an approach are discussed in Section 3.4.1 as well as in a recent review (Zhang & Guo 2016), and thus are not given here. This is usually a very demanding undertaking, as the solution of the differential equation using either time-dependent or time-independent methods is numerically costly. As a result, only small systems are amenable to such treatments. A major problem with this approach is the convergence of near-threshold resonances, particularly with the time-dependent approach. This is because the low translational energy renders the grids exceedingly large in order to accommodate the long de Broglie wavelength characteristic of the low-temperature collisions.

Quasiclassical trajectory method. A much less expensive alternative is the quasi-classical trajectory (QCT) method, which treats the nuclear dynamics in a PES with Newtonian mechanics. The classical Hamiltonian for an N-atom system is conveniently expressed in Cartesian coordinates:

$$H = \sum_{i=1}^{N} \frac{\left(p_{x_i}^2 + p_{y_i}^2 + p_{z_i}^2\right)}{2m_i} + V \tag{3.44}$$

where the potential V is only a function of the internal coordinates and m_i is the atomic mass. The propagation of the trajectories is performed by integrating Hamilton's equation. The quantization of internal energy of the reactants is included semi-classically in the initial conditions, and that for the products is also approximately enforced by various binning schemes. However, other quantum effects, such as resonances and tunneling, are not included. In addition, the semiclassical quantization of the internal degrees of freedom of the reactant can also lead to problems as zero-point energy leakage, namely the artificial energy outflow from the initially deposited vibrational energy. Interestingly, the QCT method has been shown to fare quite well with QD results, especially for highly averaged attributes such as the rate coefficient (Aoiz et al. 2006). For barrierless reactions with reaction intermediates at low temperatures, comparison with experimental results has also been quite positive, especially for reactions with a large exothermicity (Guo 2012). One potential difficulty in the QCT calculation of rate coefficients is for reactions that are near-thermoneutral, in which the rate coefficient depends sensitively on how the zero-point energy (ZPE) of the products (and reactants) are treated. In such cases, trajectories with energies below the ZPE could emerge in the asymptote due to the disregard of ZPE by classical mechanics, thus violating quantum mechanics. There have been several efforts to remedy the problem (Bowman et al. 1989; Miller et al. 1989). One such approach is to turn back the ZPE-violating trajectories by changing the sign of the atomic momenta (Paul & Hase 2016). By doing so, only trajectories above the ZPE are allowed to exit. This method was applied in a recent study of the $H + NH_2 \rightarrow H + NH_2$ exchange reaction, which is important for the ortho-to-para conversion of NH_2 in interstellar media (Le Gal et al. 2016).

Transition-state theory. A completely different approach to calculating rate coefficients is provided by the TST (Fernández-Ramos et al. 2006). For reactions with a barrier, the transition state can be readily defined at the saddle point in the PES. The rate coefficient can then be determined from the partition functions at the transition state. In the classical form, the TST rate coefficient is given by

$$k(T) = \frac{k_B T}{h} \frac{Q^{\neq}(T)}{Q^R(T)} e^{-E^{\neq}/k_B T}, \tag{3.45}$$

where $Q^R(T)$ and $Q^{\neq}(T)$ are the classical partition functions for the reactants and transition state, and h is Planck's constant. These classical partition functions can be readily replaced by their quantum mechanical counterparts. Traditionally, the transition state is often defined by a dividing surface placed at the saddle point of the PES ($E^{\neq}$) that separates reactants from products, and perpendicular to the reaction coordinate (Fernández-Ramos et al. 2006). Quantum effects, such as tunneling, are approximately taken into consideration and the vibrational partition functions are often approximated with the harmonic oscillator model. A distinct advantage of TST is that it does not require the entire PES; information near the transition state is sufficient. For barrierless complex-forming reactions, there exist bottlenecks, as described above, which are essentially free-energy barriers. The application of the TST to a barrierless reaction is based on the identification of such a free-energy barrier, which could be due either to a submerged barrier in the strong-interaction region (inner transition state), or to the long-range interaction (outer transition state) (Greenwald et al. 2005; Georgievskii & Klippenstein 2005). The latter often becomes dominant at low temperatures, due to the centrifugal potential. It is apparent that the traditional TST is based on statistical mechanics and completely ignores recrossing dynamics, which can be quite important near the transition state. In variational TST (VTST), the dividing surface is variationally shifted to search for the lower limit of the TST rate coefficient, in order to account approximately for recrossing of the transition state. The TST approach has mostly been used to study kinetics of large systems, as dynamical corrections are typically small in these cases. For example, such a method has been successfully applied to elucidate the low-temperature kinetics for O + alkene reactions (Sabbah et al. 2007), where the interplay between the inner (chemical) and outer (long-range interaction) transition states along the reaction coordinate dictates the complicated temperature dependence of the rate coefficient. It should be emphasized that TST, even in its variational form, does not consider dynamical effects, such as recrossing at the transition state, which could lead to errors, particularly for small and non-statistical systems.

Ring polymer molecular dynamics. Very recently, an approximate quantum mechanical TST based on ring-polymer molecular dynamics (RPMD) has emerged as an accurate and efficient means to compute rate coefficients (Suleimanov et al. 2016). In RPMD, quantum effects, such as ZPE and tunneling, are approximately included by taking advantage of the isomorphism between statistical properties of a quantum system and those of a classical ring polymer with harmonically linked beads, akin to the path integral concept proposed by Feynman (Habershon et al. 2013). As a result, the numerical calculation of the rate coefficients can be performed with classical trajectories and the resulting scaling laws allow for its application to large systems. Different from the traditional TST, RPMD requires the global PES to compute both the free energy and recrossing dynamics. The rate coefficient is written as a product of a static term determined solely by the free-energy barrier and a dynamic term that describes the recrossing at the transition state (ϵ):

$$k(T) = k_{\mathrm{QTST}}(T, \zeta^{\neq})\kappa(t \to \infty, \zeta^{\neq}), \tag{3.46}$$

where $\kappa(t \to \infty, \zeta^{\neq})$ is the plateau value of the transmission coefficients, and $k_{\mathrm{QTST}}(T, \zeta^{\neq})$ the centroid-density quantum TST rate coefficient. Recent tests on prototypical examples have shown that the RPMD approach gives excellent results for complex-forming reactions, such as the H_2 insertion reaction by excited atoms like $C(^1D)$, $O(^1D)$, $N(^2D)$, and $S(^1D)$ (Li et al. 2014; Suleimanov et al. 2014), some at low temperatures (Hickson et al. 2015; Kumar et al. 2018).

Statistical models. An even simpler theoretical approach is based on the capture concept, which assumes that low-temperature reactivity for barrierless reactions with large exothermicity is dominated by capture of the reactants by long-range attractive forces. Once captured, the reaction intermediate dissociates irreversibly toward the products. These capture models certainly work well with many ion–molecular reactions, but have been shown to also provide an approximate estimate of rate coefficients of many barrierless neural reactions (Clary 1990; Troe 1997; González-Lezana 2007). These capture models need to take into consideration the quantized reactant states, which typically involve the rotation of the molecules at low temperatures. For complex-forming reactions that have small exothermicity, this capture model can be extended. In these so-called statistical models, the reaction intermediate is assumed to be completely randomized in energy, and as a result, the reactivity is determined entirely by the availability of open channels in both reactant and product asymptotes. The simplest version of such a statistical approach is probably the phase space theory, which expresses the reaction probability in terms of available reactant/product channels with the energy and angular momentum constraints (Light 1964; Pechukas & Light 1965)

$$\sigma(E_c) = \frac{\pi}{2\pi E_c}\sum_J (2J+1)P_J(E_c), \tag{3.47}$$

where the reaction probability $P_J(E_c)$ is obtained by counting the numbers of open channels.

A more sophisticated statistical model relies on the capture probabilities in both reactant and product channels that are computed quantum mechanically (Rackham et al. 2003; Lin & Guo 2004). A caveat is that, for small reactive systems, as discussed above in the context of TST, there might be significant non-statistical (namely dynamical) behaviors, which could result in the failure of the capture or statistical model (Guo 2012).

Product Channels and Branching Ratios

Another important issue in interstellar chemistry is concerned with the branching ratio of the product channels. Again, a proper theoretical prediction requires knowledge of the global PES that covers not only the strongly interaction region, but also the reactant and product asymptotes. In a recent study of the $C(^3P) + H_2O$ reaction, which has been thought to be astrochemically relevant due to the large abundance of both reactant species (Hickson et al. 2016), the PES was found to have complicated topography, with multiple wells and transition states, eventually leading to several product channels (Li et al. 2017). In the statistical limit, where the energy is completely randomized, the product branching ratio can be computed using statistical models, such as that of the Rice–Ramsperger–Kassel–Marcus (RRKM) theory (Baer & Hase 1996). Specifically, the RRKM rate coefficient at a particular energy E is given as follows:

$$k(E) = \frac{N^{\neq}(E))}{h\rho^{R}(E)}, \tag{3.48}$$

where $N^{\neq}(E)$ is the numbers of states at the transition state, and $\rho^{R}(E)$ the reactant density of states per unit energy. The branching ratios are determined by the energetics of the relevant transition states or asymptotic limits. Indeed, in many large systems, the RRKM theory is quite successful in predicting the product branching ratios. However, in small systems, particularly those with H atoms, that statistical limit is often not reached. Under such circumstances, dynamics dominates. Indeed, in the case of the $C(^3P) + H_2O$ reaction, the dominant product channel is predicted to be HOC + H, which has a much higher energy than that of HCO + H (Li et al. 2017). This non-statistical effect can be attributed to the large kinetic energy gained by the dissociating H atom as it is transferred from H_2O to C.

3.5 Some Perspectives

The unprecedented signal-to-noise ratio and the high spatial resolution of the Atacama Large Millimeter Array (ALMA) and the NOrthern Extended Millimeter Array (NOEMA), are both stimulating the discovery of complex molecules in ISM, as well as giving access to the spatial distribution of these species even in distant astronomical objects. The upcoming *James Webb Space Telescope* will complement these observations in the infrared range, also with high spatial resolution. The laboratory investigation of the reactivity supported by theoretical calculations must be consolidated to keep up with these astronomical observations, in an effort to identify the key formation and destruction pathways. In particular, the quest for channel-specific rate coefficients at very low temperature should be expanded. The development of methods that can determine the precise nature of the reaction products through laser or microwave spectroscopy, or through threshold VUV single-photon ionization, appears today indispensible to the in-depth study of bimolecular reactions, and they are currently the object of various experimental developments. Furthermore, because of the low pressure prevailing in the ISM, non-local thermal equilibrium (non-LTE) populations are very common. State-to-state reactivity therefore appears highly desirable. However, this is an even more challenging difficulty than branching ratio determination. For instance, if rate coefficients have been determined for reagents in quite low vibrational states, less information is known about the influence of reagent rotational excitation on reactivity. One reason for the scarcity of such data is the celerity at which rotational energy transfer occurs during collisions in laboratory conditions, which hinders the quantitative measurement of the process. More generally, the reactivity of specific states, the assignment of the structure and internal quantum states of the products of reactive or non-reactive collisions and photodissociation are great experimental challenges. Another challenge is also to be able to address a large number of radical–radical reactions experimentally. There are also many challenges for theoretical characterization of low-temperature kinetics. Among them, the accurate determination of multidimensional PESs and advancement of methods that can accurately account for quantum effects at low-temperature collisions are most urgent. The former relies on the progress in highly accurate quantum chemical methods and high-fidelity representation of the PES. The latter requires new methods that avoid the exponential scaling law of quantum mechanics. The RPMD method appears to be quite promising. Methods that allow resolution of product channels are also needed, especially beyond the statistical limit. Finally, the inclusion of fine structure states, which might lead to non-adiabatic crossings, is necessary. This places additional burden on both PES construction and dynamics.

References

Abeysekera, C., Zack, L. N., Park, G. B., et al. 2014, JChPh, 141, 214203
Abeysekera, C., Joalland, B., Ariyasingha, N., et al. 2015, JPCL, 6, 1599
Adams, N. G. 1993, Dissociative Recombination: Theory, Experiment, and Applications, ed. B. Rowe, J. B. A. Mitchell, & A. Canosa, (Boston, MA: Springer), 99
Adams, N. G., & Smith, D. 1976, IJMSI, 21, 349
Adhikari, S., & Varandas, A. J. C. 2013, CoPhC, 184, 270
Alagia, M., Balucani, N., Cartechini, L., et al. 2000, PCCP, 2, 599
Alexander, M. H., Manolopoulos, D. E., & Werner, H.-J. 2000, JChPh, 113, 11084
Althorpe, S. C. 2001, JChPh, 114, 1601
Anicich, V. G., & Huntress, W. T. Jr 1986, ApJS, 62, 553
Aoiz, F. J., Bañares, L., & Herrero, V. J. 2006, JPCA, 110, 12546
Aquilanti, V., Cavalli, S., De Fazio, D., et al. 2005, CP, 308, 237
Aquilanti, V., Mundim, K. C., Cavalli, S., et al. 2012, CP, 398, 186
Ard, S. G., Li, A., Martinez, O. Jr, et al. 2014, JPCA, 118, 11485
Aslan, E., Bulut, N., Castillo, J. F., et al. 2012, JPCA, 116, 132
Atkinson, D. B., & Smith, M. A. 1995, RScI, 66, 4434
Baer, T., & Hase, W. L. 1996, Unimolecular Reaction Dynamics: Theory and Experiments (Oxford: Oxford Univ. Press)
Balint-Kurti, G. G., Gonzalez, I. A., Goldfield, M. E., & Gray, K. S. 1998, FaDi, 110, 169
Beck, M. H., Jäckle, A., Worth, G. A., & Meyer, H.-D. 2000, PhR, 324, 1
Bhattacharya, S., Panda, A. N., & Meyer, H.-D. 2010, JChPh, 132, 214304
Billing, G. D., & Marković, N. 1993, JChPh, 99, 2674
Bowman, J. M., Gazdy, B., & Sun, Q. 1989, JChPh, 91, 2859
Braams, B. J., & Bowman, J. M. 2009, IRPC, 28, 577
Brown, G. G., Dian, B. C., Douglass, K. O., et al. 2008, RScI, 79, 053103
Bruhns, H., Kreckel, H., Miller, K., et al. 2010, RScI, 81, 013112
Bulut, N., Zanchet, A., Honvault, P., et al. 2009, JChPh, 130, 194303
Chandler, D. W., & Houston, P. L. 1987, JChPh, 87, 1445
Che, L., Ren, Z., Wang, X., et al. 2007, Sci, 317, 1061
Chen, R., & Guo, H. 1996, JChPh, 105, 3569
Clary, D. C. 1990, ARPC, 41, 61

Crawford, J., & Parker, G. A. 2013, JChPh, 138, 054313
Cvitas, M. T., & Althorpe, S. C. 2011, JChPh, 134, 024309
Dahler, J. S., & Hirschfelder, J. O. 1959, PNAS, 45, 249
Daranlot, J., Jorfi, M., Xie, C., et al. 2011, Sci, 334, 1538
Dayou, F., Duflot, D., Rivero-Santamaría, A., & Monnerville, M. 2013, JChPh, 139, 204305
Delves, L. M. 1958, NucPh, 9, 391
Dirac, P. A. M. 1930, PCPS, 26, 376
Echave, J. 1996, JChPh, 104, 1380
Eichelberger, B., Snow, T. P., Barckholtz, C., & Bierbaum, V. M. 2007, ApJ, 667, 1283
Ely, S. C. S., Morales, S. B., Guillemin, J.-C., et al. 2013, JPCA, 117, 12155
Eppink, A. T. J. B., & Parker, D. H. 1997, RScI, 68, 3477
Evenhuis, C. R., & Manthe, U. 2011, JPCA, 115, 5992
Feit, M. D., Fleck, J. A., & Steiger, A. 1982, JCoPh, 47, 412
Feit, M. D., & Fleck, J. A. 1983, JChPh, 78, 301
Ferguson, E. E., Fehsenfeld, F. C., & Schmeltekopf, A. L. 1969, Advances in Atomic and Molecular Physics, Vol. 5, ed. D. R. Bates, & I. Estermann (New York: Academic), 1
Fernández-Ramos, A., Miller, J. A., Klippenstein, S. J., & Truhlar, D. G. 2006, ChRv, 106, 4518
Frenkel, J. 1934, Waves Mechanics (Oxford: Clarendon)
Gannon, K. L., Glowacki, D. R., Blitz, M. A., et al. 2007, JPCA, 111, 6679
Georgievskii, Y., & Klippenstein, S. J. 2005, JChPh, 122, 194103
Georgievskii, Y., & Klippenstein, S. J. 2007, JPCA, 111, 3802
Geppert, W. D., & Larsson, M. 2013, ChRv, 113, 8872
Gerin, M., Neufeld, D. A., & Goicoechea, J. R. 2016, ARA&A, 54, 181
Gerlich, D. 1993, J. Chem. Soc., Faraday Trans., 89, 2199
Gerlich, D. 1992, Inhomogeneous RF Fields: A Versatile Tool for the Study of Processes with Slow Ions (New York: Wiley)
Gerlich, D. 1995, PhST, T59, 256
Ghosh, S., Ghosh, S., Adhikari, S., et al. 2015, JPCA, 119, 12392
Gioumousis, G., & Stevenson, D. P. 1958, JChPh, 29, 294
Göğtas, F., Balint-Kurti, G. G., & Offer, A. R. 1996, JChPh, 104, 7927
Gómez-Carrasco, S., & Roncero, O. 2006, JChPh, 125, 054102
González-Lezana, T. 2007, IRPC, 26, 29
Graul, S. T., & Squires, R. R. 1988, MSRv, 7, 263
Gray, S. K., & Balint-Kurti, G. G. 1998, JChPh, 108, 950
Gray, K. S., Goldfield, M. E., Schatz, C. G., & Balint-Kurti, G. G. 1999, PCCP, 1, 1141
Greenwald, E. E., North, S. W., Georgievskii, Y., & Klippenstein, S. J. 2005, JPCA, 109, 6031
Guo, H. 2012, IRPC, 31, 1
Habershon, S., Manolopoulos, D. E., Markland, T. E., & Miller, T. F. III 2013, ARPC, 64, 387
Hankel, M., Balint-Kurti, G. G., & Gray, S. S. K. 2003, IJQC, 92, 205
Haug, K., Schwenke, D. W., Shima, Y., et al. 1986, JPhCh, 90, 6757
Herman, Z., & Futrell, J. H. 2015, IJMSp, 377, 84
Hickson, K. M., Loison, J.-C., Guo, H., & Suleimanov, Y. V. 2015, JPCL, 6, 4194
Hickson, K. M., Loison, J.-C., Nuñez-Reyes, D., & Méreau, R. 2016, JPCL, 7, 3641
Hipes, P. G., & Kuppermann, A. 1987, CPL, 133, 1
Hirschfelder, J. O., & Dahler, J. S. 1956, PNAS, 42, 363
Honvault, P., & Launay, J.-M. 2001, JChPh, 114, 1057

Honvault, P., & Launay, J.-M. 2005, Quantum Dynamics of Insertion Reactions (Dordrecht: Springer)
Huang, Y., Kouri, D. J., & Hoffman, D. K. 1994, JChPh, 101, 10493
Huang, Y., Iyengar, S. S., Kouri, D. J., & Hoffman, D. K. 1996, JChPh, 105, 927
Huarte-Larrañaga, F., & Manthe, U. 2001, JPCA, 105, 2522
Huarte-Larrañaga, F., & Manthe, U. 2000, JChPh, 113, 5115
Huarte-Larrañaga, F., & Manthe, U. 2002a, JChPh, 116, 2863
Huarte-Larrañaga, F., & Manthe, U. 2002b, JChPh, 117, 4635
Jäckle, A., & Meyer, H.-D. 1996, JChPh, 104, 7974
Jiang, B., Li, J., & Guo, H. 2016, IRPC, 35, 479
Jiménez, E., Ballesteros, B., Canosa, A., et al. 2015, RScI, 86, 045108
Joalland, B., Jamal-Eddine, N., Kłos, J., et al. 2016, JPCL, 7, 2957
Johnson, B. R. 1980, JChPh, 73, 5051
Johnson, B. R. 1983, JChPh, 79, 1906
Jorfi, M., Honvault, P., Bargueño, P., et al. 2009, JChPh, 130, 184301
Jorfi, M., & Honvault, P. 2010a, JPCA, 114, 4742
Jorfi, M., & Honvault, P. 2010b, JChPh, 133, 144315
Jorfi, M., & Honvault, P. 2009, JPCA, 113, 2316
Judson, R. S., Kouri, D. J., Neuhauser, D., & Baer, M. 1990, PhRvA, 42, 351
Kooij, D. M. 1893, ZPC, 12, 155
Kosloff, R. 1988, JPhCh, 92, 2087
Kreckel, H., Bruhns, H., Cizek, M., et al. 2010, Sci, 329, 69
Kroes, G. J., & Neuhauser, D. 1996, JChPh, 105, 8690
Kumar, S. S., Grussie, F., Suleimanov, Y. V., et al. 2018, SciA, 4, eaar3417
Kuppermann, A., Schatz, G. C., & Baer, M. 1976, JChPh, 65, 4596
Kuppermann, A., & Hipes, P. G. 1986, JChPh, 84, 5962
Kuppermann, A. 1975, CPL, 32, 374
Launay, J. M., & Lepetit, B. 1988, CPL, 144, 346
Launay, J. M., & Le Dourneuf, M. 1989, CPL, 163, 178
Launay, J.-M., & Padkjœr, S. B. 1991, CPL, 181, 95
Le Gal, R., Herbst, E., Xie, C., Li, A., & Guo, H. 2016, A&A, 596, A35
Lee, S., Hoobler, R. J., & Leone, S. R. 2000, RScI, 71, 1816
Lepetit, B., Launay, J. M., & Le Dourneuf, M. 1986, CP, 106, 103
Li, A., & Guo, H. 2014, JChPh, 140, 224313
Li, Y., Suleimanov, Y. V., & Guo, H. 2014, JPCL, 5, 700
Li, J., Xie, C., & Guo, H. 2017, PCCP, 19, 23280
Light, J. C., Hamilton, I. P., & Lill, J. V. 1985, JChPh, 82, 1400
Light, J. C. 1964, JChPh, 40, 3221
Lill, J. V., Parker, G. A., & Light, J. C. 1986, JChPh, 85, 900
Lin, S. Y., & Guo, H. 2004, JChPh, 120, 9907
Lique, F., Jorfi, M., Honvault, P., et al. 2009, JChPh, 131, 221104
Lique, F., Li, G., Werner, H.-J., & Alexander, M. H. 2011, JChPh, 134, 231101
Liu, S., Xu, X., & Zhang, D. H. 2011, JChPh, 135, 141108
Mandelshtam, V. A., & Taylor, H. S. 1995a, JChPh, 102, 7390
Mandelshtam, V. A., & Taylor, H. S. 1995b, JChPh, 103, 2903
Manolopoulos, D. E. 1986, JChPh, 85, 6425

Manthe, U. 1996, JChPh, 105, 6989
Manthe, U., & Miller, W. H. 1993, JChPh, 99, 3411
Manthe, U., Meyer, H.-D., & Cederbaum, L. S. 1992, JChPh, 97, 3199
Manzhos, S., Dawes, R., & Carrington, T. 2015, IJQC, 115, 1012
Marković, N., & Billing, G. D. 1994, JChPh, 100, 1085
McCullough, E. A. Jr, & Wyatt, R. E. 1971a, JChPh, 54, 3578
McCullough, E. A. Jr, & Wyatt, R. E. 1971b, JChPh, 54, 3592
Meyer, H.-D., Manthe, U., & Cederbaum, L. S. 1990, CPL, 165, 73
Meyer, H.-D., & Worth, G. A. 2003, ThCA, 109, 251
Meyer, H.-D., Gatti, F., & Worth, G. A. (ed) 2009, Multidimensional Quantum Dynamics: MCTDH Theory and Applications (Weinheim: Wiley)
Meyer, H.-D. 2012, ComMS, 2, 351
Mikosch, J. 2007, PhD thesis, Univ. Freiburg
Miller, G., & Light, J. C. 1971, JChPh, 54, 1635
Miller, W. H. 1974, JChPh, 61, 1823
Miller, W. H., Schwartz, S. D., & Tromp, J. W. 1983, JChPh, 79, 4889
Miller, W. H., Hase, W. L., & Darling, C. L. 1989, JChPh, 91, 2863
Monnerville, M., Péoux, G., Briquez, S., & Halvick, P. 2000, CPL, 322, 157
Morales, S. B. 2009, Thesis, Université de Rennes 1, France
Nyman, G., Harrevelt, V. R., & Manthe, U. 2007, JPCA, 111, 10331
Oldham, J. M., Abeysekera, C., Joalland, B., et al. 2014, JChPh, 141, 154202
Osborn, D. L., Zou, P., Johnsen, H., et al. 2008, RScI, 79, 104103
Pack, R. T. 1974, JChPh, 60, 633
Pack, R. T., & Parker, G. A. 1987, JChPh, 87, 3888
Pan, H., Liu, K., Caracciolo, A., & Casavecchia, P. 2017, CSRv, 46, 7517
Park, G. B., & Field, R. W. 2016, JChPh, 144, 200901
Paul, A. K., & Hase, W. L. 2016, JPCA, 120, 372
Pechukas, P., & Light, J. C. 1965, JChPh, 42, 3281
Peláez, D., & Meyer, H.-D. 2013, JChPh, 138, 014108
Peláez, D., & Meyer, H.-D. 2017, CP, 482, 100
Peng, T., Zhu, W., Wang, D., & Zhang, J. Z. H. 1998, FaDi, 110, 159
Persky, A., & Kornweitz, H. 1997, IJChK, 29, 67
Plasil, R., Mehner, T., Dohnal, P., et al. 2011, ApJ, 737, 60
Quéméner, G., Balakrishnan, N., & Kendrick, B. K. 2008, JChPh, 129, 0224309
Rackham, E. J., Gonzalez-Lezana, T., & Manolopoulos, D. E. 2003, JChPh, 119, 12895
Rajagopala Rao, T., Goswami, S., Mahapatra, S., et al. 2013, JChPh, 138, 094318
Riss, U. V., & Meyer, H. D. 1998, JPhB, 31, 2279
Rivero Santamaría, A., Dayou, F., Rubayo-Soneira, J., & Monnerville, M. 2017, JPCA, 121, 1675
Rivero-Santamaría, A., Dayou, F., Rubayo-Soneira, J., & Monnerville, M. 2014, CPL, 610, 335
Rowe, B. R., Marquette, J.-B., & Rebrion, C. 1989, FaTr2, 85, 1631
Rowe, B. R., Marquette, J. B., & Dupeyrat, G. 1985, Molecular Astrophysics (Dordrecht: Springer), 631
Sabbah, H., Biennier, L., Sims, I. R., et al. 2007, Sci, 317, 102
Saho, T., Ghosh, S., Adhikari, S., et al. 2014, JPCA, 118, 4837
Schatz, G. C., Bowman, J. M., & Kuppermann, A. 1973, JChPh, 58, 4023

Schatz, G. C., & Kuppermann, A. 1976, JChPh, 65, 4642
Schatz, G. C. 1988, CPL, 150, 92
Schiffel, G., & Manthe, U. 2010, JChPh, 133, 174124
Schröder, M., & Meyer, H.-D. 2017, JChPh, 147, 064105
Sims, I. R., Queffelec, J.-L., Travers, D., et al. 1993, CPL, 211, 461
Sims, I. R., Queffelec, J.-L., Defrance, A., et al. 1994, JChPh, 100, 4229
Skouteris, D., Castillo, J. F., & Manolopoulos, D. E. 2000, CoPhC, 133, 128
Smith, F. T. 1962, JMP, 3, 735
Smith, I. W. M. 2008, CSRv, 37, 812
Smith, M. A., & Hawley, M. 1992, in Advances in Gas Phase Ion Chemistry, Vol. 1, Ion Chemistry at Extremely Low Temperatures: a Free Jet Expansion Approach (Bingley: JAI Press), 167
Smith, D., & Spanel, P. 1994, in Advances in Atomic, Molecular, and Optical Physics, Vol. 32, ed. B. Bederson, & A. Dalgarno (New York: Academic), 307
Snow, T. P., & Bierbaum, V. M. 2008, ARAC, 1, 229
Song, H., Li, A., Guo, H., et al. 2016, PCCP, 18, 22509
Soorkia, S., Liu, C.-L., Savee, J. D., et al. 2011, RScI, 82, 124102
Spangenberg, T., Köhler, S., Hansmann, B., et al. 2004, JPCA, 108, 7527
Stone, A. 2013, The Theory of Intermolecular Forces (Oxford: Oxford Univ. Press)
Suleimanov, Y. V., Kong, W. J., Guo, H., & Green, W. H. 2014, JChPh, 141, 244103
Suleimanov, Y. V., Aoiz, F. J., & Guo, H. 2016, JPCA, 120, 8488
Sun, Q., Bowman, J. M., Schatz, G. C., et al. 1990, JChPh, 92, 1677
Sun, Z., Lin, X., Lee, S.-Y., & Zhang, D. H. 2009a, JPCA, 113, 4145
Sun, Z., Lee, S. Y., Guo, H., & Zhang, D. H. 2009b, JChPh, 130, 174102
Sundaram, P., Manivannan, V., & Padmanaban, R. 2017, PCCP, 19, 20172
Su, T., & Chesnavich, W. J. 1982, JChPh, 76, 5183
Taylor, S. E., Goddard, A., Blitz, M. A., et al. 2008, PCCP, 10, 422
Tal-Ezer, H., & Kosloff, R. 1984, JChPh, 81, 3967
Tizniti, M., Le Picard, S. D., Lique, F., et al. 2014, NatCh, 6, 141
Troe, J. 1997, AdChP, 101, 817
Truhlar, D. G., & Kuppermann, A. 1970, JChPh, 52, 3841
Varandas, A. J. C. 1988, AdChP, 74, 255
Viel, A., Krawczyk, R. P., Manthe, U., & Domcke, W. 2004, JChPh, 120, 11000
Wakelam, V., Loison, J.-C., Herbst, E., et al. 2015, ApJS, 217, 20
Warmbier, R., & Schneider, R. 2011, PCCP, 13, 10285
Webster, F., & Light, J. C. 1989, JChPh, 90, 265
Welsch, R., Huarte-Larrañaga, F., & Manthe, U. 2012, JChPh, 136, 064117
Werfelli, G., Halvick, P., Honvault, P., et al. 2015, JChPh, 143, 114304
Westermann, T., Brodbeck, R., Rozhenko, A. B., et al. 2011, JChPh, 135, 184102
Whitten, R. C. 1969, JMP, 10, 1631
Whitten, R. C., & Smith, F. T. 1968, JMP, 9, 1103
Worth, G. A., Beck, M. H., Jäckle, A., & Meyer, H.-D. 2000, The MCTDH Package, Version 8.2. Meyer, H.-D. 2002, 2007, Version 8.3, Version 8.4. See http://mctdh.uni-hd.de
Xu, C., Xie, D., Honvault, P., Lin, S., & Guo, H. 2007, JChPh, 127, 024304
Yamamoto, T. 1960, JChPh, 33, 281
Zhang, J. Z., & Miller, W. H. 1989, JChPh, 91, 1528

Zhang, D. H., & Zhang, J. Z. H. 1994, JChPh, 101, 3671
Zhang, D. H., & Zhang, J. Z. H. 1999, JChPh, 110, 7622
Zhang, J. Z. H. 1999, Theory and Application of Quantum Molecular Dynamics (1st ed; Singapore: World Scientific)
Zhang, J. Z. H., & Miller, W. H. 1987, CPL, 140, 329
Zhang, D. H., & Guo, H. 2016, ARPC, 67, 135
Zhu, W., Peng, T., & Zhang, J. Z. H. 1997, JChPh, 106, 1742
Zhu, C., Krems, R., Dalgarno, A., & Balakrishnan, N. 2009, ApJ, 703, 1176
Zickendraht, W. 1965, AnPhy, 35, 18
Zickendraht, W. 1967, PhRv, 159, 1448

Chapter 4

Radiative Processes in Astrophysical Molecules

Octavio Roncero, Alfredo Aguado and Susana Gómez-Carrasco

4.1 Introduction

Most of our information about the universe is obtained from the detection of electromagnetic waves, covering a broad spectrum from radio or micro wavelengths to the high-energy frequencies of x-rays and gamma rays. This radiation is also responsible for some transformation of the matter it passes through. The nature of the different processes involved depends upon the energy they carry. In this chapter, we will focus on the visible–ultraviolet (UV) spectrum. In this energy range, the main effect is the electronic excitation of atoms and molecules. In this last case, the excess of energy produces the fragmentation of the molecule into fragments, and eventually ionization (or both). The main subject of this chapter is the study of radiative processes in this energy range, dealing particularly with molecular fragmentation denoted by

$$AB + h\nu \rightarrow A + B,$$

where $h\nu$ denotes the energy of the absorbed photon, with h being the Planck constant and ν the radiation frequency.

In the vicinity of stars, the flux of UV radiation is high. The dense molecular cloud around young stars is washed up by the UV radiation. The irradiated part of the molecular cloud is denoted as the photon-dominated region (PDR). In these PDRs, the physical and chemical conditions are governed by the flux of the UV photons, quantified by the visual extinction, A_v (Hollenbach & Tielens 1997; Tielens 2005). The molecules that survive act as shields, reducing the flux of the absorbed frequencies. Photodissociation by UV radiation also plays an important role in other illuminated objects, such as protoplanetary disks, comets, exoplanets, etc.

UV radiation also transforms matter. When it acts on ices, the photofragments trigger a complex chemistry, giving rise to complex organic molecules. These molecules are expelled back to the gas phase, where they are detected.

doi:10.1088/2514-3433/aae1b5ch4

Radiation does not only destroy molecules. At the first steps of chemistry in the early universe and in regions with very low densities, the first molecules are formed through radiative recombination processes,

$$A + B \rightarrow AB + h\nu$$

in which the excess of energy is dissipated by the emission of photons, leading to a stable molecule, AB, formed by the colliding fragments, A and B.

All dynamical processes associated with radiative transitions need to be accounted for in radiative transfer models in order to analyze the physical and chemical conditions of different astronomical objects. The photodissociation rates, for example, are obtained by integrating the cross section with the radiation field over the entire UV spectrum. Photons of 13.6 eV or larger energy produce the ionization of hydrogen atoms, the most abundant element in space. Therefore, it is normally thought that the astronomical UV radiation field ends at this energy of 13.6 eV.

In this chapter, we shall present the theoretical framework needed to treat these phenomena. The second section starts by defining the expressions needed to quantify the absorption or emission of photons to produce any of the events discussed above. The third section will be devoted to radiationless transitions to interpret and classify the type of dynamics the molecular system may undergo. Section 4.4 briefly describes two of the most widely used quantum methods to study photodissociation, specifically the nuclear dynamics, while in Section 4.5 a brief review of the electronic structure calculations will be presented. Finally, in Section 4.6, some prototypical examples will be presented, namely the photodissociation of CO, HCN/HNC, and H_2CO, to draw some conclusions regarding the present state of the art in the theory used to treat these phenomena.

4.2 Radiative Transitions

The radiation field in the interstellar medium is of low intensity. Therefore, it can be assumed that it only induces transitions (absorption and emission) among the eigenstates of the Hamiltonian of the isolated molecular system, H. In this case, H commutes with the square of the total angular momentum operator, $\mathbf{J}$, of the system, and therefore the eigenstates of H are also eigenstates of $\mathbf{J}^2$. These eigenstates can be classified as

$$\begin{array}{llll} \Psi_i^{JM} & \text{bound} & i & \text{denoting the vibrational quantum numbers} \\ \Psi_{Ef\Omega}^{JM} & \text{dissociative} & f & \text{denoting all quantum numbers of fragments} \\ & & E & \text{being the total energy,} \end{array} \quad (4.1)$$

where M and Ω are the projection of the total angular momentum on the space-fixed and body-fixed frames, respectively. The space-fixed frame is associated with the radiation while the body-fixed frame is associated with the molecular system, typically with the z-axis parallel to the dissociative coordinate. For circularly polarized light, the space-fixed z-axis is parallel to the direction of the propagation of light. For linearly polarized light, the z-axis is considered to be parallel to the polarization vector, **e**, of the electric field, as will be the case generally considered here ($\mathbf{e}\|\mathbf{e}_z$). The dissociative wave functions are energy-normalized according to $\langle\Psi^{JM}_{Ef\Omega}|\Psi^{JM}_{E'f'\Omega'}\rangle = \delta_{\Omega\Omega'}\,\delta_{ff'}\,\delta(E - E')$, while bound wave functions are normalized as $\langle\Psi^{JM}_{i}|\Psi^{JM}_{i'}\rangle = \delta_{ii'}$.

Under low-intensity conditions of radiation, the transitions can be treated by first-order perturbation theory, in which the transition probability is proportional to the square of the matrix element between the eigenfunction of Equation (4.1) of a transition operator. Assuming electric dipole transitions, this operator is written as

$$\mathbf{e}\cdot\hat{\mathbf{d}} = |\mathbf{e}| \sum_{p=-1,0,1} (-1)^p\ \mathbf{e}_{-p}\cdot\hat{\mathbf{d}}_p \tag{4.2}$$

$$= |\mathbf{e}| \sum_{p=-1,0,1} (-1)^p\ \mathbf{e}_{-p} \sum_q D^{1*}_{pq}(\phi,\theta,0)\ \ \hat{\mathbf{d}}_q \tag{4.3}$$

where $\hat{\mathbf{d}}_X$ ($X = p$ or q) is the electric dipole operator, with components p or q in the space-fixed and body-fixed reference frames, respectively. The two frames are related by a rotation, given by the Wigner rotation matrix $D^{1*}_{pq}(\phi, \theta, 0)$ (Zare 1988), where (θ, ϕ) are the polar angles of a given vector **R** within the molecular system.

In order to treat dissociation, **R** is chosen to be parallel to the vector joining the center of mass (COM) of the A and B fragments. Accordingly, the dissociative wave functions can be expanded as

$$\Psi^{JM}_{Ef\Omega} = \sqrt{\frac{2J+1}{4\pi}} \sum_{\Omega'}\sum_{f'} \frac{\Phi^{JM,Ef\Omega}_{f'\Omega'}(R)}{R}\quad D^{J*}_{M\Omega'}(\phi,\theta,0)\quad |f'\Omega'\rangle. \tag{4.4}$$

$\mathbf{J} = \mathbf{j} + \boldsymbol{\ell}$, where **j** is the angular momentum of the fragments (including both fragments electronic and nuclear angular momenta) while ℓ is the end-over-end angular momentum between A and B. Here, ℓ is by definition perpendicular to **R,** and therefore Ω is the projection of both **J** and **j** on the body-fixed z-axis. The bound-wave functions, Ψ^{JM}_b, can be expanded according to Equation (4.4) without any loss of generality.

Here, $|f\Omega\rangle$ are the functions describing the two fragments, including electronic and nuclear motions. For diatomic molecules in Hund's case a, these functions reduce to $|\Lambda\rangle$ functions, with $\Lambda = \Omega$ being the projection of the electronic angular momentum on the diatomic axis. For triatomic systems leading to A + BC fragments in electronic closed shells, $|f\Omega\rangle$ are simply the rovibronic states of the diatomic molecule, which can be expressed in spherical coordinates as $|g\rangle\varphi_{vj}(r)\,Y_{j\omega}(\gamma,\chi)/r$, with $|g\rangle$ being the single electronic wave function.

In Equation (4.4), $\Phi^{JM,Ef\Omega}_{f'\Omega'}(R)$ are radial coefficients that are obtained by solving the set of coupled differential equations (see below), subject to the proper boundary conditions. If we consider the collision between the photofragments, A+B, the boundary conditions correspond to incoming waves with reagents in state $|f\Omega\rangle$ and outgoing waves in states $|f'\Omega'\rangle$. If we consider photodissociation as a half-collision process, its boundary conditions can be considered as the complex conjugate of the A + B collisional boundary conditions, corresponding to outgoing waves with reagents in state $|f\Omega\rangle$ and incoming waves with reagents in states $|f'\Omega'\rangle$. In the case of radiative recombination, however, the usual collisional boundary conditions are considered.

Within this representation, the electric dipole matrix elements for bound to dissociative states transitions are written as

$$\left\langle \Psi_i^{J_iM_i} |\mathbf{e}\cdot\hat{\mathbf{d}}|\, \Psi_{Ef\Omega}^{JM} \right\rangle = \sum_p (-1)^{J_i+M_i+p}(\mathbf{e})_{-p}\, \sqrt{2J_i+1} \begin{pmatrix} J_i & 1 & J \\ -M_i & p & M \end{pmatrix} \times \langle J_i;\, i||\mathbf{d}||J;\, Ef\Omega\rangle \tag{4.5}$$

where $(\cdots)$ are 3j symbols (Zare 1988) and the reduced matrix elements are given by

$$\langle J_i;\, i||\mathbf{d}||J;\, Ef\Omega\rangle = \sum_q \sum_{f_i\Omega_i} \sum_{f'\Omega'} (-1)^{-\Omega_i-J_i}\, \sqrt{2J+1} \begin{pmatrix} J_i & 1 & J \\ -\Omega_i & q & \Omega' \end{pmatrix} \times \int dR \left[\Phi_{f'\Omega'}^{JM,\, Ef\Omega}(R)\right]^* \langle f_i\Omega_i|\mathbf{d}_q|f'\Omega'\rangle \Phi_{f_i\Omega_i}^{J_iM_i}(R) \tag{4.6}$$

where $\langle\cdots\rangle$ denotes integration over all coordinates of fragments, electronic and nuclear. If the system initially is isotropically distributed, it is obtained that

$$\sum_{M_i} (2J_i+1)^{-1} \left| \left\langle \Psi_i^{J_iM_i} \,|\, \mathbf{e}\cdot\hat{\mathbf{d}} \,|\, \Psi_{Ef\Omega}^{JM} \right\rangle \right|^2 = \frac{1}{3} |\langle J_i;\, i||\mathbf{d}||J;\, Ef\Omega\rangle|^2. \tag{4.7}$$

This equation, derived for bound–dissociative matrix elements, is formally identical (with changes only in the quantum number labels) to any other matrix element, bound–bound, dissociative–bound, etc. With the expansion of the wave functions of Equation (4.4), the term depending on the Euler angles (θ, ϕ) is integrated, giving rise to a simple geometric factor that takes into account the usual $\Delta J = 0, \pm 1$ for electric dipole transitions.

4.2.1 Photodissociation

In the framework of first-order perturbation theory, the partial cross section for the photodissociation from an isotropically distributed bound state $\Psi_i^{J_iM_i}$ is given by Shapiro & Balint-Kurti (1981), Singer et al. (1985), and Schinke (1993)

$$\sigma_{J_i i}^{PD}(h\nu) = A\, h\nu \sum_{Jf\Omega} \sum_{M_i} (2J_i+1)^{-1} \left| \left\langle \Psi_i^{J_iM_i} |\mathbf{e}\cdot\hat{\mathbf{d}}|\, \Psi_{Ef\Omega}^{JM} \right\rangle \right|^2 \delta(E - E_i - h\nu) \tag{4.8}$$

$$= A\, h\nu \sum_{Jf\Omega} \frac{1}{3} |\langle J_i;\, i||\mathbf{d}||J;\, Ef\Omega\rangle|^2 \quad \delta(E - E_i - h\nu), \tag{4.9}$$

where $h\nu$ is the photon energy and $A = 1/\hbar^2\epsilon_0 c = 10.905$ au (with $\hbar = h/2\pi$). This expression gives the total absorption cross section, and dealing with dissociative states, it coincides with the total photodissociation cross section, unless the system can undergo other processes such as re-emission of photons. The partial cross section to form the fragments in state $|f'\rangle$ (irrespective of the projection Ω and *J*) is given by Equation (4.8), but restricting the sum on the right-hand side (rhs) to $f = f'$.

The final photodissociation rate constant needed in radiative transfer models is

$$K(T) = \int d\lambda \frac{F(\lambda)}{h\nu} \sigma^{PD}(h\nu, T) \qquad \text{with} \qquad \lambda = \frac{c}{\nu}, \tag{4.10}$$

and is obtained by integrating the product of the cross section, $\sigma^{PD}(h\nu)$, with the radiation field intensity, $F(\lambda)$, over the wavelength of the radiation.

The temperature, *T*, in this expression corresponds to the Boltzmann distribution of initial states $\Psi_i^{J_i M_i}$. The radiation field $F(\lambda)$ usually depends on the astronomical region under study; a typical example in the interstellar medium is that of Draine (1978), given by

$$\begin{aligned} F(\lambda) = {} & 6.362171 \times 10^7/\lambda^4 - 1.0238 \times 10^{11}/\lambda^5 \\ & + 4.081310 \times 10^{13}/\lambda^6 \text{ erg cm}^{-2}\text{ s}^{-1}\text{ Å}^{-1}, \end{aligned} \tag{4.11}$$

valid in the $912 \leqslant \lambda \leqslant 2000$Å range, shown in the left panel of Figure 4.1. In the right panel of Figure 4.1, the experimental photodissociation cross section of HCN measured by Nuth & Glicker (1982) is shown in the same wavelength range. The photodissociation rate constant of HCN obtained using Equation (4.10) with the quantities shown in Figure 4.1 is 1.9×10^{-9} s^{-1}.

A general database for photodissociation of molecules of astrophysical interest can be found in the Leiden database (https://www.universiteitleiden.nl/en/science/astronomy).

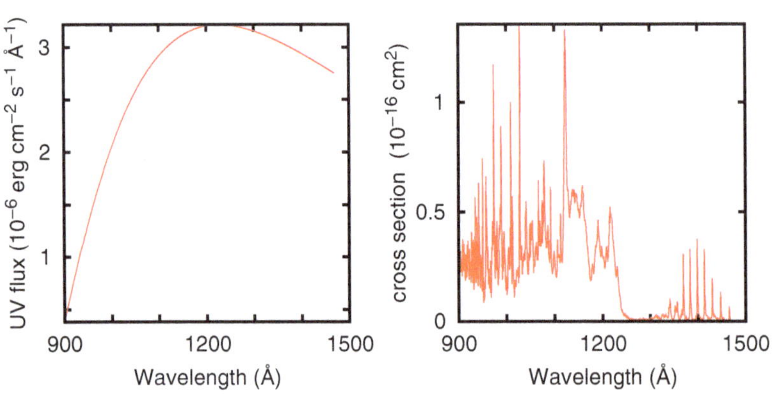

Figure 4.1. Left panel: Interstellar radiation field, $F(\lambda)$, from Draine (1978). Right panel: Experimental photodissociation cross section, $\sigma^{PD}(h\nu, T)$, of HCN by Nuth & Glicker (1982). These two quantities are integrated in Equation (4.10) to provide the photodissociation rate constant, $K(T) = 1.9 \times 10^{-9}$ s^{-1}, as discussed in the text.

4.2.2 Bound–bound Einstein Coefficients

Spontaneous and stimulated transitions among bound states are described by A and B Einstein coefficients, which are related by $A/B = 8\pi h\nu^3/c^3$. The A Einstein coefficient is the rate for the spontaneous emission of a transition from the excited (upper) $\Psi_u^{J_u M_u}$ to the final (lower) $\Psi_l^{J_l M_l}$ state, and in the first-order perturbation theory, according to Fermi's Golden rule, is given by Larsson (1983)

$$\mathrm{A}_{uJ_u;lJ_l} = \frac{B(h\nu)^3}{3} |\langle J_u;u||\mathbf{d}||J_l;l\rangle|^2 \tag{4.12}$$

where $B = 1/\pi\epsilon_0\hbar^4c^3 = 1.554 \times 10^{-6}$ au, and the reduced dipole matrix elements associated to bound–bound transitions are directly defined from Equation (4.6). The radiative lifetime is then defined as the inverse of all possible transitions as

$$\tau_{uJ_u} = \left(\sum_{lJ_l} \mathrm{A}_{uJ_u;lJ_l}\right)^{-1}. \tag{4.13}$$

Einstein A coefficients are used in radiative transfer models to account for the excitation and de-excitation of molecules induced by radiation. The typical radiative lifetime is on the order of μs–ns (10^{-6}–10^{-9} s), while photodissociation is typically faster, between ps–fs (10^{-12}–10^{-15} s). Thus, spontaneous emission is very unfavorable when competing with photodissociation, and therefore its impact is reduced to regions where photodissociation is negligible or slow. Absorption and re-emission of photons are not only important to describe the population of bound states of molecules, but also can induce interesting phenomena, such as photoisomerization, which was recently used to explain the detection of the CIS conformed of formic acid, observed for the first time in the Orion bar (Cuadrado et al. 2016).

4.2.3 Radiative Association

In radiative association, the initial state is dissociative, $\Psi_{Ei\Omega}^{JM}$, and the probability to emit to a final bound state is given by the corresponding Einstein coefficients divided by the current density of particles (Ayouz et al. 2011). Therefore, the cross section for radiative association of colliding particles at energy $E = k^2/2\mu$ (with μ being the A + B reduced mass, and k the linear momentum) emitting to a bound state $\Psi_b^{J'M'}$ is given by Bates (1951), Palmer (1967), and Babb & Dalgarno (1995) as

$$\begin{aligned}\sigma_{iJ\Omega;bJ'}^{RA}(E) &= \frac{2\pi^2}{k^2}\mathrm{A}_{iJ\Omega;bJ'} \\ &= \frac{2\pi^2}{k^2}\frac{B\,(h\nu)^3}{3} |\langle J';b||\mathbf{d}||J;Ei\Omega\rangle|^2 \quad \delta(E - E_b - h\nu),\end{aligned} \tag{4.14}$$

where the initial state $\Psi_{Ei\Omega}^{JM}$ corresponds to the collisional wave function describing A and B fragments initially in internal state i, with a total angular momentum J. This expression is very similar to that of Equation (4.8) for photodissociation, which makes possible to evaluate it as (Puy et al. 2007)

$$\sigma^{RA}_{iJ\Omega;bJ'}(E) = 2\pi\left(\frac{h\nu}{\hbar kc}\right)^2 \sigma^{PD}_{bJ',iJ\Omega}(E) \tag{4.15}$$

The total radiative association cross section is obtained after summation over the partial waves and all possible transitions as

$$\sigma_i^{RA}(E) = \sum_{J\Omega}\sum_{bJ'} \sigma^{RA}_{iJ\Omega;bJ'}(E). \tag{4.16}$$

The total radiative association rate constant is obtained after integrating over a Boltzmann distribution for the velocity, $\mathrm{v} = k/\mu$, between the reactants, and summing over the thermal distribution of initial states i. The observed radiative association rates are relatively small (Gerlich & Horning 1992; Gerlich et al. 2013), on the order of 10^{-16} cm^3 s^{-1} or smaller. Radiative association requires the formation of long-lived complexes during the collision, which increases the overlap between dissociative and bound-wave functions. Even when the rates are relatively low, radiative association plays an important role in low-density regions: in these regions, three-body collisions are very unlikely and the only way to form the first molecules is through radiative association. This is the situation found in the early universe.

The radiative association is formally similar to photodissociation. The main difficulty is found in the initial state. For photodissociation, the initial state is bound, and the rotational population can be somehow "controlled" by lowering the temperature. Thus, the number of partial waves needed to simulate the whole spectrum is relatively low, and one or few vibrational states are typically needed.

On the other hand, for radiative association, the number of partial waves needed depends on the collisional energy. If this energy is kept low, where long-lived resonances may be formed, then the number of partial waves can be relatively low as well. Otherwise, it may be very computationally demanding. In addition, the transition to many final vibrational states of the final complex adds an extra difficulty. This is the reason why most of the simulations refer to diatomic molecules (Babb & Dalgarno 1995; Martinazzo & Tantardini 2005) and only a few to triatomic systems (Ayouz et al. 2011; Stoeklin et al. 2013).

In what follows, we shall restrict our scope to photodissociation, although the same methods and arguments apply to radiative association processes.

4.3 Non-radiative Transitions

Once the initial state, Φ_0, is "prepared" by the photon excitation,

$$\Phi_0 = \mathbf{e}\cdot\mathbf{d}\,\Psi_b^{JM}, \tag{4.17}$$

the system AB evolves up to fragmentation into A + B, in the case of photodissociation. In order to interpret and classify the different processes occurring, it is interesting to factorize the total Hamiltonian according to physical arguments as

$$H = H_0 + V, \tag{4.18}$$

so that H_0 has two types of eigenfunctions, dissociative ϕ_E^α and discrete ϕ_i, that fulfill

$$\begin{aligned} H_0\phi_i &= E_i\phi_i \qquad \langle\phi_i|\phi_{i'}\rangle = \delta_{ii'} \qquad \langle\phi_i|\phi_E^\alpha\rangle = 0 \\ H_0\phi_E^\alpha &= E\phi_E^\alpha \qquad\qquad\qquad \langle\phi_E^\alpha|\phi_{E'}^{\alpha'}\rangle = \delta_{\alpha\alpha'}\delta(E - E'), \end{aligned} \tag{4.19}$$

where i represents the quantum numbers that characterize the bound states, α denotes the quantum numbers needed to specify the state of the fragments, and E is the total energy.

The term V in Equation (4.18) couples these zero-order functions, and for simplicity it will be assumed that

$$\begin{aligned} \langle\phi_i|V|\phi_{i'}\rangle &= (1 - \delta_{ii'})V_{ii'} \\ \langle\phi_i|V|\phi_E^\alpha\rangle &= V_{i,\,E\alpha} \\ \langle\phi_E^\alpha|V|\phi_{E'}^{\alpha'}\rangle &= 0. \end{aligned} \tag{4.20}$$

These functions form a complete set, in which the closure relationship can be written as

$$\begin{aligned} 1 = P + Q \quad \text{with} \quad P = \sum_i |\phi_i\rangle\langle\phi_i| \quad \text{and} \\ Q = \sum_\alpha \int dE \quad |\phi_E^\alpha\rangle\langle\phi_E^\alpha| \end{aligned} \tag{4.21}$$

where we have introduced the projector operators P and Q (Feshbach 1962; Mower 1966), which satisfy $P^2 = P$, $Q^2 = Q$ and $PQ = QP = 0$.

The initial state, Φ_0, can be expressed in terms of the zero-order eigenfunctions as (Fano 1961; Mics 1968, 1969a, 1969b)

$$|\Phi_0\rangle = \sum_i a_i(E) \quad |\phi_i\rangle + \sum_\beta \int dE' \, b_{E'\beta} \, |\phi_{E'}^\beta\rangle \tag{4.22}$$

where $a_i(E) = \langle\phi_i|\Phi_0\rangle$ and $b_{E'\beta} = \langle\phi_E^\beta|\Phi_0\rangle$.

The dissociative eigenfunctions of the total Hamiltonian, $\Psi_{E\alpha}$, (defined in the previous section with $\alpha \equiv f\Omega$ and where we have dropped the superscript JM to simplify the notation) can be expressed in terms of these zero-order functions using the Lippmann–Schwinger equation (Lippmann & Schwinger 1950; Lefebvre 1972) as

$$|\Psi_{E\alpha}\rangle = |\phi_E^\alpha\rangle + G(E)V|\phi_E^\alpha\rangle. \tag{4.23}$$

where $G(E)$ is the resolvent operator defined in Appendix 4.7.

Thus, the cross sections defined in the previous section can be rewritten as

$$\sigma(E) \propto |\langle \Phi_0 | \Psi_{E\alpha} \rangle|^2 \tag{4.24}$$

and using Equations (4.22) and (4.23) can be easily expressed in terms of the matrix elements of $G(E)$ (see Appendix 4.7).

The evolution operator can be written as (Goldberger & Watson 1964; Uzer & Miller 1991)

$$e^{-iHt/\hbar} = -\frac{1}{2\pi i} \int dE e^{-iEt/\hbar} G(E) \tag{4.25}$$

from where the evolution of the wave packet defined as

$$\Phi_t = e^{-iHt/\hbar} \Phi_0 \tag{4.26}$$

can also be written in terms of the resolvent operator matrix elements.

Below, we shall consider several limiting cases of photodissociation depending on the composition of the zero-order wave functions.

4.3.1 One Dissociative State: Direct Photodissociation

Let us consider the case of a single dissociative state and for simplicity, a diatomic molecule like HCl (see Figure 4.2). The first maxima of the continuum wave-functions appear at distances close to the classical turning point. In this region, the available kinetic energy is lower and therefore the wave length is longer and the amplitude is higher. Ergo, when this first maximum coincides in distance with the equilibrium distance in the ground electronic state, the Frank–Condon factor

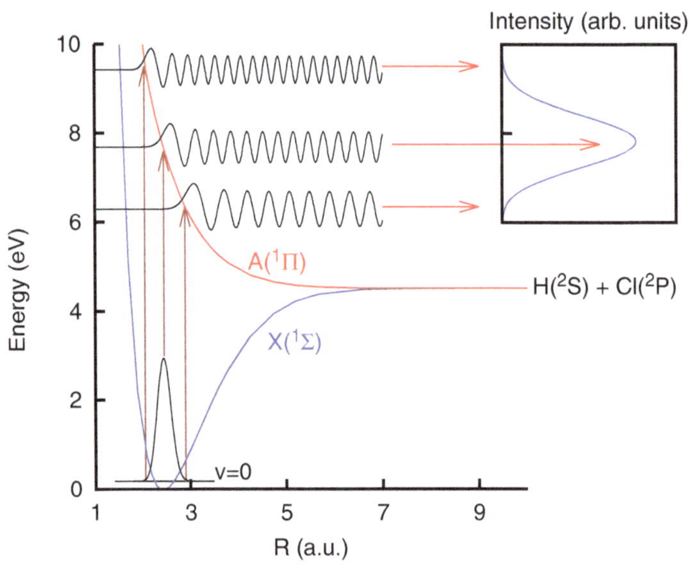

Figure 4.2. Direct photodissociation example: energy curves of HCl. The inset represents the photofragmentation cross section, $\sigma(E)$, defined in Equation (4.24), in arbitrary units.

increases, giving rise to a maximum in the photodissociation cross section, as shown in the inset of Figure 4.2.

This Frank–Condon argument naturally leads to the so-called reflection principle (Schinke 1993): tracing a vertical line at the maxima and minima of the initial state toward the dissociative state, it is possible to determine the corresponding extrema of the absorption spectra, as shown in Figure 4.2. With this simple picture, one can easily predict both the position and the width of the absorption spectrum: the larger the slope of the excited state is, the wider the spectrum shows.

Moreover, the fragmentation dynamics is also determined by the slope of the excited state potential: the increase of the slope produces an increase in the velocity, making the fragmentation dynamics much faster. Typically, the time for direct fragmentation is on the order of femtoseconds.

The situation becomes more complex for more dimensions or electronic states. In those cases, the analysis of the product's branching ratio must be used to extract more information regarding the fragmentation dynamics.

4.3.2 One Bound State Coupled to Several Dissociative States: Predissociation

Let's consider the case of excitation of a bound state, ϕ_1, of the zero-order Hamiltonian, of energy E_1. If this bound state is coupled to one or several dissociative continua, as represented in Figure 4.3, the population is going to be transferred to the continua at a speed proportional to the square of the discrete–continua couplings.

In this case, the absorption spectrum can be written as

$$\begin{aligned} \sigma(E) &\propto \sum_{\alpha} |\langle \phi_1 | G(E) | \phi_1 \rangle V^{\alpha}_{1E}|^2 \\ &= \frac{\Gamma_{11}/\pi}{(E - E_1 - \Delta_{11})^2 + \Gamma_{11}^2} \end{aligned} \tag{4.27}$$

where $\Gamma_{11} = \sum_{\alpha} \pi |V^{\alpha}_{1E_1}|^2 = \sum_{\alpha} \Gamma^{\alpha}_{11}$ is calculated at the energy E_1 according to the Fermi's Golden rule. This is the half-width at half-maximum of the Lorentzian

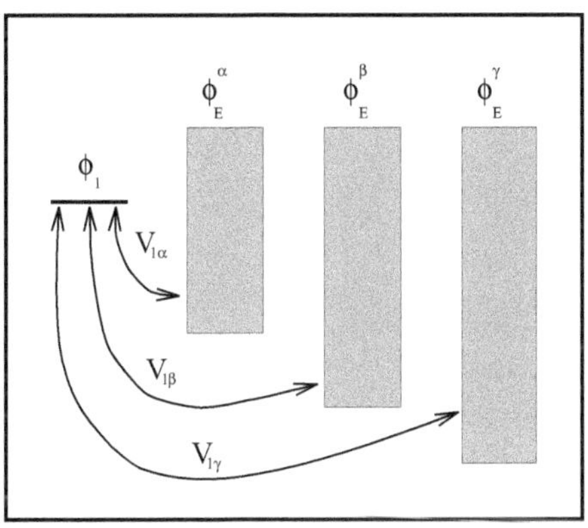

Figure 4.3. Energy diagram of one zero-order bound state coupled to several dissociation continua.

function centered at the energy $E_1 + \Delta_{11}$, where Δ_{11} is a shift of the energy of the bound state due to its coupling to the continua.

The population of the initial state decays as an exponential

$$P_1(t) = \left|\langle\phi_1|\Phi_t\rangle\right|^2 = e^{-2\Gamma_{11}t/\hbar}, \tag{4.28}$$

with a lifetime of $\tau = \hbar/2\Gamma_{11}$. This lifetime is on the order of picoseconds (10^{-12}s) or longer, about three orders of magnitudes slower than direct photodissociation.

The population of the different fragmentation continua is given by

$$P_\alpha(t) = \frac{\Gamma_{11}^\alpha}{\Gamma_{11}}(1 - e^{-2\Gamma_{11}t/\hbar}), \tag{4.29}$$

i.e., all dissociative appear at the same velocity, but with a different branching ratio given by $\Gamma_{11}^\alpha/\Gamma_{11}$.

This process is called predissociation, and it is typically classified as electronic, vibrational, or rotational, according to the quantum number which varies from the bound state to the dissociative continua.

In diatomic molecules, there is only electronic predissociation; a typical example is the CO molecule when exciting to the $B^1\Sigma^+$ state. In Tchang-Brillet et al. (1992), a model Hamiltonian was built composed of two electronic states of $^1\Sigma^+$ symmetry; one bound, the diabatic $B^1\Sigma^+$ state, and the other repulsive, the diabatic $D^1\Sigma^+$ state, as shown in the left panel of Figure 4.4. The B state has bound vibrational states that are coupled to the continuum wave functions of the dissociative D state. For $v = 0$, the overlap between the bound and dissociative states are negligible, and the bound state is long-lived. As the vibrational excitation increases, so do the overlap and the coupling. Thus, the lifetime of those vibrational states decreases with vibrational excitation, because they decay faster and faster. This is also seen in the spectrum,

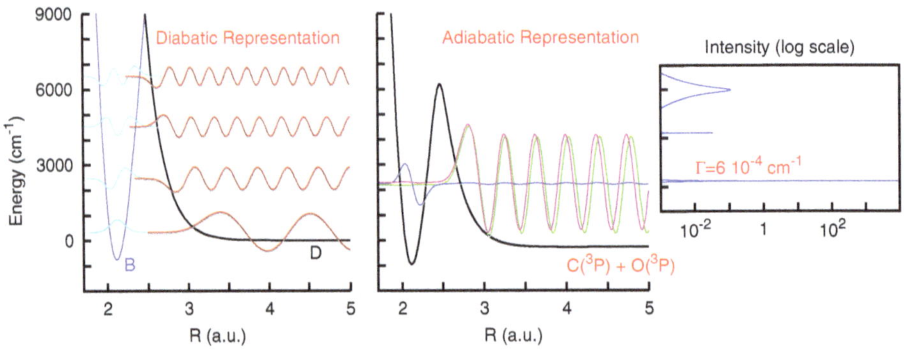

Figure 4.4. Left panel: 2x2 diabatic representation of the electronic states B and D of CO, adapted from Tchang-Brillet et al. (1992). The bound (light blue lines) and continuum (brown lines) wave functions are calculated independently in each electronic state, and correspond to the zero-order bases, ϕ_i and ϕ_E, respectively. Middle panel: lower adiabatic potential (black line) and the corresponding wave functions around the energy of $\phi_i \equiv \phi_{v=1}$ in different colors (navy blue corresponds to the energy of the resonance). Right panel: absorption spectrum calculated in the adiabatic representation with only one electronic state, using Equation (4.24)

shown in the right panel of Figure 4.4, in which the width, Γ, of the vibrational states increases with the vibrational excitation, from $\Gamma = 6 \times 10^{-4}$ cm^{-1} for v = 1–70 cm^{-1} for v = 3, whose lifetime changes from 4.5 ns to 40 fs.

This spectrum has been calculated in the adiabatic representation, taking only one electronic state, as shown in the middle panel of Figure 4.4. Such approximation is good for the lower vibrational states like v = 1, but not very good for the higher v = 3, rather close to the crossing, where the 2×2 description should be employed. In the simple adiabatic approximation with only one electronic state, this case of electronic predissociation is equivalent to the tunneling through the barrier originated by the avoided crossing. The wave function calculated on the adiabatic B state shows a rather sharp change around the v = 1 energy. The wave function amplitudes are generally negligible for $R < 2.5$ au, because the system is reflected back by the barrier. Just at the energy of the resonance, the component of the v = 1 wave function is dominant, with very small components of the continuum functions. This last contribution increases rapidly across the resonance width, becoming dominant at an energy slightly different from the resonance energy, as show in the middle panel of Figure 4.4.

For polyatomic molecules, the photon absorption can produce excitation of some vibrational modes. This vibrational energy can be transferred to other vibrational modes, which may eventually yield fragmentation, giving rise to vibrational predissociation. Van der Waals clusters provide interesting model systems for characterization, and many experimental studies exist on simple dihalogen molecules attached to rare gas atoms, in both frequency (Levy 1981; Janda 1985; Waterland et al. 1988; Evard et al. 1988; Burke & Klemperer 1993) and time (Willberg et al. 1992; Gutmann et al. 1992) domains. In many of these cases, especially in the X-BC triatomic case, the dynamics is dominated by a single resonance corresponding to a BC(B, v) excited state, which transfers a single vibrational quantum (or a few) to the weaker van der Waals modes, yielding BC (B, v'<v) + X fragments. This fragmentation process has been widely studied by quantum simulations (Beswick et al. 1979; Beswick & Jortner 1981; Halberstadt et al. 1987; Roncero et al. 1990; Halberstadt et al. 1990; Gray 1992; Roncero et al. 1997; Buchachenko et al. 2003).

4.3.3 Small-large Molecule Limits

As the number of atoms of the molecule increases, the number of vibrational modes increases as well. The energy provided by the photon to one or several vibrational modes is transferred to other modes, which may be eventually coupled to the dissociative continua. If the number of vibrational modes is high, and the bound–bound couplings are more effective than the bound–continuum couplings, the energy is randomized among the different vibrational modes and never comes back to the initial excited mode/modes (Bixon & Jortner 1969; Freed & Nitzan 1980). In such a situation, the system becomes photo-resistant, i.e., absorbs photons without dissociating. This is the case of polycyclic aromatic hydrocarbons (PAH) in the interstellar media (van Dishoeck & Visser 2015). The energy absorbed can eventually be dissipated by emitting photons of lower energy—infrared, for example.

In order to quantify the limits that may be expected, we shall here use the model displayed in Figure 4.5. Traditionally (Bixon & Jortner 1969; Lahmani et al. 1974; Freed & Nitzan 1980; Uzer & Miller 1991), a model based on an initial state, ϕ_b, coupled to an infinity number of states, ϕ_k (with energy $E_k = \epsilon + \Delta k$ and width Γ common to all of them), is used, with a constant coupling $V_{bk} = V$. Here, we shall use an analogous treatment (Roncero et al. 1997) formed by an infinite number of Lorentzian functions, describing an infinite number of equidistant bound states with the same width Γ and with energies $E_k = k\Delta$ $(-\infty \leqslant k \leqslant \infty)$. The initial state is expressed as a linear superposition of these states, with a coefficient $a_k/a_0 = V/(k\Delta - i\Gamma)$. Following the procedure of Appendix 4.7, the absorption spectrum becomes (Roncero et al. 1997)

$$\sigma(E) = \frac{B}{\pi A}\left\{\frac{\left(\Gamma^2 - |V|^2\right)/\Gamma}{E^2 + \Gamma^2} + \frac{\Gamma_{IVR}}{E^2 + 4\Gamma^2}\left[\frac{e^{\pi\Gamma/\Delta} + e^{-\pi\Gamma/\Delta}}{e^{\pi\Gamma/\Delta} - e^{-\pi\Gamma/\Delta}} + \frac{1}{E}\frac{4\Gamma\sin 2\pi E/\Delta + E\left(e^{2\pi\Gamma/\Delta} - e^{-2\pi\Gamma/\Delta}\right)}{e^{2\pi\Gamma/\Delta} + e^{-2\pi\Gamma/\Delta} - 2\cos 2\pi E/\Delta}\right]\right\} \tag{4.30}$$

where $\Gamma_{IVR} = \pi|V|^2/\Delta$, and

$$\begin{aligned} A &= \{\Gamma^2 - |V|^2\}\{e^{2\pi\Gamma/\Delta} + e^{-2\pi\Gamma/\Delta} - 2\} - \Gamma_{IVR}\Gamma\{e^{-2\pi\Gamma/\Delta} - e^{2\pi\Gamma/\Delta}\} \\ B &= \Gamma^2\{e^{2\pi\Gamma/\Delta} + e^{-2\pi\Gamma/\Delta} - 2\}. \end{aligned} \tag{4.31}$$

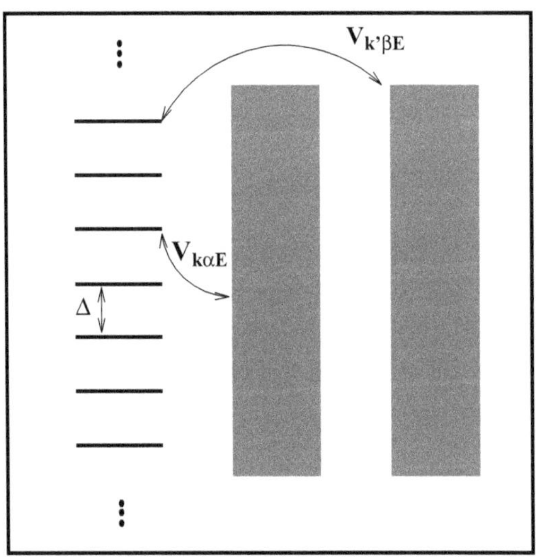

Figure 4.5. Energy diagram of a model based on an infinite number of equidistant bound states coupled to several dissociative continua.

The population of the initial state, $\phi_b = \sum_k a_k \phi_k$, becomes

$$\begin{aligned} P_b(t) = \; & \frac{B^2}{A^2\Gamma^4} \frac{e^{-2\Gamma t/\hbar}}{e^{2\pi\Gamma/\Delta} + e^{-2\pi\Gamma/\Delta} - 2} \Big\{ e^{2\pi\Gamma/\Delta} \big(\Gamma^2 - |V|^2 + \Gamma_{IVR}\Gamma e^{-\Gamma T/\hbar}\big)^2 \\ & + e^{-2\pi\Gamma/\Delta}\big(\Gamma^2 - |V|^2 - \Gamma_{IVR}\Gamma e^{\Gamma T/\hbar}\big)^2 \\ & - 2\big(\Gamma^2 - |V|^2 + \Gamma_{IVR}\Gamma e^{-\Gamma T/\hbar}\big)\big(\Gamma^2 - |V|^2 - \Gamma_{IVR}\Gamma e^{\Gamma T/\hbar}\big) \Big\} \end{aligned} \tag{4.32}$$

with $T = t - 2\pi\hbar n/\Delta$, n being the integer part of $t\Delta/2\pi\hbar$.

These expressions depend on the Γ/Δ ratio, which yields a natural classification, as previously discussed (Bixon & Jortner 1969; Freed & Nitzan 1980; Uzer & Miller 1991; Roncero et al. 1997):

- **Sparse regime,** ($\Gamma < \Delta$): In this regime, the resonances are well-separated. The initially populated bright state will usually interact with a single dark state, so quantum phenomena such as recurrences are readily apparent. However, this means that the Γ_{kj} coupling terms in Equation (4.73), neglected in this simplified model, are usually important (see Roncero et al. 1997 and references therein for a more detailed discussion). For a detailed analysis of this regime, the simplifying assumptions made above should not be used and are not necessary.
- **Intermediate regime,** ($\Gamma \approx \Delta$): In this regime, the resonances are mixed but have not completely lost their individual identity. This will be apparent in both the spectrum, in which nearby transitions will overlap with each other, and in the dynamics, which will exhibit non-exponential decay with weak recurrences in the population of the initially excited states.
- **Statistical regime,** ($\Gamma \gg \Delta$): In this regime, there are many closely spaced resonances that blend together to yield a quasi-Lorentzian excitation spectrum. As a consequence, the initial states lose their identity and decay irreversibly as a single exponential.

These three situations are shown in Figure 4.6 for the spectra. The decay of the population of the initial state is shown in Figure 4.7.

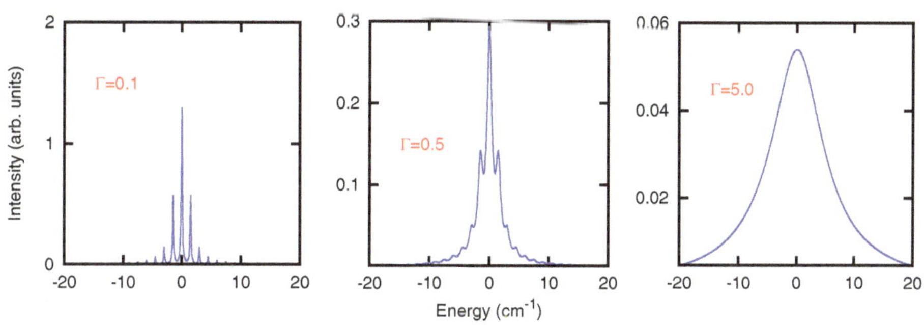

Figure 4.6. Absorption spectra for sparse ($\Gamma = 0.1 < \Delta$ in left panel), intermediate ($\Gamma = 0.5 \approx \Delta$ in middle panel) and statistical ($\Gamma = 5 > \Delta$ in right panel) limits. The values of $\Delta = 1.5\ \text{cm}^{-1}$ and $V = 1\ \text{cm}^{-1}$ have been considered in all the cases.

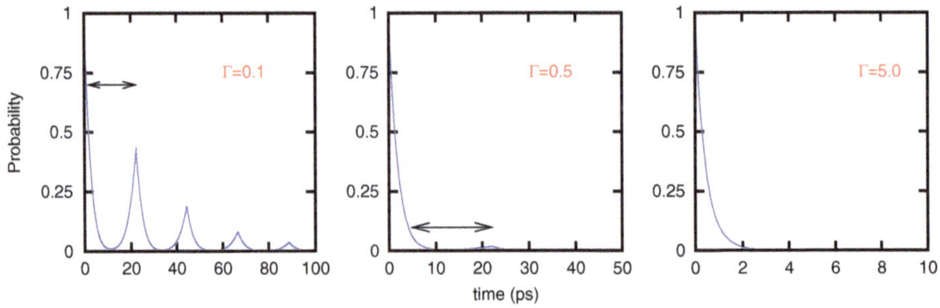

Figure 4.7. Probability of the initial state, ϕ_b, versus time for the three limits, sparse (left panel), intermediate (middle panel), and statistical (right panel). The parameters used are those of Figure 4.6. The arrows indicate the period for recurrences.

4.4 Methods

In this section, some methods to calculate the photodissociation cross section will be described. With this purpose, the total Hamiltonian of the AB system is written as

$$H = T_R + H_A + H_B + W \quad \text{with} \quad V = V_A + V_B + W \tag{4.33}$$

where the functions $|f\Omega\rangle$, already defined in Equation (4.4), are the eigenfunctions of the Hamiltonians $H_A = T_A + V_A$ and $H_B = T_B + V_B$ of the fragments, i.e.,

$$(H_A + H_B)\,|f\Omega\rangle = E_f\,|f\Omega\rangle. \tag{4.34}$$

In Equation (4.33), W is the potential describing the interaction between A and B, and T_R is the kinetic energy operator associated to the vector **R** joining the COM of A and B. This operator takes the form

$$T_R = -\frac{\hbar^2}{2\mu}\frac{\partial^2}{\partial R^2} - \frac{\hbar^2}{2\mu}\frac{2}{R}\frac{\partial}{\partial R} + \frac{\hbar^2\ell^2}{2\mu R^2}, \tag{4.35}$$

where $\mu = M_A M_B/(M_A + M_B)$ and ℓ is the end-over-end angular momentum operator. In a space-fixed frame, the spherical harmonics are eigenfunctions of ℓ^2 with eigenvalues $\ell(\ell + 1)$. In the body-fixed frame, used in the expansion of the total wave function in Equation (4.4), the matrix elements of ℓ^2 are no longer diagonal. Considering that $D^J_{M\Omega}$ are eigenfunctions of the total angular momentum and that $\ell = \mathbf{J} - \mathbf{j}$, with $\mathbf{j}$ being the total angular momentum operator of the fragments $\mathbf{j} = \mathbf{j}_A + \mathbf{j}_B$, the matrix elements of ℓ^2 also involve the $|f\Omega\rangle$ functions. Thus, the matrix elements of this operator depend on the particular choice made to represent $|f\Omega\rangle$. If $|f\Omega\rangle$ are eigenfunctions of $\mathbf{j}^2$ and $\mathbf{j}_z$ (along the body-fixed z-axis), these matrix elements take the general form

$$\begin{aligned}\langle D^J_{M\Omega}\langle f\Omega\,|\ell^2|\,f'\Omega'\rangle D^{J'}_{M'\Omega'}\rangle = \delta_{JJ'}\delta_{MM'}\delta_{ff'}\big\{\delta_{\Omega\Omega'}[J(J+1) + j(j+1) - 2\Omega^2] \\ + \delta_{\Omega\Omega'\pm1}\sqrt{j(j+1) - \Omega\Omega'}\sqrt{J(J+1) - \Omega\Omega'}\big\}.\end{aligned} \tag{4.36}$$

This term, called Coriolis coupling, couples states with different Ω values because the body-fixed frame rotates with the molecule. However, this coupling varies as R^{-2}, and may be neglected in many cases. For example, in diatomic systems, if the electronic Coriolis is not accounted for, the dynamics occurs in electronic states with fixed values of $\Omega \equiv \Lambda$ (the projection of the electronic angular momentum on the z body-fixed axis in the a and c Hund's cases). For polyatomic molecules, and when the dissociation process is fast, this term may be neglected, giving rise to the axial recoil approximation, which is equivalent to the coupled states or centrifugal sudden (CS) approach in collision dynamics.

The matrix elements of the interaction potential W are diagonal in the Ω quantum number in the body-fixed representation. This yields a considerable reduction of the matrix elements to store, as compared to those required in the space-fixed representation.

4.4.1 Time-independent Close Coupling equations

In the expansion of the total wave function of Equation (4.4), the $\Phi^{JM,\,Ef\Omega}_{f'\Omega'}(R)$ coefficients must be obtained, in general, by numerically integrating a coupled set of differential equations. These equations are obtained by first inserting Equation (4.4) into the time-independent Schrödinger equation with the Hamiltonian of Equation (4.33), then pre-multiplying by $\sqrt{2J+1/4\pi}\,D^{J}_{M\Omega''}\langle f''\Omega''|$, and finally integrating over all coordinates but the radial coordinate R. This set of equations is expressed as

$$\begin{aligned}\left\{-\frac{\hbar^2}{2\mu}\frac{d^2}{dR^2}+\frac{\hbar^2[J(J+1)+j''(j''+1)-2(\Omega'')^2]}{2\mu R^2}+E_{f''}-E\right\}&\\ \Phi^{JM,Ef\Omega}_{f''\Omega''}(R)&\\ =-\sum_{\Omega'}\delta_{\Omega\Omega'\pm1}\sqrt{j(j+1)-\Omega''\Omega'}\,\sqrt{J(J+1)-\Omega''\Omega'}\,\Phi^{JM,\,Ef\Omega}_{f''\Omega'}(R)&\\ -\sum_{f'}\langle f'\Omega''|W|f'\Omega''\rangle\Phi^{JM,\,Ef\Omega}_{f'\Omega''}(R).&\end{aligned} \tag{4.37}$$

This set of second-order differential equations is numerically solved by imposing two boundary conditions. At origin, $R\to 0$, all coefficients must be zero, i.e., $\Phi^{JM,\,Ef\Omega}_{f'\Omega''}(R=0)=0$. As $R\to\infty$, where the interaction potential W vanishes, the two fragments are described by a superposition of incoming and outgoing plane waves as

$$\begin{aligned}\Phi^{JM,Ef\Omega}_{f'\Omega'}(R\to\infty)=\sqrt{\frac{\mu}{2\pi\hbar^2}}\left\{\frac{e^{-i(k_fR-\ell\pi/2)}}{\sqrt{k_f}}\delta_{ff'}\delta_{\Omega\Omega'}\right.&\quad\text{incoming}\\ \left.-S^{J}_{f\Omega,f'\Omega'}(E)\frac{e^{-i(k_{f'}R-\ell'\pi/2)}}{\sqrt{k_{f'}}}\right\}&\quad\text{outgoing}\end{aligned} \tag{4.38}$$

where $k_f = \sqrt{2\mu(E - E_f)/\hbar^2}$ is the wave vector and $S^J_{f\Omega,f'\Omega'}(E)$ is the collision matrix, which contains all the information about the collision event. In photodissociation, the complex conjugate of this solution is imposed as commented above.

In practice, when the potential is real, it is more practical to impose real boundary conditions as

$$\hat{\Phi}^{JM,Ef\Omega}_{f'\Omega'}(R \to \infty) = \sqrt{\frac{2\mu}{\pi\hbar^2}} \left\{ \frac{\sin(k_f R - \ell\pi/2)}{\sqrt{k_f}} \delta_{ff'}\delta_{\Omega\Omega'} + K^J_{f\Omega,f'\Omega'}(E) \frac{\cos(k_{f'}R - \ell'\pi/2)}{\sqrt{k_{f'}}} \right\} \tag{4.39}$$

where $K^J_{f\Omega,f'\Omega'}(E)$ is the reaction matrix, which is related to the S-matrix as

$$\mathbf{S} = [\mathbf{1} - i\mathbf{K}]^{-1} \quad [\mathbf{1} + i\mathbf{K}] \tag{4.40}$$

where $\mathbf{S}$ is a matrix of elements $S^J_{f\Omega,f'\Omega'}(E)$, and the same holds for $\mathbf{K}$.

For collisions, all the information is extracted from the S-matrix. For photodissociation, however, we need to evaluate the matrix elements of Equation (4.6). In order to do so, the real function calculated with the boundary conditions of Equation (4.39) has to be transformed to that with the physical conditions of Equation (4.38). Arranging all independent solutions in vectors $\hat{\mathbf{\Phi}}$ and $\mathbf{\Phi}$ for the real and complex wave functions, this transformation is accomplished as

$$\mathbf{\Phi} = \hat{\mathbf{\Phi}} \times [\mathbf{1} + i\mathbf{K}]^{-1}. \tag{4.41}$$

In Appendix 4.8, the Numerov–Fox–Goodwin (or renormalized Numerov) method is described as an example for the numerical resolution of these equations.

4.4.2 Wave Packet Method

An alternative to time-independent close coupling method is to use the propagation in time of the initial state, Φ_0, defined in Equation (4.17), with

$$\Phi_t = e^{-iHt/\hbar} \quad \Phi_0. \tag{4.42}$$

The exact time-independent eigenfunctions of Equation (4.1) form a complete basis set, analogous to that of Equation (4.21), and then

$$\begin{aligned} \Phi_0 &= \sum_{f\Omega} \int dE \quad \left|\Psi^{JM}_{Ef\Omega}\right\rangle\left\langle\Psi^{JM}_{Ef\Omega}\middle|\Phi_0\right\rangle \equiv \sum_{f\Omega} \int dE \quad a^J_{f\Omega}(E)\left|\Psi^{JM}_{Ef\Omega}\right\rangle \\ \Phi_t &= \sum_{f\Omega} \int dE \quad a^J_{f\Omega}(E) \; e^{-iEt/\hbar} \quad \left|\Psi^{JM}_{Ef\Omega}\right\rangle, \end{aligned} \tag{4.43}$$

i.e., the wave packet corresponds to a linear superposition of different dissociative eigenstates $\Psi^{JM}_{Ef\Omega}$, corresponding to different energies and different final states of the fragments $f\Omega$. This last sum disappears when treating collision dynamics because a projection on a particular $f\Omega$ function is done to describe the initial state of reactants. In order to get the inverse relationship, we premultiply by $\exp(iE't/\hbar)$ both sides of the second expression of Equation (4.43) and integrate in time, so that we get

$$\frac{1}{2\pi\hbar}\int dt\ e^{-iE't/\hbar}\Phi_t = \sum_{f\Omega} a^J_{f\Omega}(E')\ \left|\Psi^{JM}_{E'f\Omega}\right\rangle, \tag{4.44}$$

where we have used the integral representation of the Dirac delta function as

$$\delta(E - E') = \frac{1}{2\pi\hbar}\int dt e^{i(E-E')t/\hbar}. \tag{4.45}$$

In order to eliminate the sum on the rhs of Equation (4.44), a projection on a particular state of the fragments, $|f'\Omega'\rangle$, must be done, obtaining

$$\frac{1}{2\pi\hbar\quad a^J_{f'\Omega'}(E')}\int dt e^{-iE't/\hbar}\langle f'\Omega'|\Phi_t\rangle|_{R\to\infty} = \left\langle f'\Omega'\middle|\Psi^{JM}_{E'f'\Omega'}\right\rangle|_{R\to\infty}, \tag{4.46}$$

where the projection has to be done at long distances R, where the interaction W becomes negligible. From this expression, the value of $a^J_{f\Omega}(E)$ can be easily obtained by using the complex conjugate of the exact function at long distances given in Equation (4.38). Similar expressions can be obtained for $R \to \infty$ when the wave packet has reached the asymptotic region. This expression provides a way to evaluate the partial cross section on a particular $|f'\Omega'\rangle$ state of the fragments through evaluation of $a^J_{f\Omega}(E)$, restricting Equation (4.8) to a single rotational transition

$$\sigma_{J_i\to J}(E) = \frac{A\,h\nu}{3}\sum_{f\Omega}\left|\left\langle\Phi_0\middle|\Psi^{JM}_{Ef\Omega}\right\rangle\right|^2 = \frac{A\,h\nu}{3}\sum_{f\Omega}\left|\ a^J_{f\Omega'}(E)\right|^2. \tag{4.47}$$

A further simplification can be made to obtain Equation (4.48), i.e., the total spectrum for a given rotational transition. By introducing Equation (4.45) into the above expression, we get

$$\begin{aligned}\sigma_{J_i\to J}(E) &= \frac{A\,h\nu}{3}\sum_{f\Omega}\int dE'\left\langle\Phi_0\middle|\Psi^{JM}_{E'f\Omega}\right\rangle\delta(E-E')\left\langle\Psi^{JM}_{E'f\Omega}\middle|\Phi_0\right\rangle\\ &= \frac{A\,h\nu}{3}\frac{1}{2\pi\hbar}\int dt e^{iEt/\hbar}\left\langle\Phi_0\right|\sum_{f\Omega}\int dE'\left|\Psi^{JM}_{E'f\Omega}\right\rangle e^{-iHt/\hbar}\left\langle\Psi^{JM}_{E'f\Omega}\middle|\Phi_0\right\rangle,\end{aligned} \tag{4.48}$$

where the closure relationship can be recognized (similar to Equation (4.21)), and it can finally be written as

$$\sigma_{J_i\to J}(E) = \frac{A\,h\nu}{3}\frac{1}{2\pi\hbar}\int dt\ e^{iEt/\hbar}\ \langle\Phi_0|\Phi_t\rangle, \tag{4.49}$$

i.e., the spectrum can be obtained by the Fourier transformation of the autocorrelation function $\langle \Phi_0 | \Phi_t \rangle$. This enormously simplifies the calculations, because this integral vanishes for long distances, where the initial wave packet Φ_0 is zero.

The time evolution of the initial wave packet can be done using many different propagation methods (Leforestier et al. 1991). These propagations in time can be done on the basis $|f\Omega\rangle$ as the time-independent close coupling calculations. However, this would require the storage of a potential matrix elements for all R distances required to converge the calculations. It is more efficient to represent the wave packet in grids, in which the potential becomes diagonal (Paniagua et al. 1999; Aguado et al. 2003; Chenel et al. 2016).

These two methods here described should give the same results, and the use of any of them is a matter of choice. Time-independent calculations are better suited for slow processes like predissociation, while time-dependent wave packets are very well-suited for fast direct photodissociation, to rapidly obtain a broad spectrum.

4.5 Electronic Structure Calculations

In order to obtain the electronic part of the $|f\Omega\rangle$ basis functions of fragments, it is convenient to factorize the total Hamiltonian of the system as:

$$H = T_{\mathbf{Q}} + H_e \qquad \text{with} \qquad H_e(\mathbf{q}, \mathbf{Q}) = T_{\mathbf{q}} + V(\mathbf{q}, \mathbf{Q}), \tag{4.50}$$

in which $\mathbf{Q}$ and $\mathbf{q}$ denote the coordinates of the nuclei and electrons, respectively. Here, $T_{\mathbf{Q}}$ and $T_{\mathbf{q}}$ are the corresponding kinetic energy operators, the particular forms of which depend on the choice of coordinates made, and $V(\mathbf{q}, \mathbf{Q})$ is the electrostatic potential among all particles of the system. In the framework of the Born–Oppenheimer approximation, the electronic functions $\phi_\alpha(\mathbf{q}; \mathbf{Q})$ are first obtained as

$$H_e(\mathbf{q}, \mathbf{Q})\, \phi_\alpha(\mathbf{q}; \mathbf{Q}) = U_\alpha(\mathbf{Q})\, \phi_\alpha(\mathbf{q}; \mathbf{Q}) \tag{4.51}$$

for each nuclear configuration denoted by $\mathbf{Q}$. The eigenvalues $U_\alpha(\mathbf{Q})$ are evaluated at each nuclear configuration, and can be considered as the adiabatic potential energy surfaces in which the nuclear motion takes place for each electronic state α. The total eigenfunctions of the system Ψ_β can be expressed using these electronic functions as

$$\Psi_\beta(\mathbf{q}, \mathbf{Q}) = \sum_\alpha \varphi_{\beta\alpha}(\mathbf{Q})\, \phi_\alpha(\mathbf{q}; \mathbf{Q}) \tag{4.52}$$

where the nuclear functions $\varphi_{\beta\alpha}(\mathbf{Q})$ satisfy

$$\begin{aligned} &\left[\left\langle \phi_\alpha(\mathbf{q}; \mathbf{Q}) | T_Q | \phi_\alpha(\mathbf{q}; \mathbf{Q}) \right\rangle + U_\alpha(\mathbf{Q}) - E\right] \varphi_{\beta\alpha}(\mathbf{Q}) \\ &\quad = - \sum_{\alpha' \neq \alpha} \left\langle \phi_\alpha(\mathbf{q}; \mathbf{Q}) | T_Q | \phi_{\alpha'}(\mathbf{q}; \mathbf{Q}) \right\rangle \varphi_{\beta\alpha'}(\mathbf{Q}). \end{aligned} \tag{4.53}$$

In the adiabatic approximation, the couplings between different electronic states are neglected, so that the nuclear dynamics is performed on each potential energy surface, $U_\alpha(\mathbf{Q})$, individually. These couplings arise from the changes of the electronic wave functions, by action of the nuclear kinetic operator. Using the Hellmann–Feynman theorem, the coupling matrix elements can be expressed as

$$\left\langle \phi_\alpha(\mathbf{q}, \mathbf{Q}) | T_Q | \phi_{\alpha'}(\mathbf{q}, \mathbf{Q}) \right\rangle = \frac{\langle \phi_\alpha(\mathbf{q}, \mathbf{Q}) | (T_Q V) | \phi_{\alpha'}(\mathbf{q}, \mathbf{Q}) \rangle}{U_{\alpha'} - U_\alpha} \quad \text{with} \quad \alpha \neq \alpha', \tag{4.54}$$

which indicates that the couplings increase when two states energetically approach each other, and therefore transitions among them must be considered in order to properly describe the dynamics of a system.

The main difficulty in treating photodissociation for astrophysical purposes is the necessity of describing the absorption spectra up to 13.6 eV. This excitation energy is close to or even higher than the ionization threshold of most molecules. Near this threshold, the bound electronic states gradually grow closer and closer, and thus many of them must be calculated. Also, these states present many crossings, which implies the necessity of going beyond the usual Born–Oppenheimer approximation.

In this section, the methods currently used to treat these problems will be briefly reviewed.

4.5.1 Excited Electronic States

Having to calculate several electronic states requires the use of multiconfigurational methods, because each electronic state is described by at least one electronic configuration.

The mono-referential methods use the Hartree–Fock (HF) solution as a reference $|\phi_{HF}>$. This is the case with the configuration interaction (CI) or coupled cluster (CC) methods.

In the CI methods, the multi-electron wavefunction for the electronic state α, $|\phi_\alpha>$, is expanded in the basis of Slater determinants obtained by excitations from the reference configuration $|\phi_{ref}>$

$$|\phi_\alpha> = (1 + \hat{C}_\alpha)|\phi_{ref}>$$

where the reference state $|\phi_{ref}>$ usually corresponds to the HF determinant, $|\phi_{HF}>$. The operator $\hat{C}_\alpha$ produces the mono-, bi-, ..., electronic excitations from occupied to virtual orbitals of the reference configuration. Usually, truncated expansions are considered in which only mono- and bi-excitations are included, giving rise to the configuration interaction with single and double excitations (CISD) method. The coefficients of the expansion are obtained using the variational principle, i.e., minimizing the energy with respect to the coefficients, that is, solving the secular equation for the Hamiltonian matrix represented in the basis of the reference and excited configurations

$$H_{\alpha,\beta} = <\phi_\alpha|H_e|\phi_\beta>.$$

According to the MacDonald–Hylleraas–Undheim theorem (Hylleraas & Undheim 1930; McDonald 1993), the αth root is an upper bound to the corresponding αth electronic state. However, using HF as a reference configuration is often a poor solution to describe excited electronic states that are energetically close or degenerate.

On the other hand, the coupled cluster method (Čížek 1966; Čížek 1969; Čížek et al. 1971) with single and double excitations (Purvis & Bartlett 1982) and including perturbatively triple excitations, CCSD(T) is commonly used only for the ground electronic state. In contrast to the CI techniques, CC methods are size-consistent. In this method, the ground electronic state is obtained using an exponential excitation operator $\hat{T}$

$$|\phi_0> = \exp(\hat{T})|\phi_{HF}> .$$

However, because a variational determination of the expansion coefficients is not feasible, projection techniques are used in the development of the CC methods. As a consequence, excited states cannot be obtained using the same technique. The equation of motion coupled cluster (EOM-CC) method is a powerful alternative to accurately describe the lower excited electronic states, as well as states of a different nature, such as interacting Rydberg and valence states.

However, when very excited states are needed, multiconfigurational methods based on a single reference configuration are not accurate enough to describe very high excitations. Multireference methods are usually required for very excited states. In these methods, several reference configurations, i.e., Slater determinants, are used to build all possible single and double excitations to generate the configuration interaction (CI) expansion of all the electronic states. There are several multireference approaches based on the coupled cluster method (Kállay & Gass 2004) (MRCC), but the most commonly used method is the multireference configuration interaction (MRCI) method (Werner & Knowles 1988).

In order to select the initial reference configurations $|\phi_{ref}>$ for the MRCI calculations, the most widely used method is the multiconfiguration one (Werner & Knowles 1984), usually using a complete active space scheme, CASSCF, in which the molecular orbitals as well as the configuration interaction coefficient (in a reduced CI matrix) are simultaneously optimized. This method is extremely dependent on the number of active orbitals and on the number of electronic states optimized and used to build the reference configurations. Because the complexity of the optimization grows rapidly with the number of active orbitals, the main difficulty is the determination of this active space and the procedure followed to optimize the orbitals. Direct minimization methods have been developed to ensure rapid convergence (Werner & Knowles 1981; Fernandez Rico et al. 1986), which allows the optimization of an energy average of several states with arbitrary weight factors.

This optimization has to be performed at many geometries of the molecule, describing regions where the molecule is bound as well as its dissociation in fragments for many electronic states. Often, the convergence of the optimization is not homogeneous all along the molecular geometries needed to describe the dissociation. This problem is particularly difficult when dealing with very excited electronic states, because the number of electronic states is very dependent on the geometries.

When the fragments are in degenerate states, it is also important to properly select the active orbitals to achieve a numerically acceptable degeneracy.

4.5.2 Transition Dipole Moments

The absorption intensity of each excited electronic state is determined by the electric dipole operator matrix elements between the ground and excited electronic states, $\langle\phi_\alpha|\mathbf{d}|\phi_\beta\rangle$ needed to evaluate the reduced matrix elements of Equation (4.6). There are, in general, three components that must be calculated.

For diatomic systems, parallel and perpendicular transitions refer to transitions in which the non-zero transition dipole moment corresponds to z, parallel to the internuclear axis, or along the x or y axes, respectively. Transitions between electronic states of the same projection of the electronic orbital angular momentum, Λ, are parallel. Transitions with $\Delta\Lambda = \pm 1$ are perpendicular, while those with $\Delta\Lambda > 1$ are forbidden. In the absence of spin–orbit couplings, transitions between states of different multiplicities are also forbidden for electric dipole transitions.

For triatomic molecules, or systems with a plane of symmetry, the electronic states are separated into two groups, A' or A'', i.e., symmetric or antisymmetric under reflection through the plane. The transitions are then in-plane, between states of same symmetry, or out-of-plane, involving transitions between states of different symmetry.

The electronic transition dipole moments $\langle\phi_\alpha|\mathbf{d}|\phi_\beta\rangle$ also depend on the nuclear coordinates $\mathbf{Q}$. Thus, in a proper description, the three components have to be calculated to properly describe the absorption intensity. In some cases, it can be approximated by those matrix elements corresponding to the equilibrium geometry in the ground electronic state. This, however, may introduce some errors, as discussed below for the HCN example.

The calculation of non-diagonal matrix elements has a problem regarding the change of the relative phase as a function of the nuclear configuration $\mathbf{Q}$. Because ϕ_α and $-\phi_\alpha$ are both eigenfunction of the electronic Hamiltonian, the solutions at different $\mathbf{Q}$ may have different relative phase. These artificial changes of sign of $\langle\phi_\alpha|\mathbf{d}|\phi_\beta\rangle$ need to be corrected because they may include artificial nodes in the initial wave packet in the excited state and thus produce wrong absorption spectra.

The avoid this problem, a careful analysis of the phase has to be performed, and the best way to correct it is to calculate the overlap of functions calculated at adjacent configuration points $\mathbf{Q}_i$ and $\mathbf{Q}_i + \Delta = \mathbf{Q}_j$, defined as

$$S_{\alpha,\beta}(\mathbf{Q}_i, \mathbf{Q}_j) = \left\langle\phi_\alpha(\mathbf{Q}_i)|\phi_\beta(\mathbf{Q}_j)\right\rangle. \tag{4.55}$$

To do so automatically, it is necessary to devise a proper grid to perform the calculations with points sufficiently close to each other that the overlap matrix is close to the unity matrix, allowing the correction of the relative sign of the dipole matrix elements.

4.5.3 Non-adiabatic Couplings

There are two non-adiabatic matrix elements associated with $\langle\phi_\alpha|T_{\mathbf{Q}}|\phi_{\alpha'}\rangle$: those of first-order derivative, $\langle\phi_\alpha|\nabla|\phi_{\alpha'}\rangle$, and of second-order derivative, $\langle\phi_\alpha|\nabla^2|\phi_{\alpha'}\rangle$. The second order are typically smaller, and when using a sufficiently large number of electronic functions, can be expressed in terms of those involving first-order derivatives.

In general, these first-order non-adiabatic matrix elements are calculated numerically, using a Taylor expansion of the wave function as

$$\phi_\alpha(\mathbf{Q}_{i\pm1}) = \phi_\alpha(\mathbf{Q}_i) \pm \boldsymbol{\Delta}\, \phi'_\alpha(\mathbf{Q}_i) + \frac{\Delta^2}{2}\, \phi''_\alpha(\mathbf{Q}_i) \quad (4.56)$$

so that, using a first-order approximation, the non-diabatic matrix elements become

$$\mathcal{F}^m_{\alpha\beta} = \left\langle \phi_\alpha \left| \frac{\partial}{\partial Q_m} \right| \phi_\beta \right\rangle = \left[S_{\alpha,\beta}(\mathbf{Q}_i, \mathbf{Q}_{i+1}) - \delta_{\alpha\beta}\right] \Big/ \Delta_m, \quad (4.57)$$

where ∇ and $\mathbf{F}_{\alpha\beta}$ are vectors of components $\partial/\partial Q_m$ and $\mathcal{F}^m_{\alpha\beta}$, respectively. It is necessary to calculate 3N − 6 matrix elements of this kind to fully describe the non-adiabatic matrix elements, which is a formidable computational task.

Moreover, these matrix elements diverge at conical intersections at which $U_\alpha - U_\beta = 0$, according to Equation (4.54). Conical intersections are always possible in polyatomic systems depending on more than two degrees of freedom (Lipkowitz & Cundari 2007).

4.5.4 Diabatization and Quasi-diabatization

To avoid the singularity of the non-adiabatic couplings at conical intersections, another electronic basis can be defined, called "diabatic," which does not depend (approximately) on the nuclear coordinates (Smith 1969; Baer 1975). In this new basis, the matrix elements of the kinetic operator T_Q, in Equation (4.54) vanish (approximately). This "diabatic" basis, $\{\tilde{\phi}_a\}$, is defined by an unitary transformation from "the adiabatic" basis, $\{\phi_a\}$, as

$$\tilde{\phi}_a = \sum_\alpha \phi_\alpha\, T_{\alpha a}(\mathbf{Q}) \quad \leftrightarrow \quad \phi_\alpha = \sum_a T^\dagger_{\alpha a}(\mathbf{Q})\, \tilde{\phi}_a. \quad (4.58)$$

If, for example, it is assumed that at a given reference geometry, $\mathbf{Q}_r$, the two bases coincide, $\tilde{\phi}_a(\mathbf{Q}_r) = \phi_a(\mathbf{Q}_r)$, for other geometries it can be considered that the unitary transformation is simply given by the overlap

$$\begin{aligned} \phi_\alpha(\mathbf{Q}) = \sum_a T^\dagger_{\alpha a}(\mathbf{Q})\, \tilde{\phi}_a \equiv \sum_a T^\dagger_{\alpha a}(\mathbf{Q})\, \phi_a(\mathbf{Q}_r) \quad \longrightarrow \quad T_{\alpha a}(\mathbf{Q}) \\ = \langle \phi_\alpha(\mathbf{Q}) | \phi_a(\mathbf{Q}_r) \rangle, \end{aligned} \quad (4.59)$$

and this implies that the electronic "adiabatic" basis should be complete for **all geometries Q**. Under these circumstances, the non-adiabatic couplings vanish, but these functions are no longer eigenfunctions of the H_e, and

$$\begin{aligned}\left\langle \tilde{\phi}_a \,|H_e|\, \tilde{\phi}_b \right\rangle &= \sum_{\alpha\beta} T^{\dagger}_{\alpha a}(\mathbf{Q}) T_{\beta b}(\mathbf{Q}) \left\langle \phi_\alpha \,|H_e|\, \phi_\beta \right\rangle \\ &= \sum_{\alpha} T^{\dagger}_{\alpha a}(\mathbf{Q}) T_{\beta b}(\mathbf{Q})\; U_\alpha(\mathbf{Q}),\end{aligned} \tag{4.60}$$

i.e., in the diabatic basis, the couplings among different electronic functions are given by potential terms.

However, the completeness condition is not fulfilled; it is only valid locally around $\mathbf{Q}_r$. An alternative is to calculate the transformation matrix as defined by Smith (1969)

$$\frac{\partial T_{\alpha a}(\mathbf{Q})}{\partial \mathbf{Q}} = \sum_{\alpha'} T_{\alpha' a}(\mathbf{Q}) \mathbf{F}_{\alpha\beta} \quad \text{with} \quad T_{\alpha a}(\mathbf{Q}_r) = \delta_{\alpha a}. \tag{4.61}$$

For a diatomic system, the vector $\mathbf{F}_{\alpha\beta}$ only has one element, the radial coupling. Equation (4.61) has to be integrated on the radial coordinate starting at the reference geometry $\mathbf{Q}_r$, and the radial non-adiabatic coupling can be eliminated in the diabatic representation.

However, this is not the situation for a general polyatomic system. First, the differential equation of Equation (4.61) may be integrated along many different paths, which indicates that the adiabatic–diabatic transformation is not unique. Second, there are several components in the non adiabatic vector $\mathbf{F}_{\alpha\beta}$, and in general it is not possible to make them all vanish (Mead & Truhlar 1982).

Therefore, for polyatomic systems, there is no strict diabatic representation, because generally the curl condition is not fulfilled (Baer 1975; Mead & Truhlar 1982) and it is not possible to eliminate all derivative couplings in all the configuration space. Nevertheless, it is convenient to transform to a quasi-diabatic representation where the derivative couplings are somehow minimized. These quasi-diabatic representations are not unique because they depend on the path chosen to solve the first-order differential equation to eliminate one/some first derivative matrices (Smith 1969; Baer 1975). To avoid this arbitrariness, it is then necessary to add to the diabatic matrix the residual derivative couplings, which are expected to be much smaller than in the adiabatic representation. When there are conical intersections, it is then crucial to eliminate the singularity of the derivative couplings at crossings. This can be accomplished by a "regularization" procedure, as shown by Thiel & Köppel (1999), producing regular diabatic states whose derivative coupling matrices do not show singularities at crossing, but instead vary smoothly.

The methods for quasi-diabatization can be classified into three groups (Köppel 2004), in order of decreasing computational effort.

1. Derivative-based methods (Smith 1969; Baer 1975, 1980, 2001) involve the calculation of non-adiabatic derivative matrix elements, which requires highly accurate electronic wavefunctions.

2. Property-based methods (Werner & Meyer 1981) require the calculation of wavefunctions to evaluate some property, such as electric dipole, which is assumed to change smoothly with the nuclear configuration and is diagonal in the diabatic representation.
3. Energy-based methods only require knowledge of the eigenvalues, are based on a good knowledge of the system, and very frequently are assumed to require some adjustable parameters.

4.6 Examples

In this section, we shall discuss the photodissociation dynamics of three representative molecules of astrophysical interest, focusing on the description of the molecular dynamics.

4.6.1 CO

CO is the second-most abundant molecule in space and the one most commonly observed to trace physical conditions of different astrophysical environments. CO is the main reservoir of carbon in the interstellar medium and is the precursor of more complex molecules through reactions in gas phase and on the surfaces of grains and interstellar grains. The major roles of this system in interstellar clouds and circumstellar disks have been reviewed by Visser et al. (2009), who state that "a key process in controlling its gas phase abundance is its photodissociation by ultraviolet photons."

CO is a very stable molecule, with a triple bond, and its first dissociation limit appears at $\approx$ 11.1 eV (or λ = 114 nm). The photodissociation of CO proceeds by predissociation of bound states. Most of these states belong to excited electronic states associated with two series of Rydberg states converging to CO^+, in either the $X^2\Sigma^+$ or the $A^2\Pi$ states (Eidelsberg & Rostas 1990; Tchang-Brillet et al. 1992; Monnerville & Robbe 1994). The higher part of the spectrum is unidentified, probably because the electronic states are highly perturbed (Eidelsberg & Rostas 1990). These bound states predissociate by coupling to the valence states, leading to $C(^3P,^1D,^1S)$ and $O(^3P,^1D)$ fragments. For some long-lived states, the dissociation competes with the fluorescence back to the ground electronic state, such that dissociation branching ratios, ν, have been also reported along with the total rate, radiative plus dissociate (Eidelsberg & Rostas 1990; Eidelsberg et al. 1991).

The different absorption bands of CO and its isotopologues have been experimentally studied in detail (Letzelter et al. 1987; Eidelsberg & Rostas 1990; Eidelsberg et al. 1991; Ubachs et al. 1994; Cacciani et al. 1998; Eidelsberg et al. 2006; Eidelsberg et al. 2012). However, a complete theoretical study treating the Rydberg and valence states was only addressed recently by Vazquez, Lefebvre-Brion, and co-workers in a series of papers (Vázquez et al. 2009; Lefebvre-Brion et al. 2010; Majumder et al. 2014; Lefebvre-Brion & Kalemos 2016).

In order to give a sense of the complexity of the electronic structure of CO, the $^1\Sigma^+$ states are shown in Figure 4.8. Ab initio calculations have been done following the CAS+MRCI-F12 calculations, using a VQZ-F12 basis set with extra diffuse

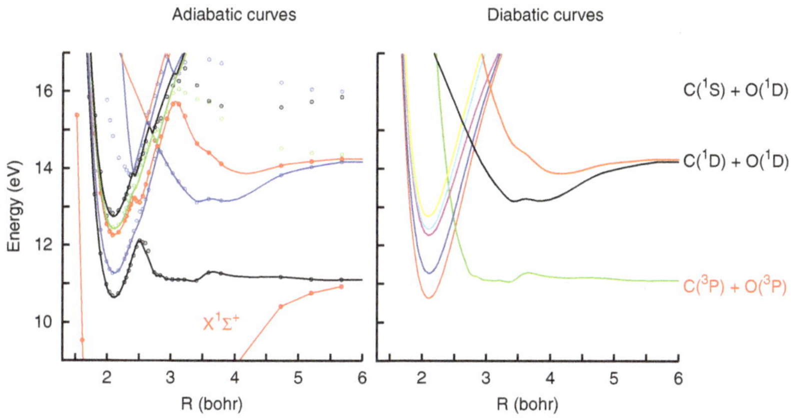

Figure 4.8. $^1\Sigma^+$ states of CO. (Left panel) Adiabatic points calculated with a MRCI-F12 method using the MOLPRO package with a VQZ-F12 basis, with extra functions to describe the excited Rydberg states. Lines are the adiabatic energies of the 8 × 8 diabatic model. (Right panel) Diabatic potentials of an 8 × 8 model, used to describe the six excited $^1\Sigma^+$ states. The final two states (up to eight) require a more complete diabatic model including more states.

functions to describe the Rydberg states. The adiabatic results of the left panel of Figure 4.8 can be compared with Figure 1 of Vázquez et al. (2009).

A 2×2 diabatic model built by Tchang-Brillet and co-workers (Tchang-Brillet et al. 1992) very satisfactorily reproduced the first two excited states: a valence dissociative state leading to $C(^3P)+O(^3P)$ fragments, the D state, and a bound Rydberg state of the $CO^+(X^2\Sigma^+)$ series, the B 3sσ state. This model reproduces rather well the first predissociative resonances of the B–D states around 11.4 eV, and was used to generate the results of Figure 4.4.

In order to describe the higher $CO(^1\Sigma^+)$ resonances, this diabatic model needs to be completed, including more states. In the right panel of Figure 4.8, we present a preliminary diabatic model including eight states. A simple energy-based diabatization method has been employed, assuming Morse potentials to describe the deep wells of the Rydberg states. This model describes approximately the ab initio points calculated up to ≈ 13.6 eV. Three dissociative valence states, correlating to $C(^3P)+O(^3P)$ and $C(^1D)+O(^1D)$ fragments have been included. To describe higher energies, more dissociative valence states are needed. For energies above 13 eV, many of the measured transitions are unassigned. In that energy region, there are many electronic states of $^1\Sigma^+$ symmetry that are difficult to converge. In this region of high density of states, tiny couplings with triplets (by spin–orbit terms) and Π states (by electronic Coriolis terms) also introduce important perturbations.

As an example, (Lefebvre-Brion & Kalemos 2016) recently studied the predissociation of the W $^1\Pi$ state of CO, including also the couplings to $^3\Pi$ states. The $^3\Pi$ state correlates asymptotically with $C(^3P)+O(^3P)$ and with $C(^3P)+O(^1D)$ and $C(^1D)+O(^3P)$ fragments. These last two asymptotes are below the $C(^1D)+O(^1D)$ limit, and clearly indicates that there are many crossings with singlet $^1\Pi$ states, at which tiny spin–orbit couplings may introduce important perturbations.

This example demonstrates that the photodissociation of diatomic molecules, apparently simple, is still a challenge for theoretical simulations.

4.6.2 HCN

HCN is one of the most abundant triatomic systems in space, probably the second after H_2O. It is characterized by two linear isomers HCN and HNC, with an energy barrier of $\approx$ 2 eV (HCN is more stable by $\approx$ 0.60 eV). These features make that, at temperatures below 5000 K, the two isomers can be considered to be independent. The two isomers are observed in different astrophysical environments and the HNC/HCN abundance ratio is commonly used to trace the physical conditions.

In cold and dense interstellar clouds, where the gas is largely shielded from the external ultraviolet radiation, the HNC/HCN abundance ratio is $\approx$ 1 (Sarrasin et al. 2010) while in regions illuminated by ultraviolet photons, HCN becomes more abundant than HNC, by a factor of $\approx$ 5 in both diffuse interstellar clouds (Liszt & Lucas 2001; Godard 2010) and photon-dominated regions such as the Orion Bar (Hogerheijde et al. 1995). In order to understand the reason for the variation of the HNC/HCN abundance ratio reactive (Loison et al. 2014) and inelastic collisions (Sarrasin et al. 2010; Dumouchel et al. 2011; Ben Abdallah et al. 2012; Hernández-Vera et al. 2017) have been investigated.

The photoabsorption and photodissociation of HCN has been studied experimentally over a wide wavelength interval, from 90 to 150 nm, by several authors (Lee 1980; Nagata et al. 1981; Nuth & Glicker 1982). There are several experimental works on individual electronic bands of HCN isomer. In Reference (Herzberg & Hines 1957) weak absorption bands associated to the $1^1A''$ and $2^1A'$ electronic states were studied. These bands are formed by narrow peaks, decaying by either electronic predissociation to the ground state or through tunneling across potential barriers due to avoided crossing with higher electronic states. These processes have been studied in detail by either experimental techniques (Herzberg & Hines 1957; Hsu et al. 1984; Meenakshi & Innes 1986; Eng et al. 1987; Jonas et al. 1990) or theoretical simulations (Peric et al. 1988; Xu et al. 2001, 2002a, 2002b, 2003). The band associated to $3^1A'$ state was measured by Mcpherson & Simons (1978) and the peaks where assigned (Mcpherson & Simons 1978; Chuljian et al. 1984; Peric et al. 1987). From the theoretical point of view, the photodissociation has been studied only for the first low lying valence states. Only very recently the higher VUV region corresponding to Rydberg states up to 13.6 eV have been studied (Chenel et al. 2016; Aguado et al. 2017).

An important aspect to perform a theoretical simulation of the HCN and HNC photodissociation is the need of determining the different destruction rate of the two isomers under UV radiation. This can not be studied easily experimentally because the higher HNC isomer can not be synthesized alone, and therefore to unravel HNC photodissociation rate is difficult. This was the aim of recent studies on the highly excited electronic states of HCN and HNC (Chenel et al. 2016; Aguado et al. 2017).

The excited electronic state energy diagrams of HCN and HNC are summarized in Figure 4.9. The electronic states of HCN are classified as A'/A'', for even/odd

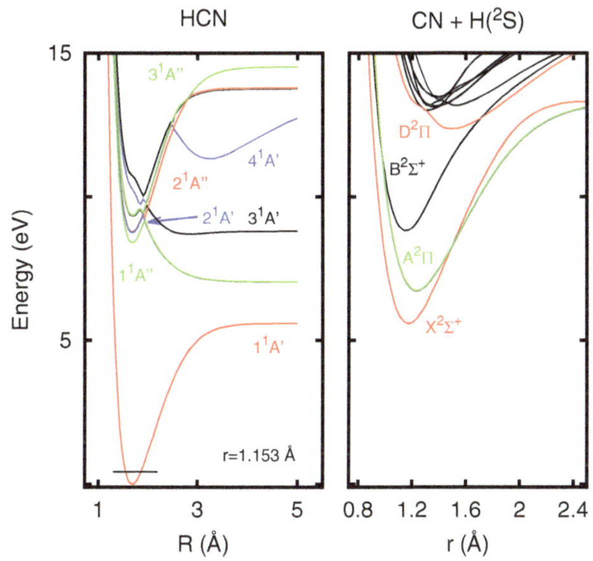

Figure 4.9. (Central panel) Potential energies of CN in different electronic states as a function of internuclear distance, r. (Left/right panels) Potential energies of the excited electronic state at the collinear configuration of HCN (left) and HNC (right), as a function of the distance between H and CN center of mass, keeping CN at the equilibrium distance r = 1.153 Å. (Credit: Chenel et al. 2016.)

parity of the electronic reflection through the plane of the molecule. CN($^2\Sigma^+$/ $^2\Sigma^-$) + H(^2S) correlate with a single $^1A'/^1A''$ state. However, the double degenerate CN($^2\Pi$ or $^2\Delta$) correlate to one $^1A'$ and $^1A''$, which are degenerate at the collinear configurations of HCN and HNC shown in the figure, because at this geometry the molecule belong to the $C_{\infty v}$ symmetry and the states are characterized as for diatomic molecules.

This symmetry breaks as the system bends, and therefore the two systems separate from each other. Moreover, the crossings at collinear geometries associated to Σ and Π states (occurring at $R \approx 2$ Å for the two isomers) constitute real conical intersections because, as long as the system bends, two of the states become of the same symmetry and separate because they interact with each other. This effect can be seen in the bottom panel of Figure 4.10 as the angle γ varies from 0 or π.

Another important aspect of the symmetry is shown in the transition dipole moments in the top panel of Figure 4.10. At collinear geometries, the selection rules are equal to those of diatomic molecules, i.e., only transitions with $\Delta\Lambda = 0, \pm 1$ are allowed and $\Sigma^+ \rightarrow \Sigma^-$ are forbidden. These conditions relax for bent geometries where $A' \rightarrow A'$ (in plane) and $A' \rightarrow A''$ (out of plane) are possible. In HCN and HNC, the rovibrational wave functions have a maximum of amplitude at $\gamma = 0$ and π, respectively, so that the selection rules are close to those of diatomic molecules (Chenel et al. 2016). Because the symmetry changes from HCN to HNC (see Figure 4.10), this introduces important different selection rules for HCN and HNC (Chenel et al. 2016) as discussed below.

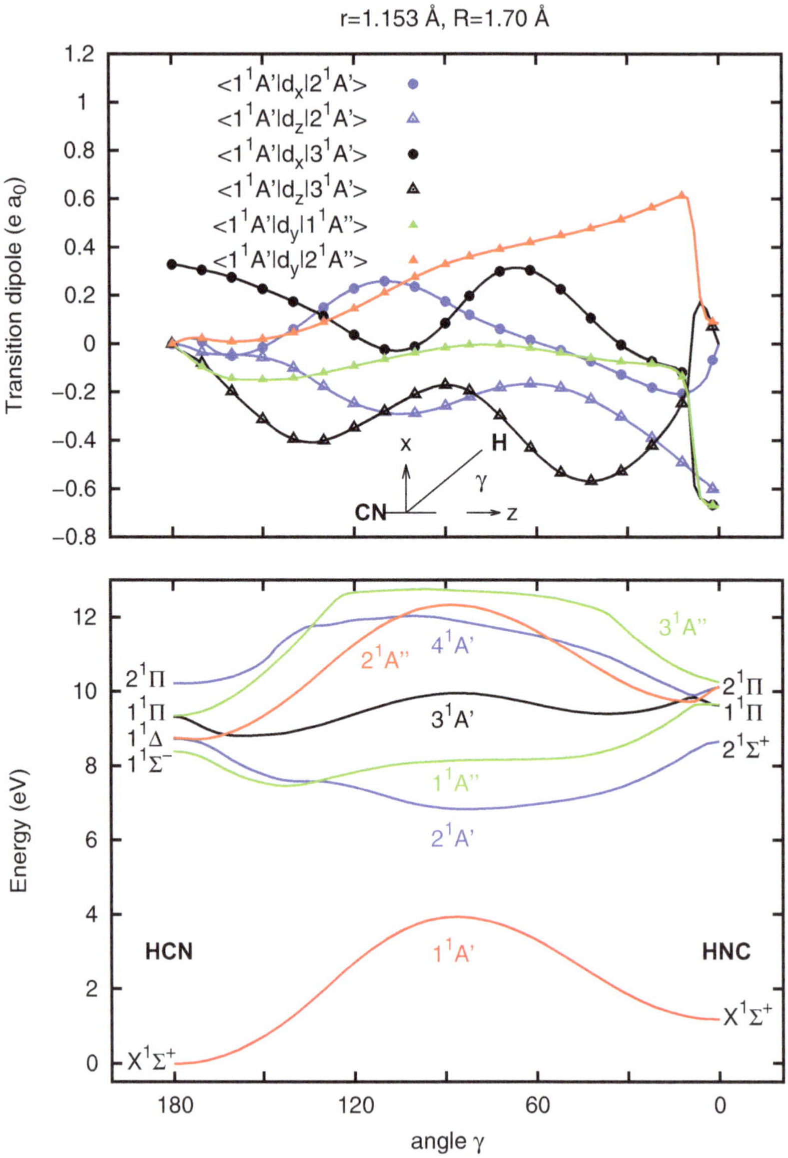

Figure 4.10. (Bottom panel) Energy diagram of the potential energy for the lower electronic state of HCN as a function of the bending angle, γ. (Top panel) Transition dipole moment from the ground to the excited electronic state as a function of the bending angle. (Credit: Chenel et al. 2016.)

The first electronic states in Figure 4.9 are dissociative valence states. The higher electronic states are bound Rydberg states, converging to a HCN^+ cation. The situation is very similar to that of CO, and in Figure 4.11 the first 21 electronic states of HCN are shown (12 of ${}^1A'$ symmetry in red and nine of ${}^1A''$ symmetry in blue) together with the first two ${}^2A'$ and ${}^2A''$ of the cation. The cation also shows two

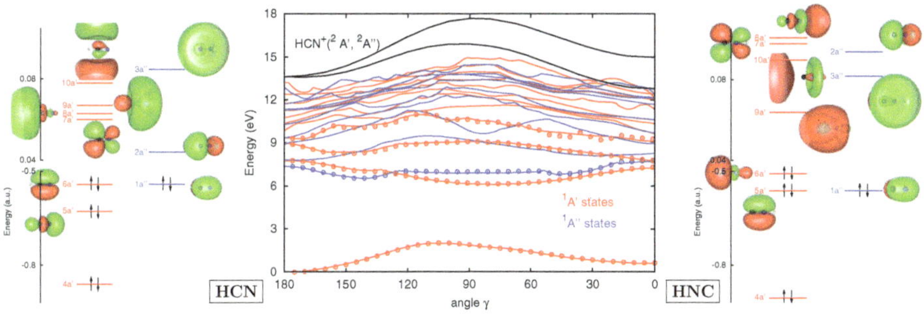

Figure 4.11. (Middle panel) Minimum energy path for the HCN/HNC isomerization for the first 12 $^1A'$ (red) and nine $^1A''$ (blue) electronic states of HCN obtained using the PESs as described in the text. In black are presented the minimum energy paths (MEPs) for the HCN^+/HNC^+ corresponding to the first $^2A'$ and $^2A''$ electronic states of the cation. The points correspond to PESs calculated previously. (Left/right panels) The molecular orbital amplitudes of HCN (left) and HNC (right). Note that the molecular orbitals are ordered by the occupation number, not their energy. (Credit: Aguado et al. 2017.)

minima at linear configurations, for HCN^+ and HNC^+ isomers, and the most stable isomer changes from A' and A''. Interestingly, all the excited Rydberg states of HCN show a similar structure and are characterized by a main configuration with very diffuse orbitals, as shown in Figure 4.11.

The adiabatic potential surfaces of Rydberg states are all bound, and the bound rovibrational states are localized on either HCN or HNC wells. Thus, the emission back to the ground state is also to the HCN or HNC wells, respectively. The radiative lifetimes are on the order of 10^{-6}–10^{-9}s (Chenel et al. 2016). The adiabatic dynamics on the dissociative valence state is either direct (in few fs) or mediated by resonances whose dissociation lifetimes are on the order of 10^{-12} s or shorter. Thus, emission from the valence dissociative states is not likely to occur.

The individual spectra for each of the 21 excited electronic states were calculated individually using a wave packet method (Chenel et al. 2016; Aguado et al. 2017). This method is well-suited to describe both dissociation and bound states. In this last case, the autocorrelation function in Equation (4.49) is multiplied by an exponential decay function, which introduces an artificial Lorentzian broadening in bound–bound transitions. The individual character of each resonance can also be analyzed by doing the Fourier transform of the wave packet at the resonance energy (Paniagua et al. 1999; Chenel et al. 2016) as in Equation (4.46). This procedure allows the assignment of the transitions, as discussed in more detail in Chenel et al. (2016) and Aguado et al. (2017). An example for the absorption band of the HCN (X $^1A'$ → $3^1A'$) transition is shown in Figure 4.12. Clearly, the peaks are associated with bending progression on the $3^1A'$ state, which is associated with a well supported by a barrier, originating from the crossing between a dissociative valence state and a bound Rydberg state, like in the case of CO. In this case, the resonances are considerably broader because the non-adiabatic couplings are larger as the system bends. In this spectrum, the peaks appearing below $E = 0$ are bound states when using the adiabatic approximation, i.e., neglecting non-adiabatic couplings toward

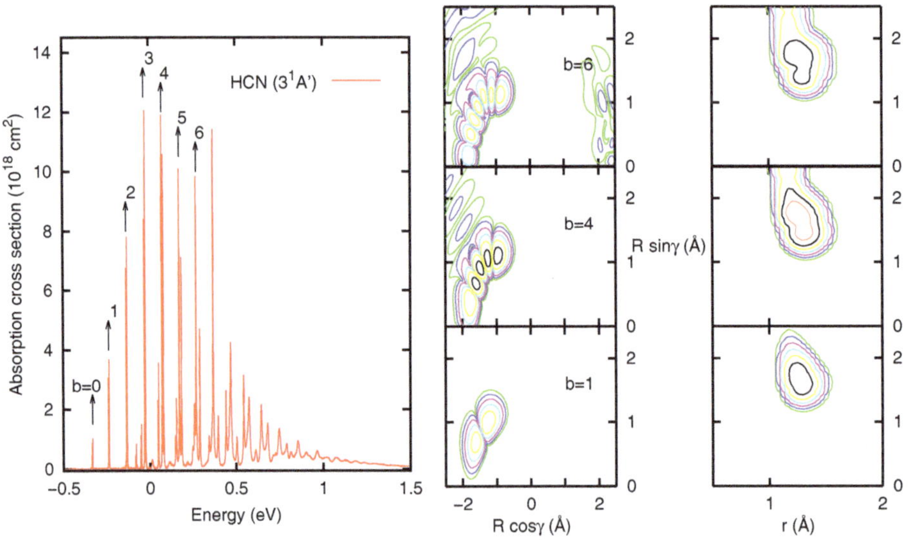

Figure 4.12. (Left panel) Absorption spectrum for the $HCN(X^1A, J = 0 \rightarrow 3^1A', J = 1)$ transition as a function of energy, referred to the $CN(B\ ^2\Sigma^+, v = 0, j = 0)$ threshold at 8.93 eV. (Right panels) Contour plots of the wave functions at the energies of some of the resonances taken from Chenel et al. (2016). (Credit: Chenel et al. 2016.)

the lower electronic states. Above $E = 0$, the peaks are resonances and the contribution of the continuum can be appreciated in the right panels for $b = 4$ and 6. The bending progression stops at approximately 0.5 eV, at the energy of the top of the barrier. Because this barrier is originated by an avoided crossing, it should continue in the upper surface. When neglecting the non-adiabatic couplings, the series in the upper surface is shifted towards higher energies. This is why, in the comparison with the experimental spectrum of Nuth & Glicker (1982) in the left panel of Figure 4.13, the peaks around photon energies of ≈ 9 eV are missing in the simulated spectrum of HCN.

This experimental spectrum is similar to others obtained for this isomer (Lee 1980). In spite of some similarities for some transitions, there is a significant difference in the simulated and observed spectra, especially in the overall intensity and density of transitions. These differences have been attributed to one reason (Aguado et al. 2017): the lack of non-adiabatic transitions among the different electronic states. The adiabatic electronic PESs cross many times, and at the conical intersections and/or avoided crossings, the non-adiabatic couplings become important. The inclusion of such terms will increase the width of the bound levels belonging to the bound Rydberg states and share the oscillators' strength. Another reason is that, in the simulations, the experimental width of the excitation pulse is not introduced and the experimental spectrum was obtained at higher temperatures that could add extra broadening of the absorption bands.

All these effects affect the two isomers in a similar way, so the simulated spectra for the two isomers can be compared. In Figure 4.13, photodissociation cross section of HNC is significantly larger than that of HCN, especially at photon energies below

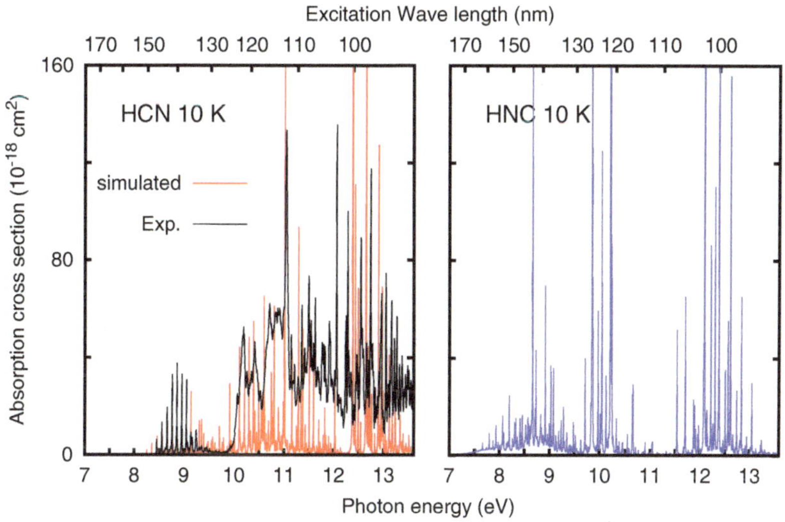

Figure 4.13. Absorption spectra of HCN (left panel) and HNC (right panel) obtained with a quantum wave packet method at 10 K, using all rotational transitions from $J_i = 0$, 1, 2, and 3. For HCN (left panel), the experimental absorption spectrum by Nuth & Glicker (1982) is shown. (Credit: Aguado et al. 2017.)

10 eV. This is due to the change of the Σ, Π or Δ character of the electronic wave function when changing from HCN to HNC linear geometries, which has an enormous impact on the amplitudes of the transition dipole moments, as explained before (Chenel et al. 2016). This fact leads to different photodissociation rates depending on the radiation field existing in different regions, which will be discussed below. The simulated spectra shown are completely ab initio and no empirical data have been introduced.

Given the discrepancy found via experiment, in order to get the photodissociation rate, $K(T)$, for a given UV radiation field, we proceed as follows: first, the ratio between HNC and HNC, $r = K_{\mathrm{HNC}}/K_{\mathrm{HCN}}$, is calculated with the simulated cross sections; second, the rate for HCN, K_{HNC}, is calculated with the experimental cross section, and then that of HNC is obtained as $K_{\mathrm{HNC}} = r\ K_{\mathrm{HNC}}$. To see the effect in diffuse clouds, the Meudon PDR code (Le Petit 2006) was used under normal conditions (Aguado et al. 2017), first using $r = 1$ (the usual choice until now), and then using $r = 2.2$ obtained via the interstellar radiation field (Draine 1978). The HNC/HCN abundance ratios obtained are 1 and 0.5–0.8 for $r = 1$ and 2.2, respectively, while the observation's is 0.21 (Liszt & Lucas 2001). Clearly, the agreement improves, showing the possible impact of using more realistic photodissociation cross sections for each isomer.

4.6.3 H_2CO

Formaldehyde was first detected in space in 1969 (Snyder et al. 1969) and it is now known to be significantly abundant among the identified interstellar molecules. It is also believed to be formed in the interstellar medium by successive hydrogenation of carbon monoxide CO on ice surfaces through a sequence of reactions, $CO \rightarrow HCO \rightarrow H_2CO$, and in the gas phase it can be formed in the reaction $CH_3 + O \rightarrow H_2CO + H$. However, despite the high number of studies on this system, its photodynamics is not yet completely understood. The photodissociation dynamics of H_2CO can be divided mainly in two energetic regions: at energies below 4.0 eV, the decomposition of formaldehyde yields molecular $H_2 + CO$ and radical $HCO + H$ products. The detailed mechanism of the photodissociation dynamics is not yet completely known, but it is clear that it involves the lowest-lying valence electronic states: the ground 1^1A_1 (S_0), the 1^1A_2 (S_1), and the 1^3A_2 (T_1) electronic states. The rupture of the CO bond is, however, not relevant for these low energies; it requires much higher energies (the CH_2+O products are formed at $\approx$ 7.6 eV, experimentally).

Most of the dynamical studies on formaldehyde, both theoretically (Schinke 1986; Chang & Miller 1990; Zhang et al. 2004, 2005; Yin et al. 2006; Farnum et al. 2007; Yin et al. 2007; Shepler et al. 2008; Araujo et al. 2008, 2009; Zhang et al. 2009) and experimentally (Chuang et al. 1987; Carleton et al. 1990; Terentis & Kable 1996; Valachovic et al. 2000; Hopkins et al. 2007; Simonsen et al. 2008), have mainly focused on the photodissociation via the lowest- lying electronic states. The formaldehyde in its S_0 ground electronic state is photoexcited to the dipole-forbidden S_1 excited state. It is believed that:

1. The molecular $H_2 + CO$ and radical $HCO + H$ products are obtained via a fluorescence or internal conversion mechanisms from the S_1 excited state to the ground S_0 state.
2. The $HCO + H$ products can be also obtained via an intersystem crossing of the S_1 state with the T_1 electronic state, which can also intersystem cross to the ground electronic state.

The $H_2 + CO$ fragments in their ground electronic states come only from the S_0 electronic state, but the $HCO + H$ products in their ground states come from both T_1 and S_0 states.

Due to the relatively high dimensionality of the problem, most of the dynamical studies for this system involving S_0, S_1 and T_1 have used quasi-classical trajectories (QCT) methods. In this regard, many QCT calculations have been performed on the S_0 ground state to study the formation of the $H_2 + CO$ products. In those calculations, the dynamics was initiated from the transition state to study the photodissociation that forms these molecular fragments (Chang & Miller 1990; Perlherbe & Hase 1996; Li et al. 2000). Later trajectories initiated from the structure of the H_2CO molecule on the S_0 state revealed *roaming* trajectories (Townsend et al. 2004). The analysis of the roaming trajectories showed the formation of an intermediate complex (H–HCO) in which a hydrogen and the radical orbit each other, until the molecular products are finally formed. Interestingly, those kinds of trajectories were not observed when starting from the transition state of the S_0 electronic state. Shepler et al. (2008) performed QCT calculations, but starting at the T_1/S_0 crossing. These calculation were performed on a modified version of a global potential energy surface for the ground state, previously reported by Zhang et al. (2004). The modifications on this surface were intended for a better description of

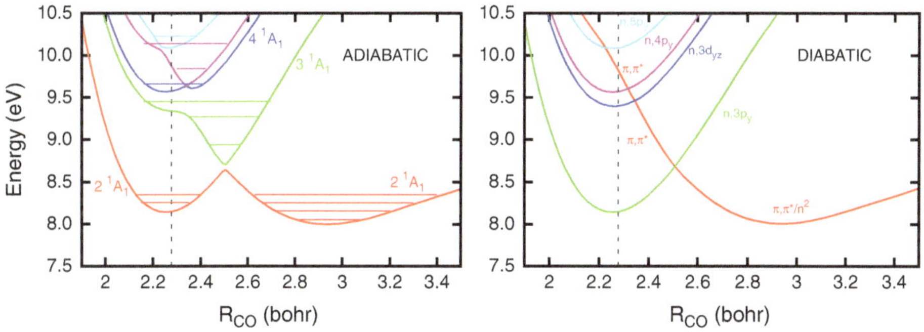

Figure 4.14. One-dimensional adiabatic (left) and diabatic (right) potential energy curves of formaldehyde of A_1 symmetry along the CO stretching coordinate, R_{CO}, at the EOM-CCSD level of theory. An internally contracted MRCI treatment is used, however, for CO distances beyond 2.4 Bohr (for details, see text). The vibrational levels are also displayed in the top panel. (Credit: Gomez-Carrasco et al. 2010.)

the crossing points. Fu et al. (2011) performed the first surface-hopping QCT calculation allowing transitions among the S_0, S_1, and T_1 states. It is clear that, in order to improve our understanding of the photodynamics of this system at energies below 4 eV, a multisurface quantum study involving at least three coupled surfaces still needs to be done.

On the other side, photoabsorption into the electronic states of formaldehyde lying above 7 eV is required to form the CH_2+O dissociative products (see Figure 4.14). The electric dipole forbidden transition ${}^1A_2(n, \pi^*) \leftarrow \tilde{X}{}^1A_1$ observed at ≈ 4.0 eV is well-studied. However, the allowed ${}^1A_1(\pi, \pi^*) \leftarrow \tilde{X}{}^1A_1$ transition that should be one of the most intense has not been experimentally detected. Likewise, the experimental assignment of the ${}^1B_1(\sigma, \pi^*) \leftarrow \tilde{X}{}^1A_1$ electronic transition also remains still uncertain. For a long time, it was believed (Langhoff et al. 1974) that the ${}^1A_1(\pi, \pi^*)$ state lies above the first ionization potential of 10.88 eV (Niu et al. 1993). Later, however, different electronic calculations predicted smaller vertical excitation energies (Gwaltney & Barlett 1995; Merchan & Roos 1995; Hachey et al. 1995; Müller & Lischka 2001; Gómez-Carrasco et al. 2010).

The reason for these non-observed states is related to the interaction between the valence and Rydberg states, mainly along the CO stretching coordinate (Grein & Hachey 1996; Perić et al. 2000). In contrast to the missing ${}^1A_1(\pi, \pi^*)$ state, the ${}^3A_1(\pi, \pi^*)$ state of formaldehyde has been observed by electron impact studies (Taylor et al. 1982). Theoretically, it has been shown (Grein & Hachey 1996) that the ${}^3A_1(\pi, \pi^*)$ electronic state lies well below the Rydberg states, and therefore it should be free of the Rydberg mixing, in contrast to its singlet counterpart.

Most of the theoretical works on this system (Whitten & Hackmeyer 1969; Buenker & Peyerimhoff 1970; Peyerimhoff et al. 1971; Harding & Goddard 1977; Fitzgerald & Schaefer 1985; Hachey et al. 1995; Merchan & Roos 1995; Gwaltney & Barlett 1995; Grein & Hachey 1996; Perić et al. 2000; Müller & Lischka 2001; Dallos et al. 2001; Gómez-Carrasco et al. 2010) have calculated vertical excitations with different ab initio methods: the CI and MRDCI calculations of Whitten &

Hackmeyer (1969), Buenker & Peyerimhoff (1970), and Peyerimhoff et al. (1971), respectively. MRCI results were given by Harding & Goddard (1977), Hachey et al. (1995), Grein & Hachey (1996), and Hachey et al. (1996). Complete active space perturbation theory (CASPT2) was applied to formaldehyde by Merchan & Roos (1995). Equation-of-motion coupled cluster including single and double excitations (EOM-CCSD) theory was used by Gwaltney & Barlett (1995). More recently, high-level MRCI and multireference averaged quadratic coupled clusters (MR-AQCC) calculations based on complete active space self-consistent field (CASSCF) wave-functions were presented by Müller & Lischka (2001) and Dallos et al. (2001), and EOM-CCSD calculations by Gómez-Carrasco et al. (2010). From the analysis of the oscillator strengths, the main contribution to the absorption spectrum comes from the $^1A_1$ and $^1B_2$ electronic states.

There are not many dynamical studies involving the high-lying electronic states of formaldehyde. It is clear that this problem requires taking into account nonadiabatic effects to account for a proper description of its photodynamics. Gómez-Carrasco et al. (2010) investigated the dynamics on five coupled $^1A_1$ and eight uncoupled $^1B_2$ potential energy surfaces, for the energy range of 7–10 eV, using a time-dependent fully quantum approach (Meyer et al. 1990; Worth et al. 2000; Beck et al. 2000; Meyer 2009). A 5×5 model Hamiltonian, $H = T_Q + \mathbf{W}$, was constructed for the diabatization of the two-dimensional $^1A_1$ potential energy surfaces, along the CO stretching (r) and the HCH bending coordinates (θ), where T_Q is the harmonic kinetic energy operator for all relevant modes and $\mathbf{W}$ is the 5×5 diabatic potential energy matrix given by

$$\mathbf{W} \equiv W^{r,\theta}_{5\times5} = \begin{pmatrix} W^{r,\theta}_{11} & W^{r,\theta}_{12} & W^{r,\theta}_{13} & W^{r,\theta}_{14} & W^{r,\theta}_{15} \\ & W^{r,\theta}_{22} & 0 & 0 & 0 \\ & & W^{r,\theta}_{33} & 0 & 0 \\ & h.c. & & W^{r,\theta}_{44} & 0 \\ & & & & W^{r,\theta}_{55} \end{pmatrix}, \tag{4.62}$$

where

$$W^{r,\theta}_{ii} = W^r_i + \Delta W^\theta_i$$
$$W^{r,\theta}_{ij} = W^r_{ij} + k_{ij}\ \Delta^\theta$$

following the ideas of the linear vibronic coupling (LVC) model (Köppel et al. 1984; Domcke et al. 2004). Both the diagonal $W^{r,\theta}_{ii}$ and the coupling $W^{r,\theta}_{ij}$ elements are written as the sum of a radial and an angular contribution.

Regarding the diagonal elements, the W^r_i are the one-dimensional diabatic CO-stretching coordinate potentials, which are estimated graphically connecting the electronic energy points of configurations with the same character. The one-dimensional diabatic surfaces are displayed in the bottom panel of Figure 4.14, so that $W^r_1 = (\pi, \pi^*)$, $W^r_2 = (n, 3p_y)$, $W^r_3 = (n, 3d_{yz})$, $W^r_4 = (n, 4p_y)$ and $W^r_5 = (n, 5p_y)$. On the other side, ΔW^θ_i are the one-dimensional $\angle$HCH (θ) bending potentials.

Regarding the off-diagonal terms of the diabatic matrix **W**, Gómez-Carrasco et al. used the simplest model for the radial contributions so that they were taken to be constants ($W_{ij}^r = W_{ij}^0$), with a value of one half of the smallest energy difference at the avoided crossing region of the adiabatic one-dimensional potential energy curves along the CO coordinate (see top panel of Figure 4.14). The off-diagonal terms in this simple model are, therefore, just θ-dependent.

Note that the avoided crossings in the one-dimensional potential energy curves along the CO coordinate in the left panel of Figure 4.14 turn into conical intersections (CoIn) when including the bending HCH degree of freedom. Because the coupling terms must vanish at the CoIns (when $\theta = \theta^{\mathrm{CoIn}}$), the interstate coupling constants k_{ij} are given by:

$$k_{ij} = -\frac{W_{ij}^0}{\Delta^{\theta=\theta_{\mathrm{CoIn}}}} \quad \text{and} \quad \Delta^\theta = \theta - \theta_{\mathrm{FC}} \tag{4.63}$$

where Δ^θ is the angular displacement with respect to its value at the FC zone ($\theta_{\mathrm{FC}} = 116.54°$).

Figure 4.15(a) and (b) shows the experimental and the calculated 2D absorption spectra, respectively, of formaldehyde after photoexcitation of the ground wave function to the several potential energy surfaces. Figure 4.15(c) and (d) shows the independent contributions to the total spectrum from the $^1A_1$ and $^1B_2$ electronic states.

Three regions can be clearly distinguished in the spectrum: at low energies, we find the transition to the $1^1B_2(n, 3s)$ Rydberg state, whose calculated excitation energy is slightly larger than the experimental one (7.20 versus 7.09 eV). The structure of this band is found to be caused by a vibrational excitation in the HCH angle, although the experimental vibrational spacing is smaller, almost half of the calculated one. Hachey et al. (1994) proposed that the vibrational structure of this Rydberg state is due to the coupling with a non-planar $2\ ^1A'$ state. This coupling can only happen under symmetry-lowering and would imply contribution of vibrational modes that are not totally symmetrical. Further calculations should be performed to clarify this situation. Note finally that the calculated intensity of the peak is smaller than the experimental one because the calculation underestimates the oscillator strength for the corresponding transition.

The next region, located between 145–160 nm, also corresponds to transitions to Rydberg states: the peak at longer wavelengths ($\approx$ 154 nm) corresponds to a transition to the $2^1B_2(n, 3p_z)$ Rydberg state, as shown in the bottom panel. At $\approx$ 152 nm, we find the vibrational progression on the HCH angle belonging to the transition to the $2^1A_1(n,3p_y)$ electronic state. The relative energetic separation between both states, 2^1B_2 and 2^1A_1, is smaller than the experimental one because, as commented above, most of the $^1B_2$ states appear shifted to higher energies. The third region corresponds to the region of strong interaction. The most noticeable feature is the intensity redistribution of the π, π^* state due to the interaction with the "surrounding" Rydberg states, showing a vibronic structure that becomes denser at higher energies. The two peaks at 139.55 and 137.20 nm in the experimental spectrum were assigned

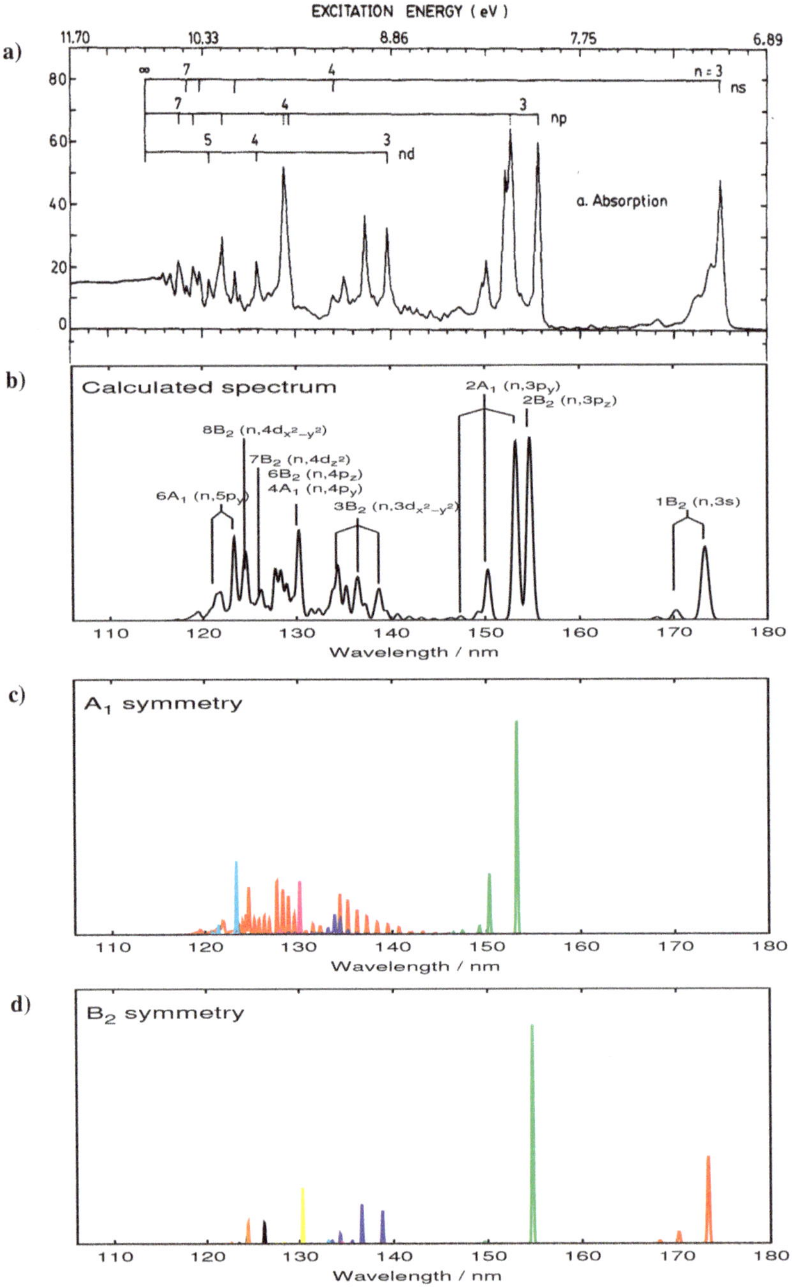

Figure 4.15. Absorption spectrum of formaldehyde: (a) Experimental spectrum, (b) calculated 2D spectrum, (c) contribution of the $^1A_1$ electronic states to the total spectrum, and (d) contribution of the $^1B_2$ electronic states to the total spectrum. The experimental spectrum is taken from Figure 1 of Suto et al. (1986). (Credit: Gomez-Carrasco et al. 2010.)

experimentally to 3D transitions (Suto et al. 1986) and theoretically are also associated to the HCH vibrational excitation of the transition to the 3 $^1B_2$ (n, $3d_{x^2-y^2}$) state, although the intensities are not well-described (see panel (d) of Figure 4.15).

At higher energies, we find two peaks: a rather intense peak that theoretically corresponds to the contribution of two states, the $4^1A_1(n,4p_y)$ and $6^1B_2(n,4p_z)$ transitions, and a second peak theoretically assigned as a transition to the $6^1A_1(n,5p_y)$ electronic state. Both are in good agreement with the experimental result.

Time-dependent wavepacket calculations were run on the five coupled $^1A_1$ electronic states using the Hamiltonian given by Equation (4.62). Figure 4.16 displays the probability of the wave packet being located in any of the five $^1A_1$ electronic states after excitation to the (π, π^*) state. The left panel displays the adiabatic populations while the right panel shows the diabatic ones. In every panel, the inset shows a detail for short propagation times. Note that the dynamics is quite similar in both cases, especially for short times, as can be seen in the insets. The dynamics of the system can be also followed by looking at Figure 4.17, where the diabatic vibrational densities are shown. Through the analysis of Figures 4.16 and 4.17, one can follow the photodynamics of the system and the consecutive transitions among the different electronic states along the time.

Regarding the inclusion of other degrees of freedom, the out-of-plane mode has been analyzed. This mode couples the $^1A_1$ and $^1B_1$ electronic states and the $^1A_2$ and $^1B_2$ ones, respectively. Dallos et al. (2001) found a CoIn between the 1^1B_1 (σ, π^*) and the 2^1A_1 (π, π^*), which was carefully analyzed in their work. Likewise, Gómez-Carrasco et al. (2010) also analyzed that CoIn, in the restricted space of higher symmetry, C_{2v}, varying the CO stretching and the symmetric HCH bending coordinates. Both results are in good agreement for CO distances larger than ≈ 2.5 Bohr. For shorter distances, however, the results are different due to the presence of the (n, $3p_y$) Rydberg electronic state. This state was not computed by Dallos et al. (2001) because they did not include diffuse functions in the atomic basis sets in the

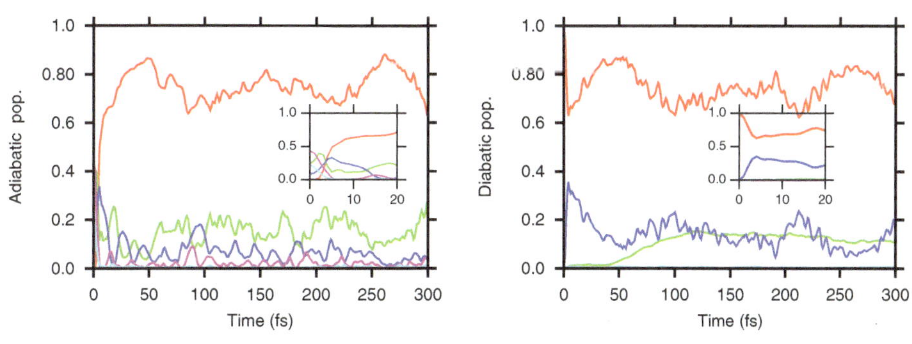

Figure 4.16. Diabatic (left) and adiabatic (right) populations as a function of the propagation time for the five $^1A_1$ potential energy curves. The initial wave packet is excited vertically to the (π, π^*) state. The inset displays a detail at short propagation times. For the line codings, see Figure 4.14. (Credit: Gomez-Carrasco et al. 2010.)

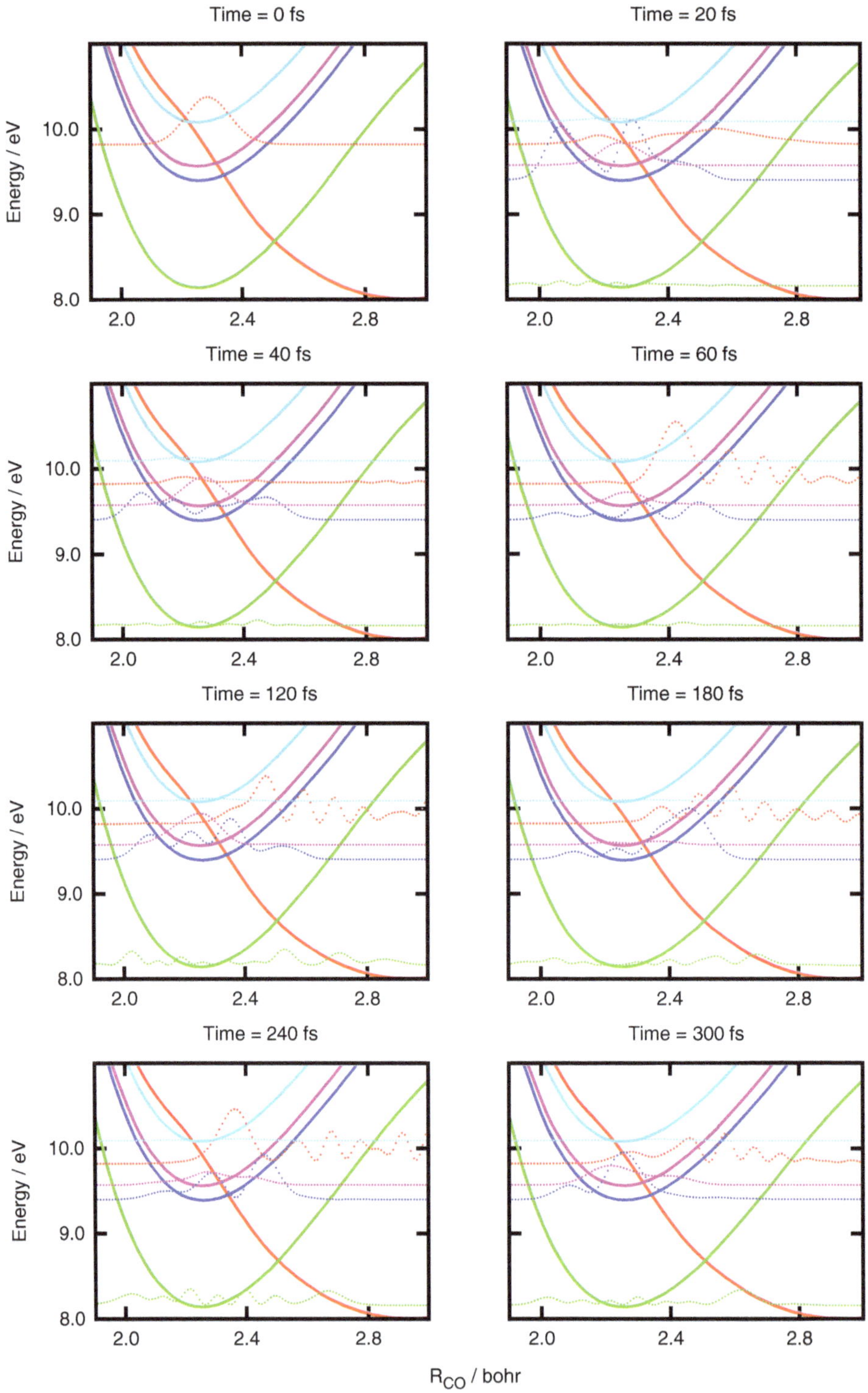

Figure 4.17. Diabatic reduced densities (dotted lines) on the different diabatic $^{1}A_{1}$ electronic states (solid lines) as a function of the propagation time (in fs) after excitation from the ground electronic state to the (π, π^{*}) state. The zero of the density is located at the minimum of the potential of each state. The PES and line codings are as in the bottom panel of Figure 4.14. (Credit: Gomez-Carrasco et al. 2010.)

electronic structure calculations. The non-adiabatic dynamics of formaldehyde, including the out-of-angle, remains to be done.

4.7 Appendix: Matrix Elements of the *G*(*E*) Operator

The resolvent operator $G(Z)$ is defined as

$$(Z - H)G(Z) = 1 \tag{4.64}$$

and is related to $G(E)$ according to

$$G(E) = \lim_{\substack{\Re(Z)\to E\\ \Im(Z)\to 0^+}} G(Z) = \lim_{\epsilon\to 0^+} G(E + i\epsilon). \tag{4.65}$$

Introducing the equalities $PP = P$ and $PQ = 0$, and the closure relationship, Equation (4.21), into the resolvent operator of Equation (4.64), the following set of equations are obtained (Feshbach 1962; Mower 1966; Shore 1967)

$$\begin{aligned} PP &= P(Z - H)(P + Q)G(Z)P = P \\ QP &= Q(Z - H)(P + Q)G(Z)P = 0. \end{aligned} \tag{4.66}$$

Using Equations (4.19) and (4.20), it is readily obtained that

$$\begin{aligned} Q(Z - H)P &= -QVP \\ P(Z - H)Q &= -PVQ \\ Q(Z - H)Q &= Q(Z - H_0)Q \\ P(Z - H)P &= P(Z - H_0)P - PVP. \end{aligned} \tag{4.67}$$

Using these expressions, Equation (4.66) becomes

$$QG(Z)P = \frac{1}{Q(Z - H_0)Q} QVP \quad PG(Z)P \tag{4.68}$$

$$PG(Z)P = \frac{1}{P(Z - H_0 - R(Z))P} \tag{4.69}$$

where $R(Z)$ is the shift operator defined as (Mower 1966)

$$R(Z) = V - VQ\frac{1}{Q(Z - H_0)Q}QV. \tag{4.70}$$

Calculating the limit $\lim_{\substack{\Re(Z)\to E\\ \Im(Z)\to 0^+}} G(Z)$, it is obtained that

$$\lim_{\substack{\Re(Z)\to E\\ \Im(Z)\to 0^+}} \langle\phi_i|Z - H_0 - R(Z)|\phi_j\rangle = (E - E_j)\delta_{ij} - V_{ij}(1 - \delta_{ij}) - \Delta_{ij} + i\Gamma_{ij} \tag{4.71}$$

where

$$\Delta_{ij} = \sum_{\alpha} \Delta_{ij}^{\alpha}; \quad \Gamma_{ij} = \sum_{\alpha} \Gamma_{ij}^{\alpha}. \tag{4.72}$$

Therefore, $PG(E)P$ in matricial form is expressed as

$$\begin{pmatrix} G_{11}(E) & G_{12}(E) & \cdots & G_{1n}(E) \\ G_{21}(E) & G_{22}(E) & \cdots & G_{2n}(E) \\ \vdots & \vdots & \ddots & \vdots \\ G_{n1}(E) & G_{n2}(E) & \cdots & G_{nn}(E) \end{pmatrix} =$$

$$\begin{pmatrix} E - E_1 - \Delta_{11} + i\Gamma_{11} & -V_{12} - \Delta_{12} + i\Gamma_{12} & \cdots & -V_{1n} - \Delta_{1n} + i\Gamma_1 \\ -V_{21} - \Delta_{21} + i\Gamma_{21} & E - E_2 - \Delta_{22} + i\Gamma_{22} & \cdots & -V_{2n} - \Delta_{2n} + i\Gamma_{2n} \\ \vdots & \vdots & \ddots & \vdots \\ -V_{n1} - \Delta_{n1} + i\Gamma_{n1} & -V_{n2} - \Delta_{n2} + i\Gamma_{n2} & \cdots & E - E_n - \Delta_{nn} + i\Gamma_{nn} \end{pmatrix}^{-1}. \tag{4.73}$$

To invert the matrix on the rhs of Equation (4.73), it is first diagonalized by an orthogonal transformation $\mathbf{T}$, according to

$$\mathbf{T}\begin{pmatrix} E_1 + \Delta_{11} - i\Gamma_{11} & V_{12} + \Delta_{12} - i\Gamma_{12} & \cdots & V_{1n} + \Delta_{1n} - i\Gamma_{1n} \\ V_{21} + \Delta_{21} - i\Gamma_{21} & E_2 + \Delta_{22} - i\Gamma_{22} & \cdots & V_{2n} + \Delta_{2n} - i\Gamma_{2n} \\ \vdots & \vdots & \ddots & \vdots \\ V_{n1} + \Delta_{n1} - i\Gamma_{n1} & V_{n2} + \Delta_{n2} - i\Gamma_{n2} & \cdots & E_n + \Delta_{nn} - i\Gamma_{nn} \end{pmatrix}\mathbf{T}^{-1} = \begin{pmatrix} \omega_1 & 0 & \cdots & 0 \\ 0 & \omega_2 & \cdots & 0 \\ \vdots & \vdots & \ddots & \vdots \\ 0 & 0 & \cdots & \omega_n \end{pmatrix}$$

where ω_k are the eigenvalues. Thus, the resolvent operator matrix elements in Equation (4.73) become

$$\begin{pmatrix} G_{11}(E) & G_{12}(E) & \cdots & G_{1n}(E) \\ G_{21}(E) & G_{22}(E) & \cdots & G_{2n}(E) \\ \vdots & \vdots & \ddots & \vdots \\ G_{n1}(E) & G_{n2}(E) & \cdots & G_{nn}(E) \end{pmatrix} = \mathbf{T}^{-1}\begin{pmatrix} 1/(E-\omega_1) & 0 & \cdots & 0 \\ 0 & 1/(E-\omega_2) & \cdots & 0 \\ \vdots & \vdots & \ddots & \vdots \\ 0 & 0 & \cdots & 1/(E-\omega_n) \end{pmatrix}\mathbf{T}. \tag{4.74}$$

With the matrix elements given by $PG(E)P$, the $QG(Z)P$ matrix elements can be evaluated using Equation (4.68)

$$\langle \phi_{E'}^{\beta} | G(E) | \phi_j \rangle = \sum_i \frac{V_{iE'}^{\beta *}}{E - E'} \langle \phi_i | G(E) | \phi_j \rangle. \tag{4.75}$$

4.8 Appendix: Numerical Method for Close Coupling Equations

To find the numerical solutions, we shall discretize the grid in the R coordinate so that the Schrödinger equation at each point i of the grid can be written as

$$\bar{\psi}_i'' = \frac{2\mu}{\hbar^2}\{\bar{V}_i - E\underline{\bar{I}}\} \quad \bar{\psi}_i \tag{4.76}$$

where we have exchanged the Energy subindex for a radial index so that $R_i = R_{ini} + (i - 1)\Delta$, Δ being the distance between two adjacent points. Here, $\bar{V}_i$ is the matrix of the potential at R_i, $\underline{\bar{I}}$ is the identity matrix, and $\bar{\psi}_i$ is the fundamental matrix at R_i.

The wavefunction and its second derivative can be expanded in a Taylor series

$$\bar{\psi}_{i\pm1} = \bar{\psi}_i \quad \pm \quad \Delta\bar{\psi}_i' + \frac{\Delta^2}{2}\bar{\psi}''_i \quad \pm \quad \frac{\Delta^3}{3!}\bar{\psi}_i''' + \ldots \tag{4.77}$$

Using these expressions as previously done for bound states, we get the three-point Numerov relation

$$\bar{\alpha}_{i-1}\bar{\psi}_{i-1} + \bar{\beta}_i\bar{\psi}_i + \bar{\gamma}_{i+1}\bar{\psi}_{i+1} = 0 \tag{4.78}$$

where

$$\begin{aligned}
\bar{\alpha}_{i-1} &= \underline{\bar{I}} - \frac{\Delta^2}{12}\{2\mu(\bar{V}_{i-1} - E\underline{\bar{I}})\} \\
\bar{\beta}_i &= -2\underline{\bar{I}} - \frac{10\Delta^2}{12}\{2\mu(\bar{V}_i - E\underline{\bar{I}})\} \\
\bar{\gamma}_{i+1} &= \underline{\bar{I}} - \frac{\Delta^2}{12}\{2\mu(\bar{V}_{i+1} - E\underline{\bar{I}})\}.
\end{aligned} \tag{4.79}$$

This expression has an error proportional to Δ^6.

Now we shall use the Fox–Goodwin algorithm, which consist of defining

$$\bar{\psi}_i = \bar{\mathcal{R}}_i\,\bar{\psi}_{i+1}. \tag{4.80}$$

As we are interested in the regular solutions for which $\psi_1 = 0$, we have

$$\begin{aligned}
\bar{\mathcal{R}}_1 &= 0 \\
\bar{\mathcal{R}}_2 &= -\bar{\beta}_2^{-1}\,\bar{\gamma}_3 \\
\bar{\mathcal{R}}_i &= -\{\alpha_{i-1}\bar{\mathcal{R}}_{i-1} + \bar{\beta}_i\}^{-1}\bar{\gamma}_{i+1}
\end{aligned} \tag{4.81}$$

which gives the recursive method to obtain $\mathcal{R}_j$.

In the asymptotic region, i.e., for large i values, we impose the stationary condition given in Equation (4.39), in order to only have to deal with real numbers, and defining the diagonal matrices

$$\begin{aligned}
\sin_i^{n,n} &= \frac{\sin k_n R_i}{\sqrt{k_n}} \quad \text{for open channels} \\
\sin_i^{n,n} &= \frac{e^{-k_n R_i}}{\sqrt{k_n}} \quad \text{for closed channels}
\end{aligned} \tag{4.82}$$

and

$$\cos_i^{n,n} = \frac{\cos k_n R_i}{\sqrt{k_n}} \quad \text{for open channels}$$
$$\cos_i^{n,n} = \frac{e^{k_n R_i}}{\sqrt{k_n}} \quad \text{for closed channels} \tag{4.83}$$

we get the matricial equation for the K transition matrix

$$\bar{K} = \{\bar{\mathcal{R}}_{i-1}\overline{\cos}_i - \overline{\cos}_{i-1}\}^{-1}\{\overline{\sin}_{i-1} - \bar{\mathcal{R}}_{i-1}\overline{\sin}_i\}. \tag{4.84}$$

Once we know the value of the wavefunction in the last point and $\mathcal{R}$ for all i values, we can regenerate the whole wavefunction by applying Equation (4.80). To obtain the outgoing solution, we then apply Equation (4.41).

Matrix Elements

The regular solution can be reconstructed once $\bar{\mathcal{R}}_i$, $i = 1, 2, \ldots N$ is known. However, it requires a back propagation, the calculation become twice as expensive and having more errors due to increase of the number of operations.

If we are interested in calculating the matrix elements of a given operator $\mathcal{O}$ between the different solutions and an initially known wavefunction Φ^g expanded in the same basis set $\{\chi_n\}$ as

$$\langle \Phi^g | \mathcal{O} | \tilde{\Phi} \rangle = \int \tilde{\psi}^g \quad \bar{\mathcal{O}} \quad \bar{\psi} \quad dR \tag{4.85}$$

where the matrix elements of $\bar{\mathcal{O}}$ are given by

$$\mathcal{O}_{nm} = \langle \chi_n | \mathcal{O} | \chi_m \rangle \tag{4.86}$$

it is then useful to use Equation (4.80) so that

$$\bar{\psi}_i = \Pi_{j=i}^{N-1} \bar{\mathcal{R}}_j \quad \bar{\psi}_N \tag{4.87}$$

so that the integrals of Equation (4.85) can be written as

$$\langle \Phi^g | \mathcal{O} | \tilde{\Phi} \rangle = \Delta \sum_{i=1}^{N} \tilde{\psi}_i^g \quad \bar{\mathcal{O}}_i \, \Pi_{j=i}^{N-1} \bar{\mathcal{R}}_j \quad \bar{\Psi}_N \tag{4.88}$$

and define another quantity that will be propagated at the same time as $\bar{\mathcal{R}}_i$ to calculate the vector of matrix elements (Gadéa et al. 1997)

$$\begin{aligned} \tilde{\mathcal{M}}_i &= \left[\tilde{\mathcal{M}}_{i-1} + \tilde{\psi}_i^g \quad \bar{\mathcal{O}}_i\right] \quad \bar{\mathcal{R}}_i \\ \tilde{\mathcal{M}}_1 &= \tilde{\psi}_1^g \quad \bar{\mathcal{O}}_1 \quad \bar{\mathcal{R}}_1 \\ \tilde{\mathcal{M}}_N &= \left[\tilde{\mathcal{M}}_{N-1} + \tilde{\psi}_N^g \quad \bar{\mathcal{O}}_N\right]\bar{\psi}_N \end{aligned} \tag{4.89}$$

so that $\tilde{\mathcal{M}}_N$ is numerically identical to the vector of integrals indicated in Equation (4.85). Finally, these quantities have to be transformed according to Equation (4.41) to get the desired matrix element with the ingoing/outgoing conditions as

$$\mathcal{M}_N^- = \tilde{\mathcal{M}}_N \quad \{\underline{\mathrm{I}} + i\bar{K}\}^{-1}. \tag{4.90}$$

References

Aguado, A., Paniagua, M., Sanz-Sanz, C., & Roncero, O. 2003, JChPh, 119, 10088

Aguado, A., Roncero, O., Zanchet, A., Agúndez, M., & Cernicharo, J. 2017, ApJ, 838, 33

Araujo, M., Lasorne, B., Bearpark, M. J., & Robb, M. A. 2008, JPCA, 112, 7489

Araujo, M., Lasorne, B., Magalhaes, A. L., et al. 2009, JChPh, 131, 144301

Ayouz, M., Lopes, R., Raoult, M., Dulieu, O., & Kokoouline, V. 2011, PhRvA, 83, 052712

Babb, J. F., & Dalgarno, A. 1995, PhRvA, 51, 3021

Baer, M. 2001, CPL, 347, 149

Baer, M. 1975, CPL, 35, 112

Baer, M. 1980, MolPh, 40, 1011

Bates, D. R. 1951, MNRAS, 111, 1951

Beck, M. H., Jäckle, A., Worth, G. A., & Meyer, H.-D. 2000, PhR, 324, 1

Ben Abdallah, D., Najar, F., Jaidane, N., Dumouchel, F., & Lique, F. 2012, MNRAS, 419, 2441

Beswick, J. A., Delgado-Barrio, G., & Jortner, J. 1979, JChPh, 70, 3895

Beswick, J. A., & Jortner, J. 1981, AdChP, 47, 363

Bixon, M., & Jortner, J. 1969, JChPh, 50, 3284

Buchachenko, A. A., Halberstadt, N., Lepetit, B., & Roncero, O. 2003, IRPC, 22, 153

Buenker, R. J., & Peyerimhoff, S. D. 1970, JChPh, 53, 1368

Burke, M. L., & Klemperer, W. 1993, JChPh, 98, 6642

Cacciani, P., Ubachs, W., Hinnen, P. C., et al. 1998, ApJ, 499, L223

Carleton, K. L., Butenhoff, T. J., & Moore, C. B. 1990, JChPh, 93, 3907

Chang, Y.-T., & Miller, W. H. 1990, JPhCh, 94, 5884

Chenel, A., Roncero, O., Aguado, A., Agúndez, M., & Cernicharo, J. 2016, JChPh, 144, 144306

Chuang, M.-C., Foltz, M. F., & Moore, C. B. 1987, JChPh, 87, 3855

Chuljian, D. T., Ozment, J., & Simons, J. 1984, JChPh, 89, 176

Čížek, J. 1966, JChPh, 45, 4256

Čížek, J. 1969, AdChP, 14, 35

Čížek, J., & Paldus, J. 1971, IJQC, 5, 359

Cuadrado, S., Goicoechea, J. R., Roncero, O., et al. 2016, A&A, 596, L1

Dallos, M., Müller, T., Lischka, H., & Shepard, R. 2001, JChPh, 114, 746

Domcke, W., Yarkony, D. R., & Köppel, H. (ed) 2004, in Conical Intersections: Electronic Structure, Dynamics and Spectroscopy (Singapore: Word Scientific)

Draine, B. T. 1978, ApJ, 37, 595

Dumouchel, F., Klos, J., & Lique, F. 2011, PCCP, 13, 8204

Eidelsberg, M., Sheffer, Y., Federman, S. R., et al. 2006, ApJ, 647, 1543

Eidelsberg, M., Lamaire, J. L., Federman, S. R., et al. 2012, A&A, 543, 69

Eidelsberg, M., Benayoun, J. J., Viala, Y., & Rostas, F. 1991, A&AS, 90, 231

Eidelsberg, M., & Rostas, F. 1990, A&A, 235, 472
Eng, R., Carrington, T., Dugan, C. H., Filseth, S. V., & Sadowski, C. M. 1987, CP, 113, 119
Evard, D. D., Bieler, C. R., Cline, J. I., Sivakumar, N., & Janda, K. C. 1988, JChPh, 89, 2829
Fano, U. 1961, PhRv, 124, 1866
Farnum, J. D., Zhang, X., & Bowman, J. M. 2007, JChPh, 126, 134305
Fernandez Rico, J., Paniagua, M., García De La Vega, J. M., Fernàndez-Alonso, J. I., & Fantucci, P. 1986, JCoCh, 7, 201
Feshbach, H. 1962, AnPhy, 19, 287
Fitzgerald, G., & Schaefer, H. F. 1985, JChPh, 83, 1162
Freed, K. F., & Nitzan, A. 1980, JChPh, 73, 4765
Fu, B., Shepler, B. C., & Bowman, J. M. 2011, JACS, 133, 7957
Gadéa, F. X., Berriche, H., Roncero, O., Villarreal, P., & Delgado-Barrio, G. 1997, JChPh, 107, 10515
Gerlich, D., Plasil, R., Zymak, I., et al. 2013, JPCA, 117, 10068
Gerlich, D., & Horning, S. 1992, ChRv, 92, 1509
Godard, B., Falgarone, E., Gerin, M., Hily-Blant, P., & de Luca, M. 2010, A&A, 520, A20
Goldberger, M. L., & Watson, K. M. 1964, Collision Theory (New York: Wiley)
Gómez-Carrasco, S., Muller, T., & Koppel, H. 2010, JPCA, 114, 11436
Gray, S. K. 1992, CPL, 197, 86
Grein, F., & Hachey, M. R. J. 1996, IJQC, 60, 1661
Gutmann, M., Willberg, D. M., & Zewail, A. H. 1992, JChPh, 97, 8037
Gwaltney, S. R., & Barlett, R. J. 1995, CPL, 241, 26
Hachey, M. R. J., Bruna, P. J., & Grein, F. 1994, FaTr, 90, 683
Hachey, M. R. J., Bruna, P. J., & Grein, F. 1995, JPhCh, 99, 8050
Hachey, M. R. J., Bruna, P. J., & Grein, F. 1996, JMoSp, 176, 375
Halberstadt, N., Beswick, J. A., & Janda, K. C. 1987, JChPh, 87, 3966
Halberstadt, N., & Janda, K. C. (ed) 1990, Dynamics of Polyatomic van der Waals Complexes (New York: Plenum)
Harding, L. B., & Goddard, W. A. 1977, JACS, 99, 677
Hernández-Vera, M., Lique, F., Dumouchel, F., Hily-Blant, P., & Faure, A. 2017, MNRAS, 468, 1084
Herzberg, G., & Hines, K. K. 1957, CaJPh, 35, 842
Hogerheijde, M. R., Jansen, D. J., & van Dishoeck, E. F. 1995, A&A, 294, 792
Hollenbach, D. J., & Tielens, A. G. G. M. 1997, ARA&A, 35, 179
Hopkins, W. S., Loock, H.-P., Cronin, B., et al. 2007, JChPh, 127, 064301
Hsu, Y. C., Smith, M. A., & Wallace, S. C. 1984, CPL, 111, 219
Hylleraas, E., & Undheim, B. 1930, ZPhy, 65, 759
Janda, K. C. 1985, AdChP, 60, 201
Jonas, D. M., Zhao, X., Yamanouchi, K., et al. 1990, JChPh, 92, 3988
Kállay, M., & Gass, J. 2004, JChPh, 121, 9257
Köppel, H., Domcke, W., & Cederbaum, L. S. 1984, AdChP, 57, 59
Köppel, H. 2004, Conical Intersections: Electronic Structure, Dynamics and Spectroscopy, ed. D. R. Yarkony, W. Domcke, & H. Köppel (Singapore: World Scientific), 175
Lahmani, F., Tramer, A., & Tric, C. 1974, JChPh, 60, 4431
Langhoff, S. R., Elbert, S. T., Jackels, C. F., & Davidson, E. R. 1974, CPL, 29, 247

Larsson, M. 1983, A&A, 128, 291
Lee, L. C. 1980, JChPh, 72, 6414
Lefebvre, R. 1972, Selected Topics in Physics, Astrophysics and Biophysics, ed. E. Abecassis de Laredo, & N. K. Jurisic (Dordrecht: Reidel), 87
Lefebvre-Brion, H., Liebermann, H. P., & Vázquez, G. J. 2010, JChPh, 132, 024311
Lefebvre-Brion, H., & Kalemos, A. 2016, JChPh, 145, 166102
Leforestier, C., Bisseling, R. H., Cerjan, C., et al. 1991, JCoPh, 94, 59
Le Petit, F., Nehmé, C., Le Bourlot, J., & Roueff, E. 2006, ApJS, 164, 506
Letzelter, C., Eidelsberg, M., Rostas, F., Breton, J., & Thieblemont, B. 1987, CP, 114, 273
Levy, D. H. 1981, AdChP, 47, 323
Li, X., Millam, J. M., & Schlegel, H. B. 2000, JChPh, 113, 10062
Lipkowitz, T. R., & Cundari, K. B. 2007, Reviews in Computational Chemistry, Conical Intersections in Molecular Systems, Vol. 23 (New York: Wiley)
Lippmann, B. A., & Schwinger, J. 1950, PhRv, 79, 469
Liszt, H., & Lucas, R. 2001, A&A, 370, 576
Loison, J.-C., Wakelam, V., & Hickson, K. M. 2014, MNRAS, 443, 398
Majumder, M., Sathyamurthy, N., Vázquez, G. J., & Lefebvre-Brion, H. 2014, JChPh, 140, 164303
Martinazzo, R., & Tantardini, G. F. 2005, JChPh, 122, 094109
McDonald, J. K. L. 1993, PhRv, 43, 830
Mcpherson, M. T., & Simons, J. P. 1978, FaTr2, 74, 1965
Mead, C. A., & Truhlar, D. G. 1982, JChPh, 77, 6090
Meenakshi, A., & Innes, K. K. 1986, JChPh, 84, 6550
Merchan, M., & Roos, B. O. 1995, AcTC, 92, 227
Meyer, H.-D. 2009, Theory and Applications Multidimensional Quantum Dynamics: MCTDH Theory and Applications, ed. H.-D. Meyer, F. Gatti, & G. A. Worth (New York: Wiley)
Meyer, H.-D., Manthe, U., & Cederbaum, L. S. 1990, CPL, 165, 73
Mies, F. H. 1968, PhRv, 175, 164
Mies, F. H. 1969a, JChPh, 51, 787
Mies, F. H. 1969b, JChPh, 51, 798
Miller, W. H. 1990, ARPC, 41, 245
Monnerville, M., & Robbe, J. M. 1994, JChPh, 101, 7580
Mower, L. 1966, PhRv, 142, 799
Müller, T., & Lischka, H. 2001, Theor. Chem. Acc., 106, 369
Nagata, T., Kondow, T., Ozaki, Y., & Kuchitsu, K. 1981, CP, 57, 45
Niu, B., Shirley, D. A., Bai, Y., & Daymo, E. 1993, CPL, 201, 212
Nuth, J. A., & Glicker, S. 1982, JQSRT, 28, 223
Palmer, H. B. 1967, JChPh, 47, 2116
Paniagua, M., Aguado, A., Lara, M., & Roncero, O. 1999, JChPh, 111, 6712
Perić, M., Grein, F., & Hachey, M. R. J. 2000, JChPh, 113, 9011
Peric, M., Buenker, J., & Peyerimhoff, S. D. 1987, MolPh, 62, 1323
Peric, M., Buenker, J., & Peyerimhoff, S. D. 1988, MolPh, 64, 843
Perlherbe, G. H., & Hase, W. L. 1996, JChPh, 104, 7882
Peyerimhoff, S. D., Buenker, R. J., Kammer, W. K., & Hsu, H. 1971, CPL, 8, 129
Purvis, G. D., & Bartlett, R. J. 1982, IJQC, 76, 1910

Puy, D., Dubrovich, V., Lipovka, A., Talbi, D., & Vonlanthen, P. 2007, A&A, 476, 685
Roncero, O., Lepetit, B., Beswick, J. A., Halberstadt, N., & Buchachenko, A. A. 2001, JChPh, 115, 6961
Roncero, O., Beswick, J. A., Halberstadt, N., Villarreal, P., & Delgado-Barrio, G. 1990, JChPh, 92, 3348
Roncero, O., Caloto, D., Janda, K. C., & Halberstadt, N. 1997, JChPh, 107, 1406
Sarrasin, E., Ben Abdallah, D., Wernli, M., et al. 2010, MNRAS, 404, 518
Schinke, R. 1986, JChPh, 84, 1487
Schinke, R. 1993, Photodissociation Dynamics, Spectroscopy and Fragmentation of Small Molecules (Cambridge: Cambridge Univ. Press)
Shapiro, M., & Balint-Kurti, G. G. 1981, CP, 61, 137
Shepler, B. C., Epifanovsky, E., Zhang, P., et al. 2008, JPCA, 112, 13267
Shore, B. W. 1967, RvMP, 39, 439
Simonsen, J. B., Rusteika, N., Jonhnson, M. S., & Sollling, T. I. 2008, PCCP, 10, 674
Singer, S. J., Freed, K. F., & Band, Y. B. 1985, AdChP, 61, 1
Smith, F. T. 1969, PhRv, 179, 111
Snyder, L. E., Bhul, D., Zuckerman, B., & Palmer, P. 1969, PhRvL, 22, 679
Stoeklin, T., Lique, F., & Hochlaf, M. 2013, PCCP, 15, 13818
Suto, M., Wang, X., & Lee, L. C. 1986, JChPh, 85, 4228
Taylor, S., Wilden, D. G., & Comer, J. 1982, JChPh, 70, 291
Tchang-Brillet, W.-Ü. L., Julienne, P. S., & Robbe, J.-M. 1992, JChPh, 96, 6735
Terentis, A. C., & Kable, S. H. 1996, CPL, 258, 626
Thiel, A., & Köppel, H. 1999, JChPh, 110, 9371
Tielens, A. G. G. M. 2005, The Physics and Chemistry of the Interstellar Medium (Cambridge: Cambridge Univ. Press)
Townsend, D., Lahankar, S. A., Leea, S. K., et al. 2004, Sci, 306, 1158
Ubachs, W., Eikema, K. S. E., Levelt, P. F., et al. 1994, ApJ, 427, L55
Uzer, T., & Miller, W. H. 1991, PhR, 199, 73
Valachovic, L. R., Tuchler, M. F., Dulligan, M., et al. 2000, JChPh, 112, 2752
van Dishoeck, E. F., & Visser, R. 2015, in Laboratory Astrophysics: from Molecules through Nanoparticles to Grains, ed. S. Schlemmer, T. Giesen, H. Mutschke, & C. Jager (New York: Wiley), 1
Vázquez, G. J., Amero, J. M., Liebermann, H. P., & Lefebvre-Brion, H. 2009, JPCA, 113, 13395
Visser, R., van Dishoeck, E. F., & Black, J. H. 2009, A&A, 503, 323
Waterland, R. L., Skene, J. M., & Lester, M. I. 1988, JChPh, 89, 7277
Werner, H.-J., & Knowles, P. J. 1981, JChPh, 74, 5794
Werner, H.-J., & Knowles, P. J. 1984, JChPh, 82, 5053
Werner, H. J., & Knowles, P. J. 1988, JChPh, 89, 5803
Werner, H. J., & Meyer, W. 1981, JChPh, 74, 5802
Whitten, J. L., & Hackmeyer, M. J. 1969, JChPh, 51, 5584
Willberg, D. M., Gutmann, M., Breen, J. J., & Zewail, A. H. 1992, JChPh, 96, 198
Worth, G. A., Beck, M. H., Jäckle, A., & Meyer, H.-D. 2000, The MCTDH Package, Version 8.2. Meyer, H.-D. 2002, 2007, Version 8.3, Version 8.4. See http://mctdh.uni-hd.de/.
Xu, D., Xie, D., & Guo, H. 2001, CPL, 345, 517
Xu, D., Xie, D., & Guo, H. 2002a, JPCA, 106, 10174

Xu, D., Xie, D., & Guo, H. 2002b, JChPh, 116, 10626
Xu, D., Guo, H., & Xie, D. 2003, JTCoCh, 2, 639
Yin, H. M., Kable, S. H., Zhang, X., & Bowman, J. M. 2006, Sci, 311, 1443
Yin, H.-M., Rowling, S. J., Büll, A., & Kable, S. H. 2007, JChPh, 127, 064302
Zare, R. N. 1988, Angular Momentum (New York: Wiley)
Zhang, X., Zhou, S., Harding, L. B., & Bowman, J. M. 2004, JPCA, 108, 8980
Zhang, X., Rheinecker, J. L., & Bowman, J. M. 2005, JChPh, 122, 114313
Zhang, P., Maeda, S., Morokuma, K., & Braams, B. J. 2009, JChPh, 130, 114304

AAS | IOP Astronomy

Gas-Phase Chemistry in Space
From elementary particles to complex organic molecules
François Lique and Alexandre Faure

Chapter 5

Electron Collision Processes

Jonathan Tennyson and Alexandre Faure

5.1 Introduction

Most of the observable universe is actually a plasma, i.e., weakly ionized matter that coexists with free electrons in some sort of dynamic equilibrium. Of course, these free electrons cannot usually be directly observed, but processes involving electron collisions with molecules can lead to observable results in a variety of astronomical environments including the interstellar medium. Examples of such collision processes are discussed below. However, first we review the various processes involving electron collisions, considering the physics behind the collision process and where the process is likely to be astronomically important. We concentrate on processes involving collisions with electrons having low-to-intermediate energies. For electron collision processes, low-energy collisions are those that do not possess sufficient energies to ionize the target species and intermediate-energy collisions are those that occur just above ionization. Taken together, the low- and intermediate-energy regimes can cover collisions with electron collisions with energies up to about 100 eV. High-energy electrons may exist in specific astronomical environments, and indeed, may be produced by energetic cosmic rays, but collisions involving these highly energetic species will not be considered further below.

It is usual for astronomical models to use rate coefficients, $k(T)$, rather than cross sections, $\sigma(E)$ (T is the electron kinetic temperature and E is the electron kinetic energy). However, most electron collision processes are not well-represented by a simple Arrhenius form ($k(T) = A\,\exp^{-Ea/(k_BT)}$, where A is a constant and E_a is an activation energy). Indeed, electron collision cross sections are often highly structured, with large enhancements occurring at so-called resonances. In the context of electron collision physics, resonances occur when a colliding electron is temporarily trapped by the target molecule to form a compound species. Some processes, notably dissociative recombination and dissociative electron attachment, can only occur through resonances, and therefore they are only observed in electron collisions with the appropriate energy. It should be noted that, unlike photon

doi:10.1088/2514-3433/aae1b5ch5

processes, which are generally observed at precise wavelengths, the effect of resonances are seen over a range of energies. This is because a resonance has a finite lifetime, and therefore, as a result of Heisenberg's Uncertainty Principle, does not have a precise energy. The energy uncertainty is generally characterized by a width, so each resonance is characterized by both an energy and a width. The width, Γ, and the lifetime, τ, are inversely related ($\tau = \frac{\hbar}{\Gamma}$) through Planck's constant divided by 2π, $\hbar$.

In this chapter, we begin by considering each low-energy electron—molecule collision process in turn. We then briefly review methods for obtaining cross sections and rate coefficients for use in astrophysical models and for interpreting astronomical observations. In the subsequent section, we consider a number of illustrative examples where electron collisions have been found to be important for processes occurring in interstellar and star-forming regions. We finish by giving some guidance on finding sources of useful electron collision data.

5.2 Fundamental Processes

When colliding with a molecule, an electron can undergo a number of distinct processes. The most important of these are listed below, where the notations ϵ, v, and J are generic quantum numbers for the electronic, vibrational, and rotational states, respectively, and AB is some generic molecule where A and B can be atoms or some molecular fragment.

Elastic scattering:

$$e^{-}+AB(\epsilon, v, J) \rightarrow AB(\epsilon, v, J) + e^{-}. \tag{5.1}$$

Electron impact rotational excitation:

$$e^{-}+AB(\epsilon, v, J) \rightarrow AB(\epsilon, v, J') + e^{-}. \tag{5.2}$$

Electron impact vibrational excitation:

$$e^{-}+AB(\epsilon, v, J) \rightarrow AB(\epsilon, v', J') + e^{-}. \tag{5.3}$$

Electron impact electronic excitation:

$$e^{-}+AB(\epsilon, v, J) \rightarrow AB(\epsilon', v', J') + e^{-}. \tag{5.4}$$

Electron impact dissociation:

$$e^{-}+AB(\epsilon, v, J) \rightarrow A + B + e^{-}. \tag{5.5}$$

Dissociative electron attachment (DEA):

$$e^{-}+AB(\epsilon, v, J) \rightarrow A + B^{-}. \tag{5.6}$$

Dissociative recombination (DR):

$$e^{-}+AB^{+}(\epsilon, v, J) \rightarrow A + B. \tag{5.7}$$

Electron impact ionization:

$$e^{-}+AB(\epsilon, v, J) \rightarrow AB^{+}(\epsilon^{+}, v^{+}, J^{+}) + 2e^{-}. \tag{5.8}$$

Electron impact dissociative ionization:

$$e^{-}+AB(\epsilon, v, J) \rightarrow A + B^{+}+2e^{-}. \tag{5.9}$$

In the following, we consider each of these processes in turn.

5.2.1 Elastic Scattering

Elastic collisions are those that result in no change in the internal state of the molecule. The elastic cross section can be defined as

$$\sigma_E(E) = 2\pi \int_0^{\pi} \frac{d\sigma}{d\theta} \sin\theta d\theta, \tag{5.10}$$

where the differential cross section $\frac{d\sigma}{d\theta}$ represents the cross section for the deflection of the colliding electron by angle θ at collision energy E. The factor of 2π comes from the fact that, as the target molecule is assumed to rotate freely in space, the electron is deflected isotropically into a cone.

Elastic collisions may seem uninteresting, but such collisions do allow for thermalization via kinetic energy exchange between the electron and the molecule. This effect is captured by the momentum transfer cross section, which is given by

$$\sigma_M(E) = 2\pi \int_0^{\pi} (1 - \cos\theta)\frac{d\sigma}{d\theta} \sin\theta d\theta. \tag{5.11}$$

At present, chemical models used for astrophysics usually assume the electrons are in local thermal equilibrium (LTE) and so do not consider the thermalization of electrons, or indeed, the thermalization of the translational motion of molecules by elastic scattering. Furthermore, it should be noted that elastic collisions between electrons and charged species are formally infinite due to the long-range properties of the Coulomb potential. Considering such cross sections in a time-dependent model would add significant complications.

Elastic collisions in general can also result in a change in the direction of the motion of the electron. This property is represented by the differential cross section. In a homogeneous environment,this change in direction is not important, as usually the net effect averages out. However, differential cross sections are important in situations where the electrons form a jet of some sort. A typical example of this is lightning. The effects of lightning are beginning to be considered in brown dwarfs and exoplanets (Ardaseva et al. 2017). Similar directional effects are also found on the aurorae of planets such as Jupiter (Gerard et al. 2013, 2016), where charged particles, including electrons, travel down magnetic field lines.

It should be noted that elastic collision cross sections are usually the easiest to both measure and calculate, and are therefore generally well-known for astrophysical species of importance. However, despite their possibly important applications, elastic collisions are generally not considered in astrophysical models and we will not consider them further here.

5.2.2 Rotational Excitation

In contrast to elastic collisions, rotational excitation by electrons can be an extremely important astrophysical process. In diffuse environments, most molecules exist at densities well below their critical density, n_c. The critical density for level j is given by

$$n_c = \frac{\sum_{i<j} A_{ji}}{\sum_{i \neq j} q_{ji}}, \tag{5.12}$$

where A_{ji} is the Einstein A coefficient and the sum runs over all possible emissions giving the emission lifetime, and q_{ji} is the rate for collisional de-population of level j, summed over all possible processes. At densities above critical, the level in question is thermalized and Boltzmann statistics can be used.

At densities well below the critical density, each process that leads to a collisional excitation can be assumed to result in the emission of a photon. As nearly all interstellar species are detected and monitored through their rotational spectrum, collisional excitation by electrons can result in directly observable effects. Of course, the main excitation processes typically involve collisions with heavy species, particularly H_2 and He. These processes are discussed extensively elsewhere in this book (see Chapter 7 by P. Dagdigian). However, although electrons are usually only trace species in astrophysical environments which contain molecules, electron-impact rotational excitation cross sections can be many orders of magnitude greater than the equivalent ones for H_2. This means rotational excitation by electrons can be important where the cross section for this process is large.

Unlike photons, collisional excitation by electrons does not obey strict selection rules, and in principle, rotational excitation can involve any change in the rotational quantum number, j. In practice, however, in situations where rotational excitation is important, $\Delta j = 1$ processes usually dominate, but consideration of higher excitations may also be important (Rabadán et al. 1998). Large rotational excitation cross sections are usually found for molecules with large dipole moments. In these cases, the magnitude of the cross section is approximately proportional to the square of molecular dipole; see below. These cross sections are usually largest at low energies near the threshold for the excitation process. Given the large (infinite) cross section for electron collisions with molecular ions, ions possessing significant dipole moments, such as HCO^+ and HOC^+, can be expected to have particularly large cross sections (Faure & Tennyson 2001). Given that molecular ions occur in regions where there are also significant densities of free electrons, it is essential to consider electron collision process in models involving ions. As discussed below, dissociative recombination also provides a major destruction process for molecular ions.

Electron impact rotational excitation cross sections are unusual in that they are relatively straightforward to compute but very difficult to measure in the laboratory. Experimentally measurement of rotationally-resolved processes in electron collisions is extremely difficult because the relatively small gaps in energies between the molecular rotational levels are usually not resolved in electron collision experiments, where the spread of energies in the beam is usually larger than the rotational excitation energies. In addition, at room temperature, many rotational states of the target molecule are usually occupied, making it even harder to interpret any change in rotational state.

There are a limited number of measurements of electron-impact rotational excitation for neutral dipolar species, but these are often difficult to interpret (Faure et al. 2004a). So far, there has only been one attempt to study rotational excitation in molecular ions experimentally, which focused on electron collisions with HD^+ (Shafir et al. 2009). This study, which actually measured deexcitation rates, provided insufficient data to directly provide cross sections. However, models based on computed cross sections were able to reproduce the measurements, giving strong support to the theoretical results.

Theoretical results are generally based on a rather simple, rigid rotor treatment of the nuclei motion and the so-called adiabatic nuclei rotation (ANR) approximation (Chang & Fano 1972; Lane 1980). At this level of treatment, the rotational excitation cross sections can be computed rather straightforwardly using a frame transformation that involves recoupling the angular momenta involved in the collision calculation from the standard body-fixed frame in which the calculations are generally performed, to a space fixed representation that allows for the rotational motion. The rotational frame transformation is performed on the body-fixed **T**-matrix as follows (for a linear molecule):

$$T^{J}_{j'l',jl} \approx \sum_{\Lambda=-l}^{l} A^{J\Lambda}_{j'l'} T^{\Lambda}_{l',l} A^{J\Lambda}_{jl} \tag{5.13}$$

with

$$A^{J\Lambda}_{j'l'} = \sqrt{\frac{2j+1}{2J+1}}\, C(jlJ;\, 0\Lambda - \Lambda), \tag{5.14}$$

where $C(.)$ is a Clebsch–Gordan coefficient. The cross section is obtained from the space-fixed **T**-matrix:

$$\sigma^{TM} = \frac{\pi}{(2j+1)k_i^2} \sum_{Jll'} (2J+1) |T^{J}_{j'l',jl}|^2. \tag{5.15}$$

In practice, the partial wave expansion is limited to some finite $l_{\max}$ value, typically $l_{\max} \leqslant 8$.

A formal treatment of the effects of the long-range dipole moment can be included at the same stage, based upon use of the Born approximation (Norcross & Padial 1982). The full integral cross section is obtained as:

$$\sigma(j \to j') = \sigma^{TM}(j \to j') + [\sigma^{CB}(j \to j') - \sigma^{PCB}(j \to j')], \tag{5.16}$$

where the second term in brackets is the long-range correction, sometimes called the "Born completion," corresponding to partial waves $l > l_{\max}$. If we adopt only the dipole term of the long-range interaction, the Born theory gives the following cross section for a neutral target (Takayanagi 1966):

$$\sigma^{CB}(j \rightarrow j+1) = \frac{8\pi}{3}\mu_D^2\frac{(j+1)}{(2j+1)}\ln\left|\frac{k+k'}{k-k'}\right|, \tag{5.17}$$

where μ_D is the dipole moment of the target molecule and k (k') is the initial (final) momentum of the electron. There are standard programs that implement the above procedure, starting from standard electron–molecule collision calculations for both neutral (Sanna & Gianturco 1998) and ionic (Rabadán & Tennyson 1998) targets. We note that, in the case of molecular ions, the long-range theory is called the "Coulomb–Born" approximation. The formula is similar to Equation (5.17) with the same dependence to the square of the dipole. However, the low-energy dependence is different with a large and finite cross section at threshold (Chu & Dalgarno 1974). Comparisons with calculations performed using a much more sophisticated procedure based on explicit close-coupling of the rotational channels suggest that the frame transformation procedures work well (Faure et al. 2006; Čurík & Greene 2017).

Although the usually dominant $\Delta j = 1$ excitation channel is often well-determined by a simple treatment of the long-range dipole moment (Chu & Dalgarno 1974), particularly for species with large permanent dipole moments, this is not true for other transitions. This means that it is important to use a good treatment of the short-range interactions in the electron-collision process, particularly to account for transitions with $\Delta j > 1$. Such transitions can be important and our experience shows that their cross sections are entirely determined by short-range effects, and as a result, show significant sensitivity to the treatment of dynamical effects such as polarization interactions in the short-range calculations.

In many environments, including the Earth, molecules occur naturally in a range of rotational levels. This means that electron impact rotational de-excitation can be important alongside the excitation process. Given cross sections for one process, such as excitation, it is reasonably straightforward to obtain the cross section for the reverse process, de-excitation, using the principle of detailed balance. This principle, which is the consequence of the invariance of the electron–molecule interaction under time reversal, states that the cross sections for the inelastic process $l \rightarrow u$ and its reverse $u \rightarrow l$ must obey the formula:

$$\sigma_{l\rightarrow u}(E_{k,l})g_l E_{k,l} = \sigma_{u\rightarrow l}(E_{k,u})g_u E_{k,u}, \tag{5.18}$$

where g_i is the statistical weight of the state i. The electron kinetic energy $E_{k,\,i}$ is related to the total energy as follows:

$$E_{\text{tot}} = E_{k,l} + E_l = E_{k,u} + E_u, \tag{5.19}$$

where E_i is the target rotational energy.

Finally, before leaving the topic of rotational excitation, it is worth noting that the vast bulk of measured electron collision cross sections are not obtained rotationally resolved. This means that, by convention, rotational excitation (and de-excitation) cross sections are included in the so-called elastic cross sections, and other processes. In fact, for systems with large dipole moments, the dominant contribution to these "elastic" cross sections are actually transitions with $\Delta j = 1$ rather than the genuinely elastic $\Delta j = 0$.

5.2.3 Vibrational Excitation

Electron impact vibrational excitation is probably not as important as rotational excitation in interstellar environments simply because, at low temperatures, the electrons do not possess enough energy to excite the vast majority of molecular vibrations. However, vibrational excitation by electrons is known to be important in other astronomical environments, such as comets or planetary ionospheres (Campbell & Brunger 2009, 2013). Although easier to study than rotational excitation, there are still only limited experimental measurements of electron impact vibrational excitation cross sections because of the relatively low energy of this process. Furthermore, it can be hard to distinguish between excitation of vibrational modes that are energetically close, and essentially all measurements are for molecules starting in their vibrational ground state. Theory suggests that electron impact vibrational excitation can lead to significantly different behavior, starting from vibrationally excited states (Laporta et al. 2015).

There are two distinct mechanisms whereby molecules undergo vibrational excitation. The standard, direct mechanism is usually by dipole interactions—meaning that, in the vibrational excitation process: (a) excitation dominantly involves a jump by one quantum in a single vibrational mode; and (b) the process obeys the same selection rules as infrared spectroscopy, i.e., modes that are infrared-active predominate in vibrational excitation. The direct mechanism can occur in collisions with electrons at any energy, provided the energy of the electron is above threshold—that is, provided the electron possesses enough energy to excite the given the vibrational mode. However, direct vibrational excitation is a relatively inefficient process, meaning that vibrational excitation cross sections are often dominated by any resonance contribution. Indeed, in the region of a resonance, vibrational excitation can become the main electron collision process, changing the shape of the total electron collision cross section (Itikawa 2006). Resonant vibrational excitation only occurs at the energy of a resonance, but can involve high levels of vibrational excitation: changes by up to 14 quanta in the vibrational mode (Allan 1985; Poparic et al. 2008) have been observed. Cross sections for resonant vibrational excitation are usually very large. In situations where there is a low-lying, broad resonance, vibrational excitation can be the dominant process in regions of the resonance. This has led to the use of rather straightforward theoretical procedures that only consider the resonance contribution in calculations of vibrational excitation (Laporta et al. 2016). This procedure can be extended to collisions involving vibrationally excited species, and at present provides the main means by which data on such processes can be obtained.

5.2.4 Electronic Excitation

Electron impact electronic excitation has a number of important features. First, electron collisions obey different selection rules to photons. This means, for example, that electron collisions obey the spin selection rule $\Delta S = 0, \pm 1$ rather than the propensity of photons to preserve the spin state ($\Delta S = 0$). As the lowest excited state of most stable molecules, which are generally spin singlets, is a triplet state, electron collisions can access this low-lying state, which is inaccessible by photon collisions. These states are often metastable, i.e., they have long lifetimes to photon decay. Electron impact electronic excitation of the hydrogen molecule produces the clearest spectral signature from the generally dark H_2 molecule, which is observed in a number of astronomical environments (Hallett et al. 2005; Egert et al. 2017).

Electronic excitation is actually the main route to electron impact dissociation (Stibbe & Tennyson 1998, 1999). As many low-lying states are dissociative, excitation to these states leads directly to the molecule falling apart. This process is illustrated in Figure 5.1.

5.2.5 Impact Dissociation

Electron impact dissociation is the main mechanism for molecular fragmentation with electrons at higher energy. The process generally relies on excitation to a (dissociative) excited electronic state as a first step, which is then followed by dissociation. As the vertical excitation to such dissociative states normally lies well above the dissociation limit, the effective threshold to electron impact dissociation usually lies considerably above the dissociation energy. For example, for H_2, the effective threshold is over 8 eV (Stibbe & Tennyson 1998), approximately twice the dissociation energy of the molecule.

The probabilities for impact dissociation are determined by the square of the overlap of the vibrational wave function of the molecule with the continuum state at the given electron energy; this term is usually referred to as a Franck–Condon factor, by analogy with the behavior of vibrational bands of the electronic spectra of molecules. The Franck–Condon factor between two states associated with the same potential energy curve is zero, which means that direct dissociation on the ground electronic state is strongly disfavored from low-lying vibrational states and this route to dissociation can generally be ignored. Instead, as mentioned, the main route to dissociation is via electronic excitation, as illustrated for the case of H_2 in Figure 5.1. Electron impact dissociation rates are quite challenging to measure in the laboratory, so the most reliable rates for dissociation of H_2, for example, come from theory (Stibbe & Tennyson 1999). However, for more complicated systems, such as methane, theory and experiment both struggle to obtain reliable results (Ziolkowski et al. 2012; Brigg et al. 2014; Song et al. 2015)

An interesting feature of this route to dissociation is that the increasing range of wave function associated with vibrationally excited states means that the impact dissociation threshold generally drops rapidly with vibrational excitation (Stibbe & Tennyson 1998; Laporta et al. 2015). This process can lead to significant

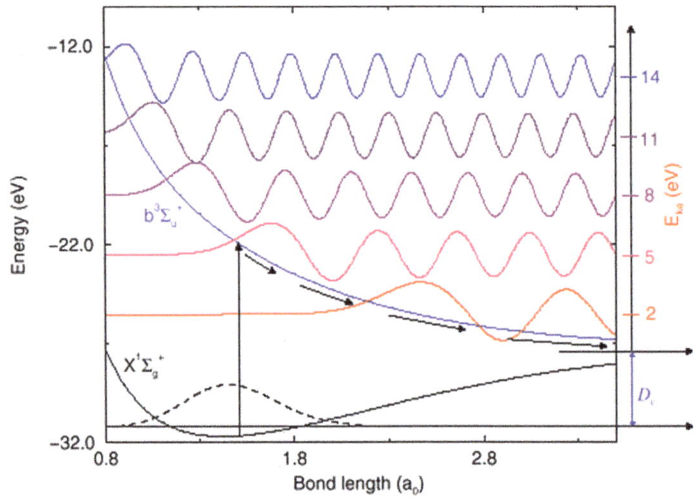

Figure 5.1. Potential energy curves and nuclear motion wave functions involved in the near-threshold electron impact dissociation of molecular hydrogen. In this process, the colliding electron excites the H_2 from the lowest ($v = 0$) vibrational state in its X $^1\Sigma_g^+$ electronic ground state (the wave function is depicted by the dashed distribution) to continuum states (given as oscillating colored curves) associated with dissociative excited b $^3\Sigma_u^+$ state. Following this state leads to dissociation. The vertical and horizontal arrows show the energy of the dissociating H atoms as given by the reflection principle. Figure adapted from Stibbe & Tennyson (1998).

modification of the vibrational distribution of molecules in certain environments (Hey et al. 2000).

5.2.6 Dissociative Electron Attachment

Dissociative electron attachment (DEA) is somewhat different from the processes discussed above, as it occurs exclusively via resonances. Figure 5.2 illustrates the general principles: an electron collides at an energy associated with a resonance. The electron plus target system forms a temporarily bound anionic state or resonance. This state can dissociate along the anion curve. As long as the resonance lives long enough to cross the neutral curve, it will lead to dissociation. In general, if the resonance decays prior to dissociation, then the main result is vibrational excitation.

DEA cross sections are relatively easy to measure, as the anionic products are easier to detect than neutral fragments. As dissociation occurs along a particular anion curve, it usually means that the result at a particular electron collision energy is usually a specific combination of neutral and anion fragments in well-defined electronic states. Different resonances normally provide a pathway to specific, well-defined fragments, which means shifts in electron collision energy can lead to entirely different DEA products. At low energy, a given resonance may not lead to DEA because the dissociation channel may not be open. This is true, for example,

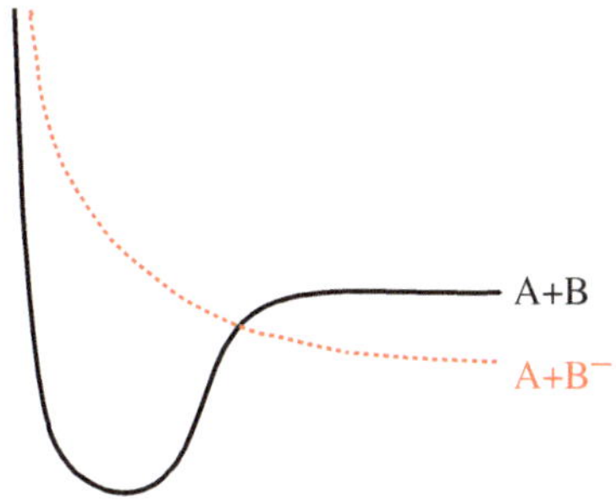

Figure 5.2. Potential energy curves involved in the dissociative electron attachment of notional molecule AB: the solid curve represents the AB ground state; the dashed curve represents the AB^- resonance state. The resonance has a finite lifetime and may decay by electron emission back to AB before the system reaches the point where the curves cross. Curve crossing leads to formation of an anion and a neutral fragment.

for the low-lying O_2^- resonance (Laporta et al. 2015). Theoretically, DEA cross sections are hard to compute for systems larger than diatomics, although there are methods of estimating such cross sections (Munro et al. 2012).

The recent detections of negative ions (anions) in space (Millar et al. 2017) has led to much discussion about the formation of these species. There are essentially two likely routes: one is as a product of DEA and the other is via direct radiative attachment of an electron:

$$e^- + AB(\epsilon, v, J) \rightarrow AB^-(\epsilon^-, v^-, J^-) + hv \tag{5.20}$$

where hv represents an omitted photon with the appropriate energy. Radiative attachment is a very inefficient process, but following a model for the process developed by Harada & Herbst (2008) is thought to provide a viable route to forming polyatomic anions in interstellar clouds. This has led to an increasing number of largely theoretical studies (Douguet et al. 2015; Gianturco et al. 2016) on species such as C_nH (n = 2, 4, 6) and C_nN (n = 1, 3, 5), whose anions have been detected in space.

Anions have also been detected in the atmosphere of Saturn's moon Titan (Coates et al. 2007; Agren et al. 2012). In the denser atmosphere of Titan, DEA provides the route to the formation of these anions (Vuitton et al. 2009).

5.2.7 Dissociative Recombination

Dissociative recombination (DR) can occur as a result of collisions between electrons and molecular ions; it is the major destroyer of molecular ions in low-temperature plasma, such as those found in the interstellar medium. DR and DEA are superficially similar, in that an electron is trapped in a resonance state, which then results in the system dissociating. However, while DEA usually involves a single curve and may only occur at certain energies, in the case of DR, many curves can be involved. For all molecular ions of astrophysical interest, the DR channel is energetically open for zero-energy electrons and can lead to a number of products.

For example, DR of H_3O^+ is considered to be an important means of producing water in cold interstellar environments. A low-energy electron will attach to H_3O^+, giving both H_2O + H and OH + H + H. As is typical when there are choices of fragmentation pathways, the pathway leading to the most fragments (OH + H + H) is the dominant one, being chosen about three time as often as the two-body (H_2O + H) pathway at low collision energies. In this case, DR leads the formation of water only about 25% of the time (Jensen et al. 2000).

The general mechanisms for DR are illustrated in Figure 5.3. Essentially, there are two competing (and sometimes interfering) routes. The direct route is similar to the mechanism for DEA and tends to be the dominant route when it is available. The indirect route involves the electron being trapped in a vibrationally excited state of a highly excited electronic state of the neutral (i.e., ion + electron) system. These nuclear excited resonances cannot dissociate directly, but the system then undergoes dissociation following crossing onto a dissociative curve.

Dissociative recombination rates are very sensitive to the precise location of the curve crossings between the neutral and the ion curve. DR rates are therefore sensitive to the vibration–rotation state of the target. This causes issues with the measurements of DR cross sections, as ions are often prepared in either ill-determined and/or excited states. The use of storage ring experiments revolutionized the measurement of DR cross sections, leading to a wealth of reliable data regardingly not only the cross sections but also the branching ratio between the various products. Comprehensive discussions of the results of these experiments has been given by Florescu-Mitchell & Mitchell (2006), and Larsson & Orel (2008). The latter work also considers sophisticated theoretical methods particularly based on the use of multichannel quantum defect theory (MQDT) (Jungen 1996, 2011) as the

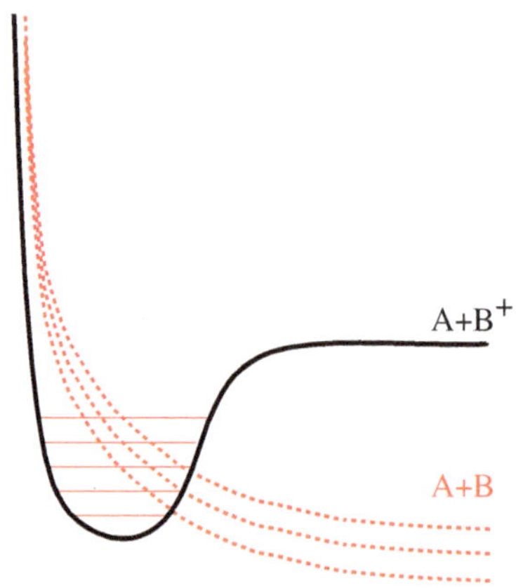

Figure 5.3. Potential energy curves involved in the dissociative recombination of the notional cation AB^+ to form neutral species A and B. The thick solid curve represents the AB^+; dashed curves are dissociative, excited states of the AB molecule. Direct dissociation occurs via the scattering electron being captured in one of these highly excited neutral states, which lead to dissociation into the neutral fragments A+B. The horizontal lines represent some of the many vibrational states associated with weakly bound (Rydberg) states of AB. Indirect dissociation occurs via the electron initially occupying one of these vibrational excited states and then undergoing a curve crossing onto a dissociative curve.

basis of the treatment of nuclear motion in the dissociation step. Given the high quality of the experimental measurements, it is hard for theory to be directly competitive. However, DR measurements for systems such as N_2^+, which are hard to cool, or for systems in vibrationally excited states are highly uncertain or not really possible, and therefore rates for these processes are best obtained theoretically (Little et al. 2014; Motapon et al. 2006).

5.2.8 Electron Impact Ionization

Cross sections for ionization by electron collisions become large at high collision energies. In many plasmas, impact ionization is the major route for production of ions in the plasma, but it does, of course, require the electrons to have sufficient energy to overcome the ionization potential (ionization energy) of the given target molecule. Ionization cross sections can be calculated in a rather straightforward manner with the semi-empirical BEB (binary encounter Born) method of Kim and co-workers (Kim & Rudd 1994) or by a number of related methods (Tanaka et al. 2016). These methods are simple to use and provide surprisingly accurate results.

5.2.9 Electron Impact Dissociative Ionization

Electron impact ionization can, if there is sufficient energy, also result in break-up of the molecular target. Calculations with the BEB method do not consider the final products and thus do not yield fragmentation patterns. However, mass spectroscopy experiments yield fragmentation patterns for a given molecular targets, but not, in general, total cross sections. A compilation of fragmentation patterns can be found on the NIST molecular database (Stein 2016). BEB and NIST data can be combined to give comprehensive fragmentation cross sections for impact ionization (Hamilton et al. 2017).

5.3 Methodology

As implied in the discussions above, electron collision data can be obtained both from laboratory measurements and from quantum mechanical calculations. In principle, measured data should be more reliable, although in practice there are issues with obtaining reliable cross sections for polar molecules (Zhang et al. 2009), and experiments are difficult to perform at the ultra-low electron energies (temperatures) encountered in interstellar space. An excellent and comprehensive review of experimental studies on electron collisions with diatomic molecules has been provided by Brunger & Buckman (2002).

Theoretically, calculations of low-energy electron collision cross sections require a quantum mechanical treatment, because quantum forces such as the exchange interaction play an important role in these collisions. There are a variety of theoretical methods available for treating such collisions (Huo & Gianturco 1995). Cross comparisons between these methods have shown in a variety of cases that, for a given theoretical model, the numerical procedures usually give very similar results; see Faure & Tennyson (2002), for example. It should be noted that results that agree between different calculations using the same model should not be confused with the correct answer for the problem, as the uncertainty in most theoretical calculations is dominated by the choice of model (Chung et al. 2016). Nonetheless, first-principles quantum mechanical calculations are increasingly becoming the main provider of electron collision data for plasma models (Bartschat & Kushner 2016; Bartschat et al. 2017).

A full treatment of electron collisions with molecules, particularly at low energies, requires consideration of both the electronic and the nuclear motion. Such a treatment is essential for many of the processes considered in the previous section. Standard treatments of molecular structure are normally based on the Born–Oppenheimer approximation, which separates the electronic and nuclear motion. Crudely explained, the Born–Oppenheimer approximation assumes that the electrons move so fast that they relax instantly to any change in the position of the nuclei. This allows one to solve the electronic structure problem for fixed (or frozen) nuclei and thus build up potential energy curves, such as the ones displayed in the figures above, upon which the nuclei move. The trouble with the electron–molecule collision problem is caused by resonances, which means that there are two sets of potential curves upon which the nuclei might move: one set belonging to the original target molecule (AB above) and the other belong to the compound (AB^-) system (Tennyson 1996b). For example, the two mechanisms for electron-impact vibrational excitation discussed above differ in terms of the curves upon which the nuclei move: the direct mechanism can be treated by extracting vibrationally resolved cross sections by averaging at an appropriate point in the calculations (actually the so-called T-matrices; Mazevet et al. 1999) using target vibrational wave functions. Conversely, the treatment of the resonance-enhanced vibrational excitation involves consideration of nuclear motion on the resonance curves (Laporta et al. 2012). Simultaneous treatment of both mechanisms requires going beyond the Born–Oppenheimer approximation. Such calculations have been performed for the explicit case of electron-impact vibrational (and rotational) excitation of diatomic molecules (Domcke 1991; Horacek et al. 1998). However, there is no general non-adiabatic (non-Born–Oppenheimer) treatment available that gives cross sections for all the processes considered in the previous section using a single unified approach. Furthermore, there exist a few practical methods for treating polyatomic systems without significant further simplifications, see Munro et al. (2012) and Laporta et al. (2016) for examples. This means that the effects of nuclear motion, which are often crucial (Stibbe & Tennyson 1998), are in practice treated on a case-by-case basis. Indications of how this is achieved in practice are given in the discussion of the individual processes above.

There are a number of theoretical methods available for treating the electronic motion in low- and intermediate-energy electron collision calculations within the Born–Oppenheimer approximation (Huo & Gianturco 1995). Of these, the R-matrix method is the major theoretical method used for treating electron collisions with atomic ions, as witnessed by a series of papers on “atomic data for astrophysics” (Del Zanna & Badnell 2016). Our method of choice for solving the electronic structure problems involved in electron–molecule collisions is also the R-matrix method, as implemented in the UK Molecular R-matrix (UKRMol) codes (Carr et al. 2012).

The R-matrix method has been comprehensively reviewed by Burke (2011), who led its development for treating a variety of problems over several decades. A more technical review of the implementation of the R-matrix method to electron–molecule collisions has been provided by one of us (Tennyson 2010); readers wishing to get a full understanding of the methodology are to referred to this previous work. Here, we only give broad outlines.

The basic idea of the R-matrix method, as illustrated for electron—H_2 collisions in Figure 5.4, is the division of space into two regions. The inner region is where most of the physics occurs, and once the continuum is discretized by enclosing the inner region in a sphere, the inner-region problem has strong similarities with the molecular electronic structure problem as implemented in a variety of different quantum chemistry codes. In the outer region, the scattering electron moves only in the long-range potential due to the target molecule. At very low electron energies, such as those found in the interstellar medium, these long-range potentials can be particularly important; indeed, if the molecular target possess a dipole moment, it can be the dominant effect.

The UKRMol code offers a variety of implementations of the inner region electronic structure problem. The current standard implementation uses Gaussian-type orbitals (GTOs) to represent both the target and the continuum wave function (Morgan et al. 1997; Faure et al. 2002). The new UKRMol+ implementation (Darby-Lewis et al. 2017) uses a mixed B-spline/GTO basis for the continuum, which should extend both the energy range and size of target molecule for which the codes can be used. The computational bottleneck in these codes is the Hamiltonian matrix construction and diagonalization. The UKRMol codes employ an algorithm that has been especially adapted to efficiently deal with the standard structure of the inner-region scattering wave functions (Tennyson 1996a); this too has recently been updated to take advantage of modern computer architectures (Al-Refaie & Tennyson 2017). Although not discussed in this chapter, the UKRMol codes can also treat photoionization with a particular emphasis on photon energies where post-photoionization interaction between the ionized target molecule and the outgoing

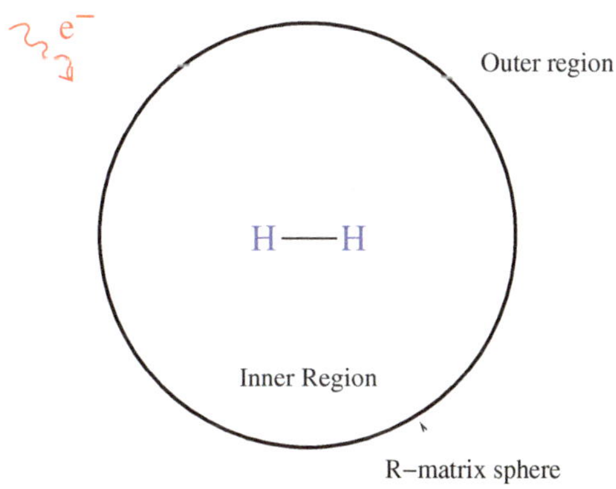

Figure 5.4. Schematic representation of the R-matrix method, illustrated for a calculation regarding electron collisions with molecular hydrogen.

electron are significant (Harvey et al. 2014). As these codes are far from straightforward to use, they can be run using the Quantemol expert system for both electron-scattering (Tennyson et al. 2007) and photoionization (Brigg et al. 2015). Both the UKRMol/UKRMol+ and Quantemol codes are making increasing use of the MOLPRO molecular electronic structure code (Werner et al. 2012) to provide an accurate representation of the target and other related information.

5.4 Astrophysical Examples

5.4.1 Dissociative Recombination of H_3^+

Protonated molecular hydrogen, H_3^+, is the simplest triatomic molecule. H_3^+ is also a strong acid (i.e., proton donor), which initiates a rich ion–molecule chemistry in the interstellar medium (Oka 2006). The ubiquity of this ion in interstellar space was revealed in the 1990s thanks to infrared absorption observations (Geballe & Oka 1996; McCall et al. 1998). The simple chemistry of H_3^+ makes this ion a powerful astrophysical probe, particularly to measure the cosmic-ray ionization rate ξ (in s^{-1}) in the interstellar medium. In diffuse clouds, H_3^+ is destroyed by recombination with electrons (DR) released from the photoionization of carbon atoms. The electron fraction, $x_e = n(e^-)/n(\mathrm{H})$, is therefore $x(e^-) \sim n(\mathrm{C}^+)/n(\mathrm{H}) \sim 10^{-4}$. Equating the production rate $\xi n(\mathrm{H}_2)$ and the destruction rate $k_e n(\mathrm{H}_3^+)n(e^-)$, the H_3^+ number densities at steady state can be written as (McCall et al. 2003):

$$n(\mathrm{H}_3^+) = \frac{\xi}{k_e} \times \frac{n(\mathrm{H}_2)}{n(e^-)} \tag{5.21}$$

where k_e is the H_3^+ DR rate coefficient (in $cm^3\,s^{-1}$). This rate coefficient has been controversial for many decades because both different experimental techniques and theoretical calculations have yielded results differing by several orders of magnitude. In the last 15 years, however, there has been a great advance in theory, thanks to the MQDT calculations of Greene and co-workers (Kokoouline et al. 2001; dos Santos et al. 2007). Experimentally, excellent agreement has been reached between different storage ring measurements, at the cross section level. Calculations now give a rate coefficient k_e, which agrees well with the latest experimental values, i.e., $k_e = 1 - 5 \times 10^{-7}\ \mathrm{cm^3\,s^{-1}}$ in the range $T_k = 10 - 100$ K, as shown in Figure 5.5. Using Equation (5.21) above with $k_e(23\ \mathrm{K}) = 2.6 \times 10^{-7}\ \mathrm{cm^3\,s^{-1}}$, McCall et al. (2003) have shown that the cosmic-ray ionization rate ξ in a diffuse cloud (towards the star ξ Persei) is $\sim 1.2 \times 10^{-15} \mathrm{s}^{-1}$, i.e., 40 times faster than previously assumed. This important result, which has now been confirmed toward other sources, relies on a unique laboratory parameter: the DR rate coefficient $k_e(T)$.

Despite the theoretical and experimental progress, significant disagreements remain between calculations and measurements, e.g., in the resonance structure of the cross sections, as discussed by Oka (2015). Among the possible sources of discrepancies, the rotational temperature of H_3^+ in the storage ring experiments is an important issue. It was revealed by Petrignani et al. (2011) that the stored H_3^+ ions are much hotter than originally thought, as supported by theory (Kreckel et al. 2002). Indeed, for electron energies above 10 meV (i.e., 100 K), excitation by the

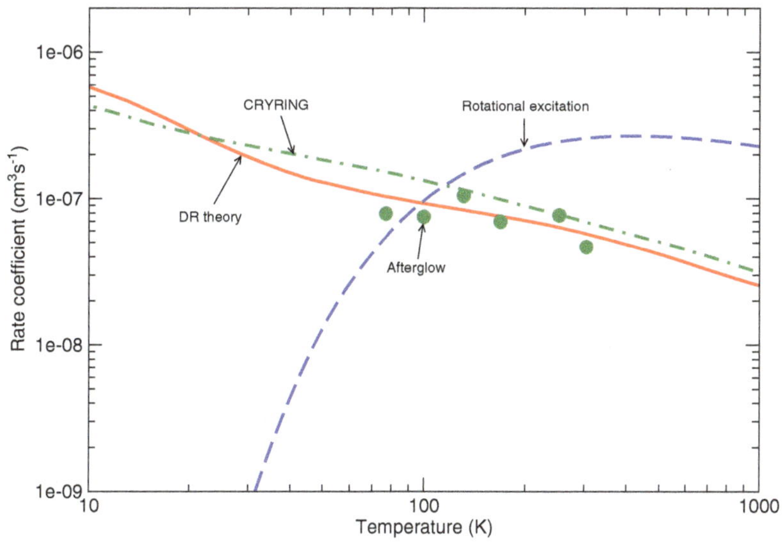

Figure 5.5. Rate coefficients as a function of temperature for electron-H_3^+ collisions. The solid line denotes the MQDT calculations of dos Santos et al. (2007) for the DR of H_3^+, assuming the ions and electrons are in common thermal equilibrium. The dot-dashed line gives the storage ring DR CRYRING experimental results of McCall et al. (2004). The filled circles correspond to the DR afterglow experimental results of Glosík et al. (2009). The dashed line gives the MQDT rate coefficient for the rotational excitation $j_k = 1_1 \rightarrow 2_1$ in para-H_3^+, as computed by Kokoouline et al. (2010).

electrons themselves becomes competitive with (or even dominant over) the DR process (Faure et al., 2006, 2009). This is illustrated in Figure 5.5 for the ground-state excitation $j_k = 1_1 \rightarrow 2_1$ of para-H_3^+. An extensive set of MQDT cross sections and rate coefficients for the electron-impact rovibrational excitation of H_3^+ have now been provided by Kokoouline et al. (2010). These data, combined with rotationally-resolved DR cross sections, might help to clarify the remaining discrepancies between the experimental and theoretical DR of H_3^+, which is of crucial astrophysical importance.

5.4.2 Electron-impact Excitation of Water

Water vapor has been detected in a great variety of astronomical environments using both spacecraft and Earth-based observations. The *ISO*, *Spitzer*, and *Herschel* space telescopes, in particular, have revealed the ubiquity of water in the interstellar medium, from star-forming regions to envelopes of evolved stars (van Dishoeck et al. 2013). Water has also been observed on planets, including extra-solar planets, and it is the main constituent of cometary comae. In environments where the electron-fraction x_e is larger than about 10^{-5}, electron collisions can dominate water excitation because, due to the large H_2O dipole (1.8 D), electron-impact collisional rates exceed those for excitation by neutral species by typically five orders of magnitude. It is thus recognized that the rotational temperature of H_2O in cometary comae is controlled by a competition between electron collisions and solar photon pumping, depending on the cometocentric distance (Xie & Mumma 1992; Bensch et al. 2006).

Electron-impact excitation of H_2O has been widely studied, both experimentally and theoretically, for many years. The first accurate rotational calculations were published in 2004 using the **R**-matrix theory combined with the ANR approximation (Faure et al. 2004a, 2004b). A few years later, Čurík et al. (2006) measured ultra-low-energy elastic cross sections, down to 18 meV. These authors also provided rotational cross sections using a semi-empirical quantum-defect theory. As shown by Zhang et al. (2009), the agreement between the **R**-matrix calculations of Faure et al. (2004b) and the measurements of Čurík et al. (2006) is very good. Excellent agreement between these calculations and the latest experimental elastic differential cross sections was also observed at energies above 1 eV (Zhang et al. 2009). To our knowledge, the only experimental attempt to measure rotationally inelastic cross sections in electron–water collisions was reported by Jung et al. (1982), for collision energies of 2.14 eV and 6 eV. The agreement with the **R**-matrix calculations of Faure et al. (2004b) for transitions with $\Delta j = 0$ and $|\Delta j| = 1$ was found to be satisfactory (see Figure 5.6); the small discrepancies were attributed to the contribution of $\Delta j = 2$ transitions, which were neglected in the experimental fitting procedure (Gorfinkiel et al. 2005).

A set of rate coefficients for the rotational excitation of water was published in Faure et al. (2004a). This data set has been employed to model water excitation in various astrophysical objects, such as comets, proto-planetary disks, and evolved stars. Significant effects (with respect to old data based on the Born approximation)

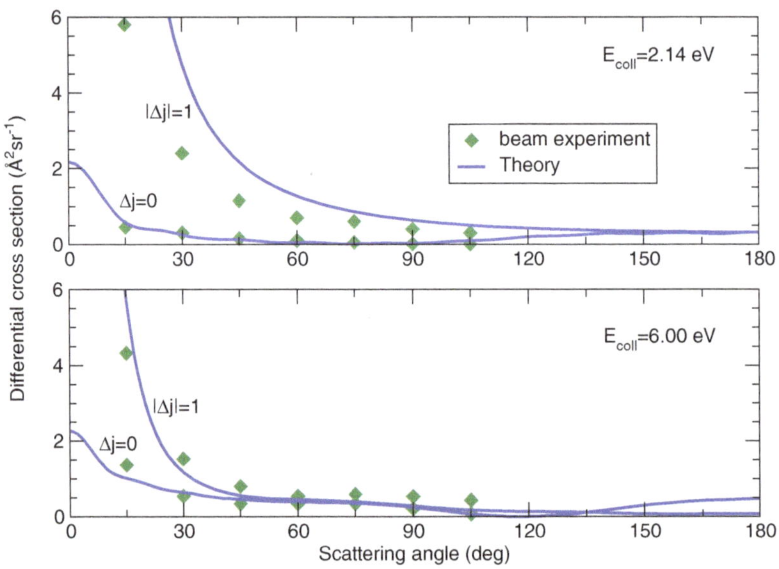

Figure 5.6. Absolute differential cross sections for pure elastic scattering ($\Delta j = 0$) and rotational (de-)excitation ($|\Delta j = 1|$) of H_2O at incident energies of 2.14 eV (upper panel) and 6.0 eV (lower panel). The symbols denote the beam experiment of Jung et al. (1982), while the solid lines correspond to the **R**-matrix calculations of Faure et al. (2004b). See Gorfinkiel et al. (2005) for more details on the calculations.

were reported, particularly for transitions among rotationally excited levels (Bensch et al. 2006). We note that these rotational rate coefficients were also combined with the vibrational calculations of Nishimura & Gianturco (2004) to provide an extensive set of ro-vibrational state-to-state rate coefficients (Faure & Josselin 2008).

5.4.3 Electron-impact Rotational Excitation of Molecular Ions: HCO^+ and ArH^+

Protonated CO, HCO^+, is the most abundant ion in the interstellar medium. Owing to its strong dipole (3.9 D), the rotational transitions of HCO^+ are widely observed. The strong dipole also means that the rate coefficients for electron-impact rotational excitation of HCO^+ are extremely large. The **R**-matrix/ANR calculations of Faure & Tennyson (2001) thus give $k \gtrsim 10^{-5}$ cm^3 s^{-1} for dipolar transitions ($\Delta j = 1$) at very low temperature. This is five orders of magnitude larger than the rate coefficients for collisions with H_2 and two orders of magnitude larger than the HCO^+ DR rate coefficient (Faure et al. 2009). At high electron fractions ($x_e \sim 10^{-4}$), free electrons are therefore expected to entirely control the HCO^+ excitation. Evidence for an electron density enhancement in the magnetic precursor of a C-type shock was actually provided by the observation of an over-excitation of $H^{13}CO^+$ with respect to $H^{13}CN$ and $HN^{13}C$ toward the L1448-mm outflow (Jimenez-Serra et al. 2006). The dipoles of HCN and HNC (~3 D) being similar to that of HCO^+, the over-excitation of the latter is, in fact, consistent with the different threshold behavior of electron–ion collisions with respect to electron–neutral collisions: for neutrals, electron-impact cross sections go to zero at threshold, while for ions they are large and finite (Jimenez-Serra et al. 2006). On the other hand, we note that because the rate coefficients for collisions of HCN with H_2 are about a factor of 10 lower than for HCO^+, HCN may provide a better probe of electron excitation in environments with lower electron fractional abundance ($x_e < 10^{-4}$), as discussed recently by Goldsmith & Kauffmann (2017).

The argonium cation, ArH^+, is another strongly polar molecular ion (the dipole is 2.1 D) recently discovered in the interstellar medium. This is the first noble gas molecular ion detected in space, thanks to the *Herschel* space telescope. The first identification of ArH^+ was reported by Barlow et al. (2013) toward the Crab Nebula, a supernova remnant in the constellation of Taurus. The $j = 1 \rightarrow 0$ and $j = 2 \rightarrow 1$ rotational lines were observed in emission at several points in the nebula. As the electron abundance in this source can reach very high fractions ($x_e \gg 10^{-4}$), this detection has motivated the calculation of electron-impact excitation rate coefficients. Cross sections for the rotational excitation of ArH^+ by electrons were thus computed by Hamilton et al. (2015) using the R-matrix theory. The three isotopes of argon (^{36}Ar, ^{38}Ar and ^{40}Ar) were considered, but astronomically the major isotope is ^{36}Ar. These rate coefficients were combined with the radiative rates to model ArH^+ excitation in the nebula through radiative transfer calculations (Hamilton et al. 2015). The large velocity gradient approximation (see Chapter 7 by P. Dagdigian in the present book) for an expanding sphere was employed, assuming the typical physical conditions inferred for the molecular gas associated with the filaments and knots in the Crab Nebula: the dominant colliders are free electrons,

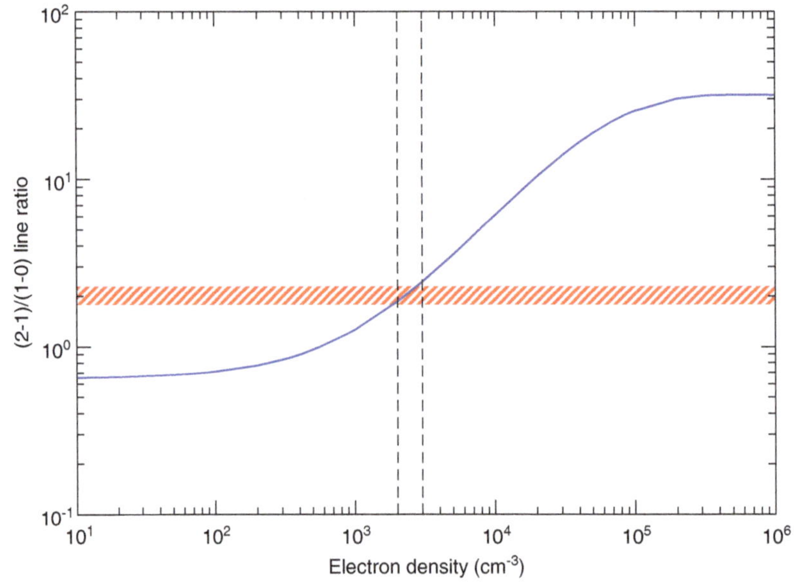

Figure 5.7. Plot of the $(2 \rightarrow 1)(1 \rightarrow 0)$ line ratio predicted for an ArH^+ column density of 2×10^{12} cm^{-2}. The red hatched zone shows the *Herschel* observational ratio (Barlow et al. 2013) at a particular position in the nebula. Adapted from Hamilton et al. (2015).

and the kinetic temperature is in the range 2000–3000 K. In these conditions, the emission line ratio $(2 \rightarrow 1)/(1 \rightarrow 0)$ was shown to be extremely sensitive to the electron abundance in the nebula. This is illustrated in Figure 5.7, where the line ratio is plotted as function of the electron density. The observational ratio is reproduced for $n(e^-) \sim 2 - 3 \times 10^3$ cm^{-3}, which was found not to depend significantly on the ArH^+ column density. Along with HCO^+, the rotational emission of ArH^+ thus provides a powerful probe of the electronic density in highly ionized interstellar region.

5.4.4 Electron-impact Rotational Excitation of CN

In diffuse interstellar clouds, the rotational temperature of optically detected polar molecules such as CN is generally close to that of the cosmic microwave background (CMB) radiation temperature ($T_{CMB} = 2.73$ K). Thus, optical absorption-line measurements of interstellar CN have long been used to estimate the temperature of CMB radiation at 2.6 and 1.3 mm, the wavelengths of the two lowest CN rotational transitions $N = 1 \rightarrow 0$ and $N = 2 \rightarrow 1$, respectively (Thaddeus 1972). It was soon realized, however, that the accuracy of this indirect method is limited by local collisional effects. Now that the CMB temperature has been measured with high precision by the *COBE* satellite, with the latest value at $T_{CMB} = 2.72548 \pm 0.00057$ K (Fixsen 2009), CN absorption line observations can be used to probe the rotational excitation of CN in excess of T_{CMB}, i.e., the local excitation. This probe, however, requires good collisional rate coefficients for the rotational excitation of CN. These data, including hyperfine-resolved transitions, have recently become available for both CN–H_2 (Kalugina et al. 2012; Kalugina & Lique 2015) and CN–electron (Harrison et al. 2013) collisions. In particular, the **R**-matrix theory was employed by Harrison et al. (2012) to compute, within the ANR approximation, the spin-coupled cross sections, which were subsequently employed to provide hyperfine-resolved rate coefficients. These data were used in a radiative transfer code to investigate the influence of varying the electron density on the local CN excitation, as described in Semenov et al. (2017). Such analysis was performed in the past using approximate collisional data (Black & van Dishoeck 1991).

Observationally, the most recent CN optical absorption line measurements have provided a weighted mean value of T_{01}(CN) = 2.754 ± 0.002 K (Ritchey et al. 2011), where T_{01} is the excitation temperature between the levels $N = 0$ and $N = 1$. This implies an excess over the temperature of the CMB of $T_{\text{loc}} = 29 \pm 0.3$ mK. The radiative transfer calculations of Harrison et al. (2013) have shown that this local excitation can be reproduced for a rather small range of electron density $n(e^-) = 0.01 - 0.06$ cm^{-3} for typical CN column densities and total hydrogen densities. These electron densities correspond to electron fractions in the range $10^{-5} - 6 \times 10^{-4}$, consistent with the abundance of interstellar C^+, which is the main source of electrons in diffuse clouds.

In the case of the diffuse cloud toward the star HD 154368, in addition to the optical absorption lines, the weak CN rotational emission $N = 1 \rightarrow 0$ at 2.6 mm was also observed. Because the physical conditions in this source are well-constrained, it was possible to study the sensitivity of this emission line as a function of the electron density. Harrison et al. (2012) found that the observed intensity could be reproduced for an electron density of ~0.03 cm^{-3} (see Figure 5.8), corresponding to an electron fraction of $\sim 2 \times 10^{-4}$, as expected from the carbon abundance. In addition, the associated excitation temperature was found to be $T_{01} = 2.75$ K, in very good agreement with the weighted mean value of 2.754 K determined in Ritchey et al. (2011).

5.4.5 Electron-impact Rotational Excitation of CH^+

The methylidyne ion CH^+ was the first molecular ion to be identified in the diffuse interstellar medium in the 1940s. Since then, CH^+ optical absorption has been observed toward many background stars, demonstrating the ubiquity of this simple carbon hydride in diffuse clouds. Emission lines are less commonly observed, but they have been detected toward several dense sources ($n_H > 10^4$ cm^{-3}) illuminated by a strong far-ultraviolet (FUV) radiation field. In such dense "photon-dominated regions" (PDR), a reservoir of rovibrationally excited (by the FUV fluorescence) H_2 is expected to provide a chemical route to CH^+ through the normally endothermic reaction $C^+ + H_2$. The formation pathway of CH^+ is important because, this ion being highly reactive with e^-, H, and H_2, it never reaches full equilibration through inelastic collisions. CH^+ thus belongs to the class of "reactive" ions whose emission spectrum should retain some memory of their formation process, as suggested by Black (1998). As a result, when solving the coupled equations of statistical equilibrium and radiative transfer, it is necessary to include the *state-resolved*

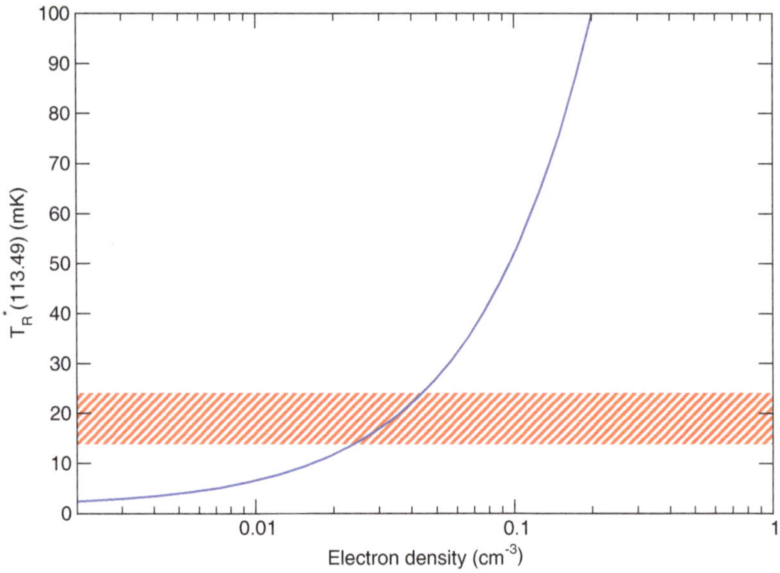

Figure 5.8. Plot of the intensity of the CN rotational emission line $N = 1 \rightarrow 0$ at 2.6 mm (113.49 GHz) as a function of electron density towards the star HD 154368. The red hatched zone shows the observed intensity (Palazzi et al. 1990). Adapted from Harrison et al. (2013).

(here rotationally-resolved or *j*-dependent) formation and destruction rates in addition to the usual radiative and inelastic rates.

An extensive set of state-resolved data for C^+–H_2, CH^+–H, and CH^+–electron collisions has recently become available, as summarized in Faure et al. (2017). In particular, the *j*-dependent rate coefficients for the electron-impact rotational excitation and DR of CH^+ were computed using the **R**-matrix and MQDT theory, respectively. The full set of collisional data was employed in Faure et al. (2017) to model the rotational emission of CH^+ in the young planetary nebula NGC 7027. As shown in Figure 5.9, the observational line fluxes suggest an increase of the electron fraction from the standard value $x_e = 10^{-4}$ to $x_e = 10^{-3}$. This result is, in fact, consistent with the higher elemental abundance of carbon in this "carbon-rich" circumstellar envelope. Faure et al. (2017) have shown that electron collisions contribute up to ~40% of the line intensities in this source, due to the high electron fraction. We finally note that, in this model, H_2 was assumed to lie exclusively in its first vibrationally excited state $v = 1$. An even better agreement with observations can be obtained if a significant fraction of $H_2(v = 2)$ exist in this source Faure et al. (2017).

5.5 Sources of Data

There are a number of databases that provide cross sections and/or rates for electron collision processes. The main ones are listed in Table 5.1. Many of these databases are accessible simultaneously using the portal provided by the Virtual Atomic and Molecular Data Centre (VAMDC) (Dubernet et al. 2010, 2016).

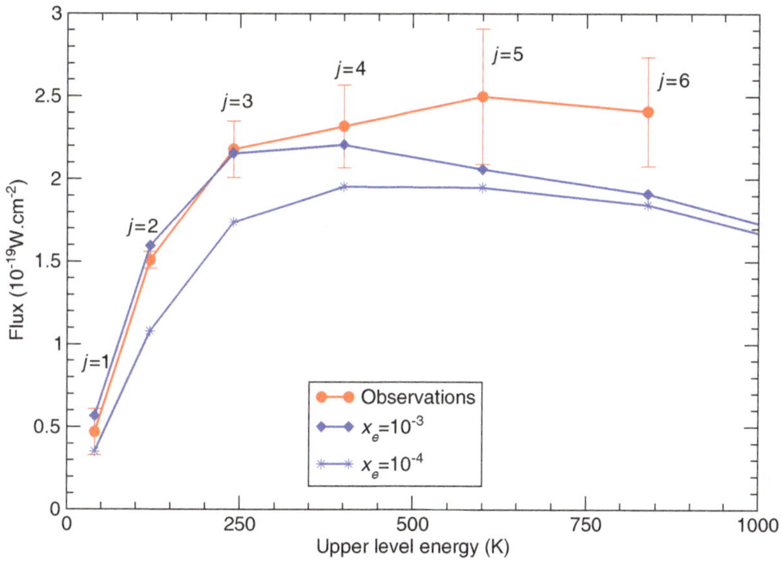

Figure 5.9. CH^+ line fluxes of $j \rightarrow j - 1$ rotational transitions as functions of the upper-level energy, as predicted by the non-LTE calculations of Faure et al. (2017). The CH^+ column density (2×10^{15} cm^{-2}) was adjusted to best reproduce the *ISO* and *Herschel* observations of Cernicharo et al. (1997) and Wesson et al. (2010). In this calculation, the formation of CH^+ is due to collisions between C^+ and H_2 in its first vibrationally excited state $\nu = 1$. See Faure et al. (2017) for full details.

Table 5.1. Key Databases Providing Electron–molecule Collision Data

Database	Reference	Electron-collision data	Target audience	Other data included
BASECOL	(Dubernet et al. 2013)	Rotational excitation	Astrophysics	Heavy particle inelastic cross sections
KIDA	(Wakelam et al. 2015)	Dissociative recombination	Astrophysics	Reaction rates
LXCat	(Pitchford et al. 2017)	Excitation processes	Plasma physics	Atomic cross sections
Phys4Entry	(Celiberto et al. 2016)	Vibrational excitation	Atmospheric re-entry	Heavy particle inelastic cross sections
QDB	(Tennyson et al. 2017)	Excitation processes	Technological plasmas	Reaction rates

Note: Only recent or currently maintained databases are considered.

In addition to the databases listed, there are an increasing number of studies providing critically assessed cross section sets for electron collisions with individual molecules. Recent compilations for astronomically important molecules include those for electron collisions with methane (Song et al. 2015), acetylene (Song et al. 2017), O_2 (Itikawa 2009), water (Itikawa & Mason 2005) and H_2 (Yoon et al. 2008).

We note that models of terrestrial plasma do not generally use rate coefficients for electron collisions, but instead consider the underlying cross sections. This is because, for such plasmas, the electron energy distribution function (EEDF), which can often be explicitly measured, is rarely found to be thermal; this invalidates the assumption of a thermal rate coefficient. For most astronomical environments, electron collisions are not the dominant or driving collision process. This has meant that the explicit use of electron collision cross sections has not generally been deemed necessary, and rate coefficients based on the notion of a thermalized electron distribution are generally used to represent processes of interest. It is unlikely the EEDFs in low-density astronomical environments are actually thermal; however, they remain difficult to determine directly. Nonetheless, at some point it is likely that astronomical EEDFs will have to be explicitly considered. At this stage, it will be necessary to explicitly employ electron collision cross sections.

References

Agren, K., Edberg, N. J. T., & Wahlund, J. E. 2012, GeoRL, 39, L10201

Al-Refaie, A. F., & Tennyson, J. 2017, CoPhC, 214, 216

Allan, M. 1985, JPhB, 18, 4511

Ardaseva, A., Rimmer, P. B., Waldmann, I., et al. 2017, MNRAS, 470, 187

Barlow, M. J., Swinyard, B. M., Owen, P. J., et al. 2013, Sci., 342, 1343

Bartschat, K., Tennyson, J., & Zatsarinny, O. 2017, Plasma Processes Polymers, 14, 1600093

Bartschat, K., & Kushner, M. J. 2016, PNAS, 113, 7026

Bensch, F., Melnick, G. J., Neufeld, D. A., et al. 2006, Icar, 184, 602

Black, J. H. 1998, FaDi, 109, 257

Black, J. H., & van Dishoeck, E. F. 1991, ApJL, 369, L9

Brigg, W. J., Harvey, A. G., Dzarasova, A., et al. 2015, JaJAP, 54, 06GA02

Brigg, W. J., Tennyson, J., & Plummer, M. 2014, JPhB, 47, 185203

Brunger, M. J., & Buckman, S. J. 2002, PhR, 357, 215

Burke, P. G. 2011, R-Matrix Theory of Atomic Collisions: Application to Atomic, Molecular and Optical Processes (Springer: Berlin)

Campbell, L., & Brunger, M. J. 2009, PhyS, 80, 058101

Campbell, L., & Brunger, M. J. 2013, PSST, 22, 013002

Carr, J. M., Galiatsatos, P. G., Gorfinkiel, J. D., et al. 2012, EPJD, 66, 58

Celiberto, R., Armenise, I., Cacciatore, M., et al. 2016, PSST, 26, 033004

Cernicharo, J., Liu, X.-W., González-Alfonso, E., et al. 1997, ApJL, 483, L65

Chang, E. S., & Fano, U. 1972, PhRvA, 6, 173

Chu, S.-I., & Dalgarno, A. 1974, PhRvA, 10, 788

Chung, H.-K., Braams, B. J., Bartschat, K., et al. 2016, JPhD, 49, 363002

Coates, A. J., Crary, F. J., Lewis, G. R., et al. 2007, GeoRL, 34, L22103

Čurík, R., & Greene, C. H. 2017, JChPh, 147, 054307

Čurík, R., Ziesel, J. P., Jones, N. C., Field, T. A., & Field, D. 2006, PhRvL, 97, 123202
Darby-Lewis, D., Masin, Z., & Tennyson, J. 2017, JPhB, 50, 175201
Del Zanna, G., & Badnell, N. R. 2016, A&A, 585, A118
Domcke, W. 1991, PhRv, 208, 97
dos Santos, S. F., Kokoouline, V., & Greene, C. H. 2007, JChPh, 127, 124309
Douguet, N., Fonseca dos Santos, S., Raoult, M., et al. 2015, JChPh, 142, 234309
Dubernet, M.-L., Alexander, M. H., Ba, Y. A., et al. 2013, A&A, 553, A50
Dubernet, M.-L., Antony, B. K., Ba, Y. A., et al. 2016, JPhB, 49, 074003
Dubernet, M. L., Boudon, V., Culhane, J. L., et al. 2010, JQSRT, 111, 2151
Egert, A., Waite, J. H. Jr., & Bell, J. 2017, JGRA, 122, 2210
Faure, A., Gorfinkiel, J. D., Morgan, L. A., & Tennyson, J. 2002, CoPhC, 144, 224
Faure, A., Gorfinkiel, J. D., & Tennyson, J. 2004a, MNRAS, 347, 323
Faure, A., Gorfinkiel, J. D., & Tennyson, J. 2004b, JPhB, 37, 801
Faure, A., Halvick, P., Stoecklin, T., et al. 2017, MNRAS, 469, 612
Faure, A., & Josselin, E. 2008, A&A, 492, 257
Faure, A., Kokoouline, V., Greene, C. H., & Tennyson, J. 2006, JPhB, 39, 4261
Faure, A., Kokoouline, V., Greene, C. H., & Tennyson, J. 2009, J. Phys. Conf. Ser., 192, 012016
Faure, A., & Tennyson, J. 2001, MNRAS, 325, 443
Faure, A., & Tennyson, J. 2002, JPhB, 35, 1865
Fixsen, D. J. 2009, ApJ, 707, 916
Florescu-Mitchell, A. I., & Mitchell, J. B. A. 2006, PhR, 430, 277
Geballe, T. R., & Oka, T. 1996, Natur, 384, 334
Gerard, J. C., Bonfond, B., Grodent, D., & Radioti, A. 2016, P&SS, 131, 14
Gerard, J. C., Grodent, D., Radioti, A., Bonfond, B., & Clarke, J. T. 2013, Icar, 226, 1559
Gianturco, F. A., Grassi, T., & Wester, R. 2016, JPhB, 49, 204003
Glosík, J., Plašil, R., Korolov, I., et al. 2009, PhRvA, 79, 052707
Goldsmith, P. F., & Kauffmann, J. 2017, ApJ, 841, 25
Gorfinkiel, J. D., Faure, A., Taioli, S., et al. 2005, EPJD, 35, 231
Hallett, J. T., Shemansky, D. E., & Liu, X. 2005, ApJ, 624, 448
Hamilton, J. R., Faure, A., & Tennyson, J. 2015, MNRAS, 455, 3281
Hamilton, J. R., Tennyson, J., Huang, S., & Kushner, M. J. 2017, PSST, 26, 065010
Harada, N., & Herbst, E. 2008, ApJ, 685, 272
Harrison, S., Faure, A., & Tennyson, J. 2013, MNRAS, 435, 3541
Harrison, S., Tennyson, J., & Faure, A. 2012, JPhB, 45, 175202
Harvey, A. G., Brambila, D. S., Morales, F., & Smirnova, O. 2014, JPhB, 47, 215005
Hey, J. D., Chu, C. C., & Hintz, E. 2000, CoPP, 40, 9
Horacek, J., Cizek, M., & Domcke, W. 1998, AcTC, 100, 31
Huo, W. M., & Gianturco, F. A. (ed) 1995, Computational Methods for Electron Molecule Collisions (New York: Plenum)
Itikawa, Y. 2006, JPCRD, 35, 31
Itikawa, Y. 2009, JPCRD, 38, 1
Itikawa, Y., & Mason, N. 2005, JPCRD, 34, 1
Jensen, M. J., Bilodeau, R. C., Safvan, C. P., et al. 2000, ApJ, 543, 764
Jimenez-Serra, I., Martin-Pintado, J., Viti, S., et al. 2006, ApJ, 650, L135
Jung, K., Antoni, T., Mueller, R., Kochem, K.-H., & Ehrhardt, H. 1982, JPhB, 15, 3535

Jungen, C. (ed) 1996, Molecular Applications of Quantum Defect Theory (London: Taylor and Francis)
Jungen, Ch. 2011, Elements of Quantum Defect Theory (Chichester: Wiley)
Kalugina, Y., & Lique, F. 2015, MNRAS, 446, L21
Kalugina, Y., Lique, F., & Kłos, J. 2012, MNRAS, 422, 812
Kim, Y. K., & Rudd, M. E. 1994, PhRvA, 50, 3945
Kokoouline, V., Faure, A., Tennyson, J., & Greene, C. H. 2010, MNRAS, 405, 1195
Kokoouline, V., Greene, C. H., & Esry, B. D. 2001, Natur, 412, 891
Kreckel, H., Krohn, S., Lammich, L., et al. 2002, PhRvA, 66, 052509
Lane, N. F. 1980, RvMP, 52, 29
Laporta, V., Cassidy, C. M., Tennyson, J., & Celiberto, R. 2012, PSST, 21, 045005
Laporta, V., Celiberto, R., & Tennyson, J. 2015, PhRvA, 91, 012701
Laporta, V., Tennyson, J., & Celiberto, R. 2016, PSST, 25, 06LT02
Larsson, M., & Orel, A. E. 2008, Dissociative Recombination of Molecular Ions (Cambridge: Cambridge Univ. Press)
Little, D. A., Chakrabarti, K., Mezei, J. Z., Schneider, I. F., & Tennyson, J. 2014, PhRvA, 90, 052705
Mazevet, S., Morrison, M. A., Boydstun, O., & Nesbet, R. K. 1999, JPhB, 32, 1269
McCall, B. J., Geballe, T. R., Hinkle, K. H., & Oka, T. 1998, Sci., 279, 1910
McCall, B. J., Huneycutt, A. J., Saykally, R. J., et al. 2004, PhRvA, 70, 052716
McCall, B. J., Huneycutt, A. J., Saykally, R. J., et al. 2003, Natur, 422, 500
Millar, T. J., Walsh, C., & Field, T. A. 2017, ChRv, 117, 1765
Morgan, L. A., Gillan, C. J., Tennyson, J., & Chen, X. 1997, JPhB, 30, 4087
Motapon, O., Fifirig, M., Florescu, A., et al. 2006, PSST, 15, 23
Munro, J. J., Harrison, S., Tennyson, J., & Fujimoto, M. M. 2012, J. Phys. Conf. Ser., 388, 012013
Nishimura, T., & Gianturco, F. A. 2004, EPL, 65, 179
Norcross, D. W., & Padial, N. T. 1982, PhRvA, 25, 226
Oka, T. 2006, PNAS, 103, 12235
Oka, T. 2015, EPJWC, 84, 06001
Palazzi, E., Mandolesi, N., Crane, P., et al. 1990, ApJ, 357, 14
Petrignani, A., Altevogt, S., Berg, M. H., et al. 2011, PhRvA, 83, 032711
Pitchford, L. C., Alves, L. L., Bartschat, K., et al. 2017, Plasma Proc. Polymers, 14, 1600098
Poparic, G. B., Ristic, M., & Belic, D. S. 2008, JPCA, 112, 3816
Rabadán, I., Sarpal, B. K., & Tennyson, J. 1998, MNRAS, 299, 171
Rabadán, I., & Tennyson, J. 1998, CoPhC, 114, 129
Ritchey, A. M., Federman, S. R., & Lambert, D. L. 2011, ApJ, 728, 36
Sanna, N., & Gianturco, F. A. 1998, CoPhC, 114, 142
Semenov, M., Yurchenko, S. N., & Tennyson, J. 2017, JMoSp, 330, 57
Shafir, D., Novotny, S., Buhr, H., et al. 2009, PhRvL, 102, 223202
Song, M.-Y., Yoon, J. S., Cho, H., et al. 2015, JPCRD, 44, 023101
Song, M.-Y., Yoon, J.-S., Cho, H., et al. 2017, JPCRD, 46, 013106
Stein, S. E. 2016, in NIST Chemistry WebBook, NIST Standard Reference Database Number 69, ed. P. J. Linstrom, & W. G. Mallard (Gaithersburg, MD: National Institute of Standards and Technology), 20899
Stibbe, D. T., & Tennyson, J. 1998, NJPh, 1, 2

Stibbe, D. T., & Tennyson, J. 1999, ApJ, 513, L147
Takayanagi, K. 1966, JPSJ, 21, 507
Tanaka, H., Brunger, M. J., Campbell, L., et al. 2016, RvMP, 88, 025004
Tennyson, J. 1996a, JPhB, 29, 1817
Tennyson, J. 1996b, CoAMP, 32, 209
Tennyson, J. 2010, PhR, 491, 29
Tennyson, J., Brown, D. B., Munro, J. J., et al. 2007, J. Phys. Conf. Ser., 86, 012001
Tennyson, J., Rahimi, S., Hill, C., et al. 2017, PSST, 26, 055014
Thaddeus, P. 1972, ARA&A, 10, 305
van Dishoeck, E. F., Herbst, E., & Neufeld, D. A. 2013, ChRv, 113, 9043
Vuitton, V., Lavvas, P., Yelle, R. V., et al. 2009, P&SS, 57, 1558
Wakelam, V., Loison, J.-C., Herbst, E., et al. 2015, ApJS, 217, 20
Werner, H.-J., Knowles, P. J., Knizia, G., Manby, F. R., & Schütz, M. 2012, WIREs Comput. Mol. Sci., 2, 242
Wesson, R., Cernicharo, J., Barlow, M. J., et al. 2010, A&A, 518, L144
Xie, X., & Mumma, M. J. 1992, ApJ, 386, 720
Yoon, J.-S., Song, M.-Y., Han, J.-M., et al. 2008, JPCRD, 37, 913
Zhang, R., Faure, A., & Tennyson, J. 2009, PhyS, 80, 015301
Ziolkowski, M., Vikar, A., Mayes, M. L., et al. 2012, JChPh, 137, 22A510

Chapter 6

Molecular Spectroscopy of Astrophysical Molecules

Stephan Schlemmer

6.1 Introduction

Spectroscopy is one of the most powerful tools in astrophysics because of the wealth of data contained in each spectrum. The field of astro-spectroscopy started in the early 19th century with Joseph Fraunhofer's discovery of the dark lines in the spectrum of the Sun. He used these lines to control the quality of the lenses he fabricated for microscopes and telescopes. The famous Fraunhofer lines are associated with atomic transitions, which were only found in 1859 thanks to Kirchhoff and Bunsen, who built advanced prism spectrometers to investigate the emission and absorption of vapors. The quantized transitions of atoms and molecules are the fingerprints we receive of very distant places in the universe. They tell us not only about the presence of these species but also about their abundance, the ambient temperature, and the relative velocity of the observed object. Therefore, molecular spectroscopy, the topic of this chapter, is of indispensable value for astrophysicists. However, this treasure can only be obtained if the spectra of the molecules are known in great detail from terrestrial laboratories. This is in fact, the topic of this chapter.

Today's telescopes and laboratory spectroscopy instruments are extremely powerful in terms of angular and spectral resolution, and thus reveal a richness that can be coined as the boon and bane of spectroscopy. When pointing the telescope toward different regions in the galactic or extragalactic environments, several very general spectral distinctions become apparent and should be inspected to introduce and highlight the main spectral characteristics. As an example, for more than one hundred years, astronomers observed absorption features throughout the visible and near-infrared (IR) regime when the light of distant stars passed through the interstellar medium. These so-called diffuse interstellar bands (DIBs) are a series

doi:10.1088/2514-3433/aae1b5ch6

of absorption bands, which appear in many observations but apparently also vary depending on the nature of the environment.

Figure 6.1 shows such an observation. Like for the Fraunhofer lines, the DIBs are spectral features with reduced intensity associated with the absorption of molecules. This is derived from general trends concerning the variation of these features, e.g., the width of each feature, when comparing different regions in space and thus different physical conditions, like temperature. In fact, the spectral features are thought to arise from excitations of the electronic states of a molecule. Due to the quantum mechanical nature of molecules, these transitions are accompanied by a potential change of the vibrational and the rotational state of the molecule. Therefore, the spectra can be rather complex and varied. Almost none of these many features could be associated with any known molecule, and thus this absorption phenomenon remains elusive even after a century of investigation. The recent discovery associating two of the prominent bands in the near-IR regime to the C_{60}^{+} molecule came as a surprise and will be discussed at the end of this chapter. It took very sophisticated spectroscopy methods to obtain the spectra in the laboratory, which is another aspect of this chapter.

Another interesting spectral region is the infrared (IR) range, where many molecules have their characteristic vibrational features. They appear as bands in low-resolution spectra but turn into rovibrational spectra with extreme richness when the resolution is high enough, as will be laid out for very selective examples later in this chapter. Before going into more detail, it is interesting to take a look at the IR to far-IR (FIR) spectral region. Figure 6.2 shows a spectral observation ranging from 1–100000 μm. This becomes possible thanks to high-altitude or space-borne telescopes for which Earth's atmosphere becomes transparent and thus the window into the terahertz (THz) region is opened. The composite spectrum reveals sharp spectral features as well as very broad ones, all of which are observed in emission. Other than for the Fraunhofer lines here, dust particles and molecules emit their radiation against a cold background. The broad background radiation peaking around 100 microns stems from cold emitting dust, which by itself is an interesting subject of research. The IR bands, here just indicated qualitatively in the range 3–12 μm, are associated with a whole set of carbonaceous molecules, namely the

Figure 6.1. DIBs observed in the visible spectrum, as this overlay of the absorption features onto the colors of the visible range of the spectrum illustrates. These absorption bands, over one hundred in all, are prominent features viewed in the diffuse interstellar medium toward many lines of sight when the star light looks redder than it should. This is a clear sign of dust and gas in the line of sight. These DIBs are associated with the electronic bands of molecules that are excited by the visible photons coming from the observed star. Only a few of these bands could be assigned to particular molecules. (Figure courtesy of Ben McCall, University of Illinois.) (Credit: Bergin et al. 2010.)

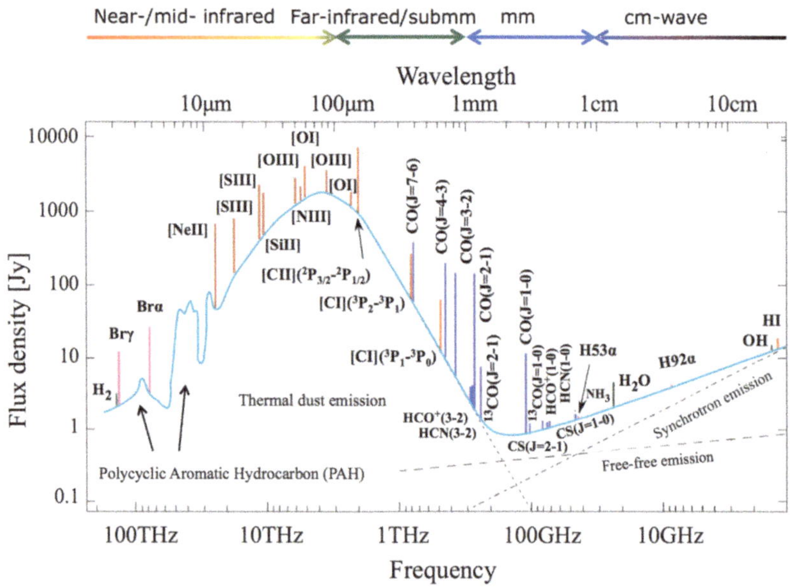

Figure 6.2. Composite infrared-to-radio spectrum of the starburst galaxy M82. The cyan line shows the continuum emission and broad emission features, and vertical lines show various types of emission lines (hydrogen recombination lines in magenta, fine structure lines in red, molecular rotational lines in blue, and masers in green). (Figure courtesy of Kotaro Kohno, Institute of Astronomy, The University of Tokyo.)

polycyclic aromatic hydrocarbons (PAHs). The fact that these emission features appear at much shorter wavelengths, i.e., higher energies, than the emission from the cold dust immediately shows that these molecules are highly excited. In fact, these molecules are not in a thermodynamic equilibrium, but they are electronically excited and then undergo a process where the electronic energy is converted into vibrational excitation. As a result, the vibrational "hot" molecules emit the bands, which are often called aromatic IR bands (AIR) or also unidentified IR bands (UIR). The latter denomination points at the fact that no single feature can be associated to just one molecule, instead, the bands from this class of molecules overlap. This makes the interpretation of these observations much more complex. This is the subject of many detailed studies, review articles, and books. Therefore, the reader is referred to these publications, e.g., Tielens (2005), for more details.

Further inspection of Figure 6.2 shows some sharp emission features on top of the broad background. The two most prominent ones are emission lines from the atoms C^+ ([CII], 157 μm) and O ([OI], 88 μm). These are the dominant cooling lines of molecular clouds. This cooling allows the clouds to contract without heating, as it would be required by the ideal gas law. Thus, the cooling is a prerequisite for star and planet formation, the subject of current astrophysical research. From this rather preliminary view of spectroscopy, very detailed data on the astrophysical environments are obtained. Apart from the atomic lines, some selective emissions from molecules are also depicted in Figure 6.2. Among those are the CO lines, which correspond to the change in the rotational state of this simple but very abundant molecule in space. The still-prominent isotopic variants of CO are also shown. More complex molecules like hydrogen cyanide (HCN), water (H_2O), and ammonia (NH_3) are other prominent molecules. With today's telescopes these species are observed not only in our astrophysical neighborhood, such as in the Orion molecular cloud (see below), but also toward the galactic center. Even maps of CO and HCN are taken from other galaxies. These observations will help to compare star formation in our Milky Way with that of more distant places in the universe. Molecules are excellent tools for these studies.

The richness of molecular spectra multiplies by a large factor when increasing the resolution in the FIR regime. This is the kingdom of radio telescopes. Spectra are recorded at a resolution of 10^{-6} or better. The spectral coverage with a telescope like the Atacama Large Millimeter Array (ALMA) ranges from 80 GHz up to almost 1 THz (i.e., 1000 GHz). Thus, the spectral information explodes, as shown by Figure 6.3, in just two small sections of the spectrum. The two windows shown span a bandwidth of ≈50 GHz, one slightly above 1 THz and the other above 500 GHz. These emission spectra, (Bergin et al. 2010) have been observed with the Heterodyne Instrument in the Far Infrared (HIFI) spectrometer onboard the Herschel Space

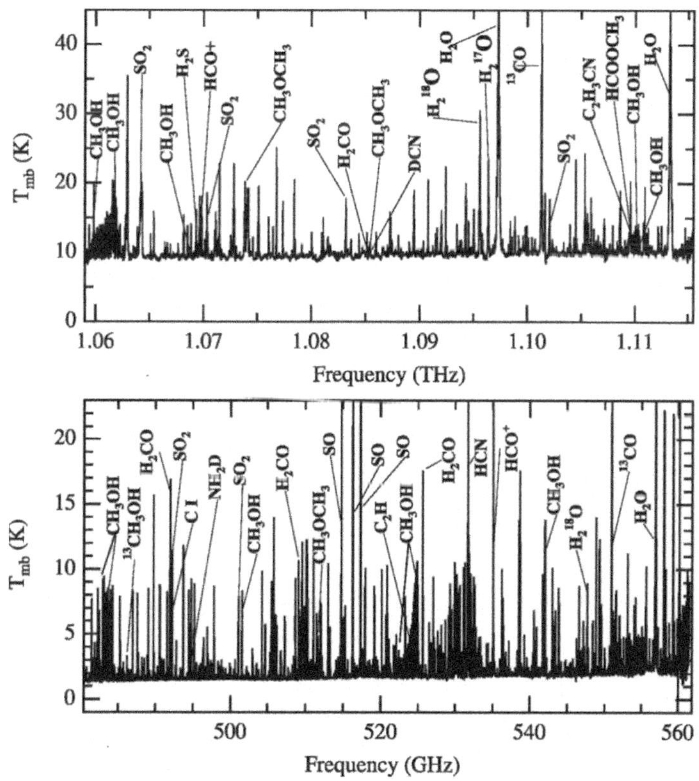

Figure 6.3. Part of the emission spectrum of the interstellar hot-core region Orion KL, observed with the HIFI spectrometer onboard the Herschel Space Observatory. Strong lines in both spectra are identified as belonging to some prominent molecules and their isotopic species. The figure is taken from Bergin et al. (2010).

Observatory. Many strong lines in both spectra are identified. They belong mostly to the rotational spectra of known molecules. As can be seen, a significant fraction of these spectral lines correspond to only a few prominent molecules, such as methanol (CH_3OH), sulfur dioxide (SO_2), formaldehyde (H_2CO), water (H_2O), or dimethyl ether (CH_3OCH_3).

Like for the Fraunhofer lines, each of these lines is indicative of a single molecule, for which the transition frequencies have to be known to high accuracy—as seen already from inspection of these rather limited spectra. Laboratory spectra for each molecule are needed for the interpretation of such spectra. It is another beauty of molecular spectroscopy that these spectra can be predicted to rather high accuracy, as the transition frequencies follow those of rather simple, yet very detailed quantum mechanical models. It is the subject of this chapter to elucidate these models and their power.

The spectral signatures of typical molecules will be introduced in this chapter, to teach the reader about the common structures of molecular spectra. Today, computer programs like the popular pgopher program (Western 2017) are used for these analyses. Such programs incorporate a lot of the physics behind the analysis that is introduced and explained in this chapter. These programs can not only be used to identify spectral features, but very detailed molecular parameters can be derived with incredible accuracy. This gives these programs a predictive power to simulate experimentally available spectra, but also to predict spectral ranges that might not be available in the laboratory. More importantly, it allows us to predict those spectra at different densities and temperatures, and to predict spectra for so-called isotopologues, i.e., molecules where the abundant main isotopes are replaced by rare isotopes of very small abundance. Examples of these species are molecules like DCN present in the spectra shown in Figure 6.3 near 1.09 THz.

Just from this simple inspection of astronomical observations in the range from the visible throughout the IR and FIR regime, molecular transitions between their electronic (UV–vis), vibrational (IR), and rotational states (FIR) are clearly prominent features of distant astrophysical objects. Understanding the observed spectra on the basis of comparing laboratory spectra to quantum mechanical models is a prerequisite to deriving astrophysical quantities for any further analysis of the astrophysical evolution, e.g., the process of star and planet formation.

Molecular spectroscopy is the subject of many very good text books; see, e.g., Bernath (1995) and Hollas (1996). It is impossible to cover the topics of all aspects of molecular spectroscopy in just one chapter. This could not even be done in another complete book. Instead, the scope of this chapter is to introduce some basic concepts of spectroscopy from a physical point of view. Basic knowledge in quantum mechanics and atomic physics, e.g., the spectroscopy of atomic hydrogen or the treatment of the harmonic oscillator, is presumed. In many instances, spectroscopy models derived for general cases can only be discussed for one example, i.e., for one molecule. For more rigorous derivations, the reader will be referred to the literature, especially when details of microwave molecular spectra (Gordy & Cook 1984; Townes & Schawlow 1975), molecular vibrations (Wilson et al. 1980), molecular symmetry (Bunker & Jensen 1998, 2005; Bishop 1993), or laser spectroscopy

(Demtröder 1996) are concerned.[1] In this chapter, some general reasoning on the basis of concepts in physics (mainly atomic physics) is presented to motivate the reader to go deeper into the details of spectroscopy by consulting the appropriate literature. The discussion begins with a simple model of a diatomic molecule. Many concepts found in this section of "spectroscopy in a nutshell" will reappear for more complex cases discussed in later sections. The chapter will conclude by showing a few selected cases where molecules have been found recently in space based on their spectra from the laboratory.

6.2 Molecular Spectroscopy in a Nutshell: Diatomic Molecules

6.2.1 Vibrational States

Molecules are highly complex systems consisting of atomic nuclei and electrons, which form the chemical bonds that put the nuclei at more or less fixed positions in space. As a result, most molecules have a clear structure that we visualize in stick and ball pictures, such as the very well-known ones for the bend H–O–H water molecule or the linear H–C ≡ N hydrogen cyanide molecule. In fact, this structure is the result of arranging the light electrons and their wave-functions around the nuclei while the heavy nuclei—more or less—stand still. This principle was first recognized by Born and Oppenheimer. It allows to separate the dynamics of the electrons from that of the nuclei. The total energy of all electrons is calculated by solving the Schrödinger equation for the electrons in the Coulomb field of the nuclei. This results in a potential energy surface (PES) that is depicted most easily for a diatomic AB molecule. In Figure 6.4, the total energy of the electrons in their lowest energy state plus the potential energy of the two nuclei is plotted as a function of the internuclear distance A–B. This so-called adiabatic potential describes the chemical bond between the atoms A and B with masses m_A and m_B, respectively.

At long distances, the two atoms are separated to (A + B) and their respective energies are given as the sum of the energies of the electrons in the respective atom. In Figure 6.4, this energy state is the $E = 0$ reference state. At shorter distances, the two atoms experience a net attraction that leads to chemical bonding. At even shorter distances, the interaction turns into a net repulsive force due to the repulsion of the electrons. As a result, the potential exhibits a minimum at the equilibrium distance R_e, as depicted in Figure 6.4. At this position, the potential energy has a corresponding well depth $-D_e$ with respect to the dissociated molecule A + B.

In the vicinity of the minimum, the potential is well-described by a harmonic oscillator potential of the generic form

$$V(R) = 1/2k(R - R_e)^2 \tag{6.1}$$

with a harmonic force field $F = -k(R - R_e)$ having k as the respective force constant. This finding leads to another common picture that shows a diatomic molecule as two masses connected via a spring with k as the respective force constant

[1] Often, it is also very instructive to consult the books of Herzberg on the various aspects of spectroscopy (Herzberg & Spinks 1945; Herzberg 1989, 1991a, 1991b).

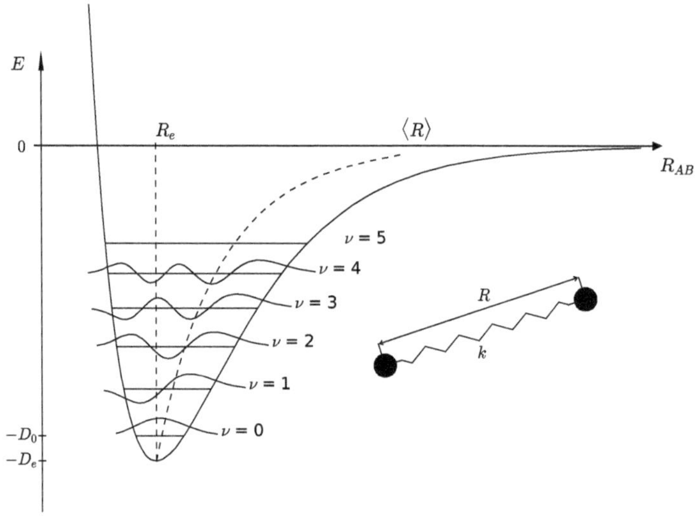

Figure 6.4. Potential energy curve of a diatomic molecule as a function of the internuclear separation. The vibrational energy levels and their respective wave functions are depicted. Near the potential minimum, the potential can be approximated by that of a harmonic oscillator with an average bond distance, $\langle R \rangle = R_e$. For higher vibrational excitation, $\langle R \rangle$ increases up to the bond dissociation. Inset: Model of a harmonic oscillator represented by two balls (nuclei) connected via a spring with force constant k.

or spring constant. Consequently, the two nuclei experience an oscillatory motion about the equilibrium distance $R = R_e$. This is the viewpoint of classical mechanics.

For the quantum mechanical harmonic oscillator, the energy eigenstates are quantized with the energies

$$E_n = \hbar\omega(n + 1/2), \tag{6.2}$$

with the quantum numbers $n = 0, 1, 2, \ldots$, as can be found in any elementary text book on quantum mechanics. The vibrational frequency is given by $\omega = \sqrt{k/\mu}$, with the force constant k as derived from the potential (Figure 6.4) and the reduced mass $\mu = \frac{m_A \cdot m_B}{m_A + m_B}$. This solution to the vibrational problem of a diatomic molecule, however, is only approximate, for a number of reasons. First of all, the true potential is anharmonic, to the extent that the two atoms can be separated, i.e., the molecule dissociates. In fact, the energy between consecutive vibrational states n and n+1 is no longer constant, as for the harmonic oscillator, $\Delta E = \hbar\omega$, but shrinks with increasing n. Thus, there is only a limited number of bound vibrational states, as expected. An improved model for the diatomic potential taking this into account is the Morse potential, as described in several textbooks

$$V(R) = D_e \cdot (1 - \exp(-\beta \cdot (R - R_e)))^2 \tag{6.3}$$

with $k = 2\beta^2 D_e$ and ω as defined above. The corresponding energy eigenstates are given by

$$E_n = \hbar\omega(n + 1/2) - \hbar\omega x_e(n + 1/2)^2. \tag{6.4}$$

Here, the additional parameter x_e is a quantity smaller than unity, which shows that the second term in Equation (6.4) can be considered as a small but considerable correction to the approximate solution of the harmonic oscillator given above. Depending on the exact potential of the diatomic molecule, additional correction terms of decreasing importance can be expressed as a power series in $(n+1/2)$, with terms in the form $x_{e,\,l}\,(n+1/2)^l$ appearing in the energy expression E_n, describing the true energy term for molecular vibration with very high accuracy as will be seen below. As a result, the vibrational state of a diatomic molecule is described very well by the quantum mechanical harmonic oscillator and some corrections. The diatomic molecule is therefore an application showcase for the harmonic oscillator in fundamental quantum mechanics.

Already in this "spectroscopy in a nutshell" description, some fundamental aspects of molecular physics and its application to astrophysics become apparent. The energy of a chemical bond is described by the bond dissociation energy, D_e. In the lowest vibrational energy state ($n = 0$), the interatomic distance cannot be exactly equal to R_e due to Heisenberg's uncertainty principle ($\Delta x \cdot \Delta p \geqslant \frac{\hbar}{2}$). Instead, a zero-point vibrational motion is associated with this state. It raises the energy of this state to about 1/2 $\hbar\omega$ with respect to the $-D_e$ minimum potential energy, and thus effectively lowers the bond dissociation energy to $-D_0$ as depicted in Figure 6.4.

In any case, this bond energy is of the same order of magnitude as the energy of electronic states in atoms, i.e., electron volt, eV. It is thus a natural energy scale for the molecular bond. Molecular vibrations have to be excited to rather high quantum numbers n before the molecule dissociates. Therefore, the natural energy scale for molecular vibration, $\hbar\omega$, is only a fraction of an eV. We will see below that the natural rotational energy scale is much lower, by a factor of 100–1000. These energy scales should be compared to thermal energies, $E = k_B T$, where k_B is the Boltzmann constant and T the temperature of the environment (laboratory or astrophysical). As a consequence, 1 eV corresponds to about 11,600 K, meaning that at ambient temperatures, molecules reside almost 100% in their electronic ground state: this is why excited states and their respective potential energy curves have not been considered in Figure 6.4. Furthermore, vibrational energies, $E_{\text{vib}} = \hbar\omega$, correspond to thousands of Kelvin in ordinary molecules. Therefore, it is often safe to assume that molecules reside in their vibrational ground state, or at least that only soft vibrations will be excited under ambient conditions. In contrast, the energy stored in molecular rotation, E_{rot}, is much smaller and therefore many rotational states will be excited in thermal environments—such as cold molecular clouds, just to refer to the coldest places where molecules are found. The ruling relation for the populations of states is the Boltzmann expression

$$P(E) \sim e^{-\frac{E}{k_B T}} \tag{6.5}$$

with E being the energy term of the electronic, vibrational, and rotational state of the molecule.

Transitions between states of molecular rotation are observed by radio telescopes operating in the range from centimeter to sub-millimeter wavelengths (λ), which are related to the frequencies (ν) by c = $\lambda \cdot \nu$. The natural energy scale for molecular rotation is therefore the unit cm^{-1}, i.e., counting the number of nodes in an electromagnetic wave over a length of one centimeter. For comparison, 1 eV, as given above as a natural energy scale for molecular bonds, amounts to 8066 cm^{-1}, and 1.4 K corresponds to 1 cm^{-1}. As discussed above, vibrational frequencies are typical for the chemical bond and for the atomic masses involved in the diatomic molecule. As an example, characteristic X–H bonds with X=C, N, or O exhibit vibrational frequencies in the 2800–3800 cm^{-1} range. As a consequence, transitions between vibrational states fall into the infrared regime, $\lambda = \frac{c}{\nu}$, i.e., around 3 μm. For the excitation of electronic transitions, radiation roughly ten times shorter in wavelength (UV) is required. These differences in energies therefore require rather different observational platforms to record molecular rotation (far-infrared), vibration (infrared), or electronic transitions (visible–UV), as well as observational strategies that will be explained later.

6.2.2 Rotational States

Very rich rotational spectra are observed with radio telescopes. Therefore, the last point in this introductory section is concerned with the energy states of molecular rotation of a diatomic or linear molecule. The Schrödinger equation describing the internal motion of the diatomic molecule AB can be written as

$$\hat{H}(R, \theta, \phi)\Psi(R, \theta, \phi) = E\Psi(R, \theta, \phi) \tag{6.6}$$

with

$$\hat{H}(R, \theta, \phi) = \hat{T}(R, \theta, \phi) + V(R). \tag{6.7}$$

Here, the Hamilton operator $\hat{H}$ is split into a kinetic energy operator $\hat{T}$ and the potential V(R). The latter is the potential energy, which only depends on the internuclear separation A–B as shown in Figure 6.4. This potential does not depend on the relative orientation of the molecule in space, which is described by the polar coordinates θ and ϕ. Therefore, the Schrödinger equation of the diatomic molecule is that of a centrosymmetric problem, which is conceptually identical to that of the hydrogen atom, except in that the interaction potential there is the Coulomb potential, and here it is that of an anharmonic oscillator as discussed above. Thanks to this similarity, we do not need to derive the solution explicitly; we already made use of this when discussing the solution to the radial problem, the anharmonic oscillator. The solution to the angular part of the Schrödinger equation will, however, yield the same spherical harmonics for both cases as solutions. We therefore refer the reader to any textbook on quantum mechanics to look up those solutions for the angular part of the electronic state of the hydrogen atom. A closer inspection of the angular part of the kinetic energy operator reveals the difference in the energy of the rotational states of the hydrogen atom and the diatomic molecule.

$$\hat{T}(\theta, \phi) = \frac{\hat{J}^2}{2I} \tag{6.8}$$

Here, $\hat{J}$ is the operator for the angular momentum of the rotating molecule, and I is the moment of inertia of the diatomic molecule. The expectation value of the $\hat{J}^2$ operator is simply given by

$$\left\langle \hat{J}^2 \right\rangle = J \cdot (J + 1)\hbar^2 \tag{6.9}$$

which does not differ for the two cases discussed. The moment of inertia, $I = \mu \cdot R^2$, however, is vastly different. In case of the hydrogen atom, it is the mass of the electron circulating the proton.[2] Here, it is the reduced mass of the AB molecule as given above.[3] R is the internuclear distance which can be approximated to high accuracy as R_e. When introducing a new constant

$$B = \frac{\hbar^2}{2I} \tag{6.10}$$

with $I = \mu \cdot R_e^2$ and $\mu = \frac{m_A \cdot m_B}{m_A + m_B}$, the energy eigenstate of molecular rotation may be written as

$$E_{\text{rot}} = BJ \cdot (J + 1). \tag{6.11}$$

Calculating values for μ and R_e then leads to values for B in the order of a cm^{-1}.

Thus, without any cumbersome derivation, the energy states of diatomic molecules are easily obtained by translating the results from an introductory quantum mechanics course. In this chapter, we will continue the discussion on the basis of these results and extend the solutions to more complex situations, as the line of reasoning follows the ideas already presented for the molecular vibration where, e.g., deviations from the simple harmonic oscillator are introduced in a qualitative fashion in order to help understanding the physical concepts. For a more rigorous mathematical description, the reader is referred to standard literature on the subject that goes beyond the fundamental quantum mechanics description.

In order to complete the "spectroscopy in a nutshell," transitions between molecular energy states are treated before more details on the rotation and vibration of molecules are given. This will allow us to immediately compare the results of this treatment with experimental data from the laboratory and space. Molecular spectroscopy describes the interaction of the molecule with a photon. In fact, the absorption or emission of a photon may be described as a reaction just like a chemical reaction or a reaction in nuclear physics (e.g., β-decay). For the diatomic molecule AB, discussed in this section, the reaction reads

$$AB(n'', J'') + h\nu \rightarrow AB(n', J') \tag{6.12}$$

for absorption and

---

[2] More precisely, it is given by the electron and proton circulating about their center of mass, i.e., the reduced mass of the system circulating the center of mass.

[3] The two-body problem is split into the translation of the center of mass (CM) with total mass $m_A + m_B$ and the relative motion of the two atoms with the reduced mass, the internuclear separation R, and the angles θ and ϕ in the reference frame of the CM.

$$AB(n'', J'') \rightarrow AB(n', J') + h\nu \tag{6.13}$$

for the emission of a photon with energy, hν. Here, n and J describe the vibrational and rotational quantum numbers of the molecules before ($''$) and after ($'$) the collision with the photon. Writing such processes as reactions implies that conservation laws have to be respected in the course of the annihilation or creation of a photon. First, the total energy before and after the collision is conserved, meaning that energy for the excitation of the molecule is provided by the photon or the photon created is carrying away the energy release of the process. As a consequence of this, the transitions described by Equations (6.13) and (6.12) occur at sharp energies, which are observed as narrow lines in a frequency spectrum. Momentum is also conserved; this is not of concern in this chapter, but is relevant for laser cooling of atoms, etc. As the molecule carries an angular momentum, so too does the photon: it carries an angular momentum of one times the elementary angular momentum, i.e., $1\hbar$. As a result, the angular momentum of the molecule has to change by $\pm\hbar$. This simply means that the angular momentum quantum number J changes by ±1. This fundamental result reduces the number of energetically possible transitions, and is therefore called the selection rule.

Figure 6.5 shows the energy-term diagram for a diatomic molecule with rotational constant B. The levels follow the quadratic behavior $E_{\rm rot} = BJ \cdot (J + 1) \sim J^2$. Arrows pointing downward depict transitions to be observed in emission with $\Delta J = -1$. Those pointing upward correspond to transitions observed in absorption with $\Delta J = +1$. These are the rotational transitions to be expected in a spectrum of a diatomic—or

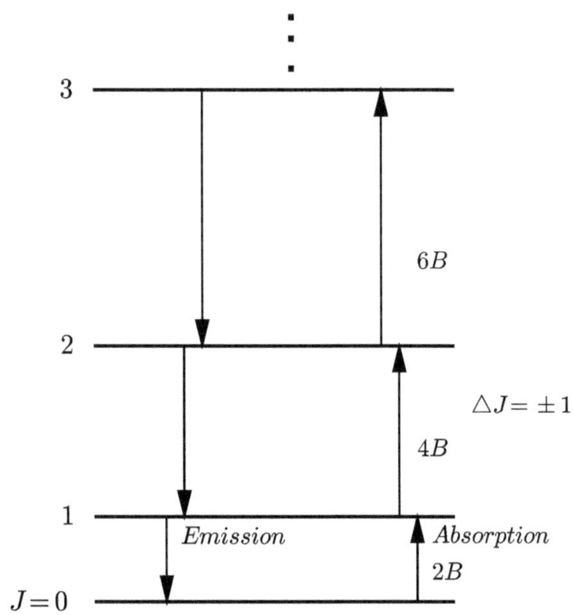

Figure 6.5. Energy level diagram for a diatomic (linear) rotor. Transitions observed in emission (absorption) are indicated by arrows pointing down (up). Due to the quadratic dependence of the rotational quantum number J, the energy spacings of allowed transitions follow a harmonic series, $2B \cdot (J + 1)$.

more general, linear—molecule. Calculating the energy differences for the $\Delta J = \pm 1$ transitions reveals the lengths of the arrows to be

$$\Delta E_{\text{rot}} = 2B(J + 1) = h\nu. \tag{6.14}$$

Therefore, the energy spacings follow a harmonic progression, $2B$, $4B$, etc, which should be visible in a spectrum.

There are several computer programs, with different levels of sophistication, that can be used to simulate and also fit molecular spectra. These can be used to predict spectra in order to compare to laboratory spectra or astrophysical observations. Pgopher is an extremely user-friendly and therefore very popular program (Western 2017). It is easy to operate and yet allows the user to simulate rather complex spectra, as we will see later on. Here, we use this program to just show how well this "spectroscopy in a nutshell" section has prepared you to interpret your first spectrum. The example presented in Figure 6.6 shows the spectrum of CO, the second-most abundant molecule in space after hydrogen, H_2. The upper part shows a stick diagram

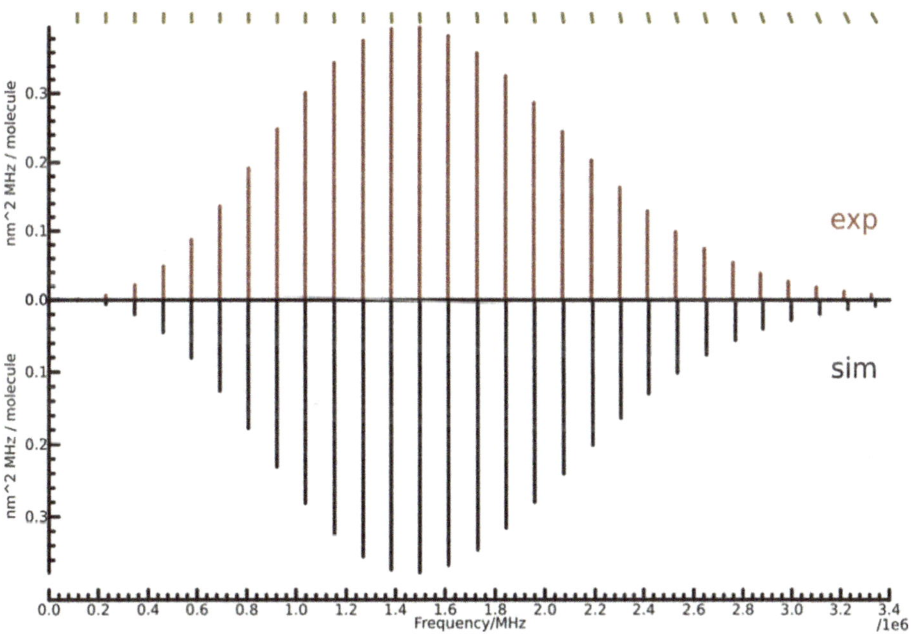

Figure 6.6. Rotational spectrum of CO at $T = 300$ K in the range 0–3.4 THz as output of the publicly available pgopher program (Western 2017). The top series of lines represents the best-fit simulated spectrum for the rotational transitions $J + 1 \leftarrow J$. The two transitions in the maximum correspond to the $12 \leftarrow 11$ and $13 \leftarrow 12$ transitions. The top series does not deviate from the experimental data (on the scale of this plot), therefore it is marked as experimental. The lower series of lines is a fit to the upper series with just the rotational constant B as one parameter, marked as simulation. Above 1 THz, frequency deviations become apparent. Using the centrifugal distortion constant D as described in the text, the resulting spectrum is very close to the top series. This shows how well the spectrum is already characterized by the rigid rotor model (lower series) and how well corrections treated in this section lead to an (almost) perfect fit. This behavior is also observed for spectra of non-linear, more complex molecules discussed later.

indicating the resonant transition frequencies observed in the range 0-3400 GHz or 0 to 113 cm^{-1}.[4] The lower part is a stick diagram of a predicted spectrum on the basis of a given B constant, $B = 57.635968$ GHz, as taken from the Cologne Data Base for Molecular Spectroscopy (CDMS). The spectrum consists of many transitions, which are about equally spaced by $2B$ as expected from our derivation with the fundamental transition from $J = 0$ to 1 at around 115 GHz. The experimental resolution in laboratory experiments is so high that errors in the frequency are not visible on the scale of the plot in Figure 6.6. A comparison between the upper and lower panel shows that the harmonic progression of $2B$ in the spectrum is a reasonable description of the experimental values. However, all transition frequencies lie below the expected value with increasing deviations for transitions of rotationally excited states. Like for the discussion of the anharmonicity for the molecular vibration, there are systematic deviations from the simple rigid rotor model with just the B constant.

6.2.3 The Non-rigid Rotor

In our current, simple interpretation, the rotational constant is just inversely proportional to the moment of inertia of the diatomic molecule at the equilibrium distance R_e; see Equation (6.10). This picture is rationalized by the much faster vibrational motion (higher in energy) compared to the slow rotational motion where the distance between the two nuclei is approximated by R_e, or more precisely, where it should be approximated by the expectation value of R, $\langle R \rangle$, loosely speaking—i.e., the quantum mechanical average. The value of the latter will change due to the anharmonicity of the potential. See how $\langle R \rangle$ increases for higher energies as depicted in Figure 6.4. This is the reason why the rotational constant decreases with the vibrational quantum number n, as will be discussed later.

More importantly, the internuclear distance is not constant, but instead like two balls connected via a spring[5] spinning at higher rotational speeds, i.e., at higher rotational quantum numbers J, the distance will increase due to a balance between the centrifugal force acting outward and the restoring spring force.

$$F_c = -\frac{d}{dR}\frac{\hat{J}^2}{2\mu R^2} = \frac{2\hat{J}^2}{2\mu R^3} = k(R - R_e) = kR_e \cdot x = F_{\text{osc}} \tag{6.15}$$

Here, $x = (R - R_e)/R_e$ is a small correction between $\langle R \rangle$ and R_e, and calculating its value for F_c at $\langle R \rangle = R_e$ from this equation the original expression for E_{rot} thus modifies to

$$E_{\text{rot}} = \frac{\langle \hat{J}^2 \rangle}{2\mu R^2} + \frac{1}{2}k(R - R_e)^2 = BJ \cdot (J + 1) \cdot \frac{1}{(1 + x)^2} + \frac{1}{2}kR_e^2 \cdot x^2. \tag{6.16}$$

[4] 1 cm^{-1} is approximately equivalent to a frequency of 30 GHz.
[5] This was the model picture for the interaction described in Figure 6.4.

Note that the vibrational energy $\frac{1}{2}k(R - R_e)^2$ also has to be accounted for in E_{rot}, as E_{vib} is no longer a constant but instead depends on R or x, which is to say that there is a coupling of rotation and vibration. Because x is much smaller than unity, Equation (6.16) may be expanded in a power series of x for the first part of the Equation, to obtain

$$E_{\text{rot}} = BJ \cdot (J + 1)(1 - 2x + 3x^2 + \ldots) + \frac{1}{2}kR_e^2 \cdot x^2. \tag{6.17}$$

Solving the left part of Equation (6.15) for

$$x = \frac{2\hat{J}^2}{2k\mu R_e^4} = 2J \cdot (J + 1)\frac{B^2}{\omega^2}. \tag{6.18}$$

the result can be used in Equation (6.17) to replace x. When dropping higher-order terms in x, the final expansion gives

$$\begin{aligned} E_{\text{rot}} &= BJ \cdot (J + 1)\left(1 - 4\frac{B^2}{\omega^2}J \cdot (J + 1) + \ldots\right) \\ &= BJ \cdot (J + 1) - DJ^2 \cdot (J + 1)^2 + \ldots \end{aligned} \tag{6.19}$$

Due to the nature of the correction, the new rotational constant D is called the centrifugal distortion constant. According to its derivation, D, B, and ω, the vibrational frequency,[6] are related by

$$D = 4B\frac{B^2}{\omega^2}. \tag{6.20}$$

In fact, the correction to B is just given by the factor $4\frac{B^2}{\omega^2}$. Recalling that B is about 100–1000 times smaller than ω, D is smaller than B by many orders of magnitude. Only when J is rather large does the correction to the term energy become significant. We find this exactly by inspection of the experimental spectrum of CO and its prediction using just B. In order to check the validity of the relation between B, D, and ω, Equation (6.20) is solved for ω:

$$\omega = 2B\sqrt{\frac{B}{D}}. \tag{6.21}$$

Using the rotational parameters $B = 1.93128087\ \text{cm}^{-1}$ and $D = 6.12147 \times 10^{-6}\ \text{cm}^{-1}$ from another valuable source of molecular information, the NIST Chemistry WebBook (Huber & Herzberg 2000), the vibrational frequency is calculated to be $\omega = 2169.553135\ \text{cm}^{-1}$, which deviates from the listed value of $\omega = 2169.81358\ \text{cm}^{-1}$ by only about 10^{-4}.

[6] D, B, and ω are all expressed as energies, such that ratios are unitless for simplicity. This also holds for Equation (6.18).

Despite this very gratifying result, higher-order distortion constants can be considered when even higher rotational corrections become detectable. The upper panel of Figure 6.6 shows the predicted spectrum for CO (termed as experimental as explained in the figure caption), including the first- and second-order centrifugal correction terms. Discrepancies between the observed and simulated spectrum reach deviations scattering around zero in the range of statistical experimental errors. Again, the lower part of Figure 6.6 shows the high-quality prediction of the rigid rotor model, where R_e (and thus the moment of inertia) is assumed to be constant. Furthermore, the harmonic oscillator described above is a reasonable approximation to the energy spectrum of molecules like CO.

6.2.4 Intensities, Populations, and Transition Probabilities

Before closing the "spectroscopy in a nutshell" section, let us consider intensities, as they carry fundamental information—especially in cases where intensities in astrophysical observations are used to retrieve the number of molecules that emit the detected radiation. In the spirit of this chapter, a rough reasoning of the intensities displayed in Figure 6.6 is given. As discussed in textbooks on atomic physics and quantum mechanics, intensities of dipole allowed transitions are given by the product of the population of the respective state from which the transition starts and the transition probability between the two states of consideration. Selection rules state which transitions are possible, but do not yield their intensities. The level population is given by the Boltzmann relation of Equation (6.5). This basic law of physics leads to the result that intensities for transitions of increasing energy level off exponentially. This is readily observable in the rotational spectrum of CO in Figure 6.6 for higher-energy transitions, i.e., at higher frequencies.

For degenerate states, i.e., states with different quantum numbers but the same energy term, the population given by Equation (6.5) has to be multiplied by the degeneracy, g, i.e., the number of elementary states with this energy, in order to evaluate the population. For the rotating molecules, so far, we have only considered the rotational quantum number J, which determined the energy of the rotor. However, for each rotational state J, there are $g_J = 2J + 1$ individual orientations of the angular momentum vector $\hat{J}$ in space, each orientation described by the additional quantum number $M = m_J$.

This general property of a quantum mechanical angular momentum should be well-known from the orbital angular momentum, $\hat{L}$, in the hydrogen atom, a vector of well-defined length. We say l is a "good" quantum number. Of this vector, $\hat{L}$, there is also one component, $\hat{L}_z$, that has a well-defined value, $m_l \cdot \hbar$. This component, and thus the m_l quantum number, can take values between $-l$ and $+l$, which are $2l+1$ values. In the same line of reasoning for the diatomic molecule, the degeneracy g_J increases linearly with growing J. Therefore, the intensities in the rotational spectrum first increase, starting from the non-degenerate ground state $J = 0$, before the exponential behavior of the Boltzmann population dominates. In total, the overall behavior of the line intensities depicted for CO in Figure 6.6 is well-characterized by these points of discussion.

In contrast to the influence of the level population, the transition probabilities do not seem to play a very important role. However, they do have a role to play, and their values are given by the well-known Einstein coefficients for emission, A, and for absorption, B.[7] A describes the rate at which an excited state decays, and the product of B $\cdot$ $\rho(\nu)$ describes the rate of absorption when the molecules are subject to a radiation field with a spectral energy density $\rho(\nu)$. The latter is related to the intensity, $I(\nu)$, of a radiation source by $\rho(\nu) = I(\nu)/c$, where c is the speed of light. The Einstein B coefficient is proportional to the square of the matrix element of the transition dipole moment, which is defined as

$$\mu_{if} = \langle \Psi_f | \hat{\mu} | \Psi_i \rangle. \tag{6.22}$$

Here, Ψ_i and Ψ_f are the wavefunctions of the initial and final state of the transition. For the angular part, i.e., the rotational problem in the coordinates θ and ϕ, the wavefunctions are given by the spherical harmonics Y_{JM} as they are for the angular part in the hydrogen atom. These functions form an orthonormal basis set of functions for the molecular rotation. Furthermore, they follow the addition theorem, which means a product of two Y_{JM} yields a sum of other $Y_{J'M'}$. The dipole moment operator $\hat{\mu}$ is proportional to the position operator. Therefore, it can be expressed as a linear combination of $Y_{1M'}$ with $M' = -1, 0, 1$. As a result of these properties of the spherical harmonics and the nature of the dipole moment operator, the matrix element of the transition dipole moment leads to only a few contributions. The reader is referred to textbooks on quantum mechanics for the proper calculation of the values. As a result, integrals of most triple products in Equation (6.22) vanish and only those with $J' - J' = \pm 1$ contribute to the matrix element. Closer inspection shows that this finding is identical to the selection rule $\Delta J = \pm 1$ as derived from the conservation law for the angular momentum in a collision of a molecule with a photon. From now on, we shall assume that the dipole moment matrix elements are calculated and used in programs like the pgopher program to simulate the spectra.

In summary, the Einstein B coefficient is simply given by Demtröder (1996)

$$\mathrm{B}_{if} = \frac{\pi \mu_{if}^2}{3\epsilon_0 \hbar^2} \tag{6.23}$$

where ϵ_0 is the permittivity of vacuum. It only depends on the matrix element μ_{if}. In a thermal radiation environment, the spectral energy density $\rho(\nu)$ is given by Planck's radiation law. Emission and absorption of light will not change the level population from the Boltzmann distribution of Equation (6.5). Thus, the rate of absorption and emission are in a steady state, which puts the Einstein A coefficient in relation to the Einstein B coefficient

$$\mathrm{A} = \mathrm{B}\frac{8\pi h \nu^3}{c^3}. \tag{6.24}$$

[7] The Einstein B coefficient and the rotational constant B should not be confused.

The spontaneous emission rate therefore depends not only on the matrix element μ_{if}, but also very sensitively on the transition frequency ν. Again, it is not the purpose of this section to present rigorous derivations, but rather to point at some key aspects of spectroscopy that become apparent even without detailed mathematical derivations. In fact, Planck's radiation law results from a thermal environment that is derived in detail in respective text books.[8] Here, we just need to obtain a rough understanding of the dependencies of the Einstein coefficients on properties of the molecule (μ_{if}) and the associated radiation field. Throughout the next sections, only selection rules are discussed; in combination with populations, they are sufficient to interpret the shown spectra. Accurate intensities are, however, needed when it comes to deriving state populations and thus temperatures from observed spectra. In Figure 6.6, the intensities of the upper spectrum represent the absorption line strengths as given by the CDMS catalog; in the lower part, the spectrum simulated with pgopher uses the same intensity units. Apart from the differences in line positions discussed above, the spectra agree in line positions and intensities to a very high degree. This demonstrates that spectroscopy is a very powerful tool to unravel the physical conditions (temperature and number density) as well as the chemical composition of near and distant places in the universe.

6.3 Laboratory Rotational Absorption Spectroscopy

6.3.1 Experimental Setup

Before any molecules can be identified unambiguously by observation of rotational lines with radio telescopes, their spectra have to be recorded with sufficient spectral resolution and accuracy in the laboratory. There are many different ways to record such spectra. Throughout this chapter, some widely used methods will be briefly introduced. The techniques used are largely dependent on the desired frequency range. Rotational spectroscopy lies in the far-infrared regime where the ALMA telescope presents a very wide coverage, namely from 80 GHz up to almost 1 THz. For this reason, laboratory instruments should cover a similar range. We have seen the very high-quality predictions of rotational spectra of the diatomic CO molecule just using the B rotational constant. Recording a few lines in the series displayed in Figure 6.6 would be sufficient to predict all other lines with high precision. This finding is also true for many spectra of more complex molecules. Despite the marvelous predictions of molecular spectra just from fractions of the complete spectrum, a wide frequency coverage is desirable.

Absorption spectrometer like the one depicted in Figure 6.7 are used in many laboratories to record such spectra. The molecules of interest are kept in the central gas cell shown at the top of the figure. Light from the radiation source, shown at the upper right part, is directed toward the cell via a standard gain antenna, passes through a polarizer grid, is focused by a lens, and then passes through the cell window. At the other end of the cell, the beam is redirected using a rooftop-shaped

[8] Einstein introduced the process of induced emission in addition to spontaneous emission as the counterpart of absorption. Only this leads to a proper derivation of Planck's radiation law.

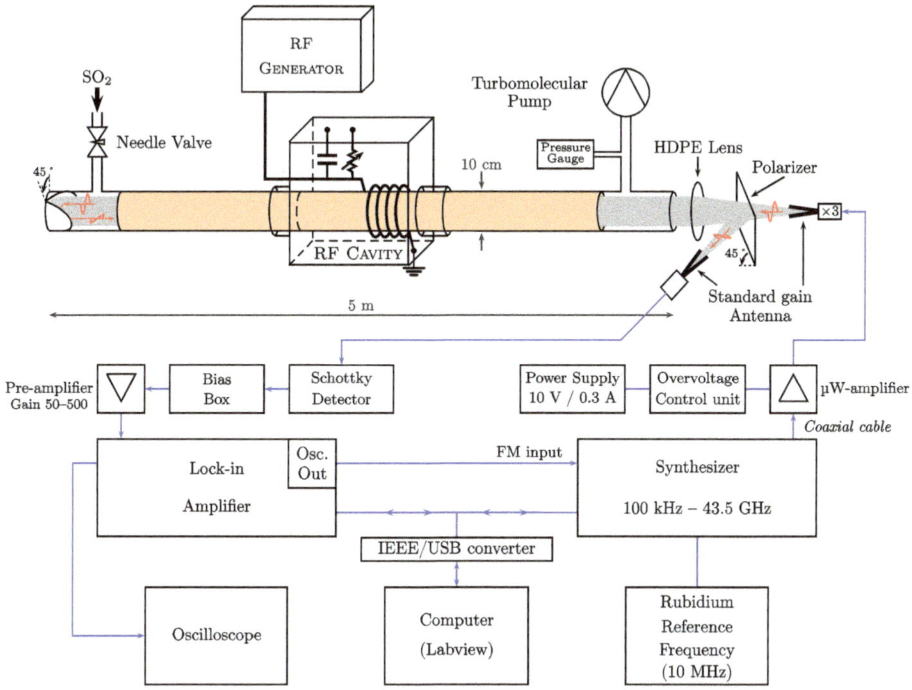

Figure 6.7. Schematic diagram of a traditional absorption spectrometer operating in the range 0.1–2 THz. The upper part shows a glass cell under vacuum containing a sample of molecules at pressures below 10^{-2} mbar. The lower part displays the frequency modulation (FM) electronics for improved instrument sensitivity. Details are given in the text. A radio-frequency discharge source on the glass cell is used in some experiments to create unstable molecules or synthesize molecules of astrophysical interest. More details on these aspects can be found in a current publication (Martin-Drumel et al. 2015), from which this figure has been reproduced.

mirror that also changes the direction of polarization of the light. After leaving the cell, the radiation is reflected by the polarizer grid because of the earlier rotation of the polarization. Past this reflection, the beam is collimated onto a detector, which in most cases is a Schottky diode. This diode rectifies the electrical field of the radiation, leading to an electrical signal that can be amplified and recorded using an analog-to-digital conversion. Very widely tunable and narrow bandwidth ($\Delta\nu/\nu < 10^{-11}$) radiation is used for these experiments. Upon scanning the frequency of radiation, transition frequencies are hit in this process and the radiation level hitting the detector is reduced significantly. This is the general concept of the absorption spectrometer, described here in brief.

One practical limitation of this approach is to discriminate a very small signal change due to molecular absorption on a tremendous "background," similar to seeing a fly when looking into the Sun. The sensitivity of this instrument is enlarged by many orders of magnitude using the so-called frequency modulation (FM) technique, the components of which are shown schematically in the lower part of Figure 6.7. In this FM approach, an on/off resonance signal is created by a periodic frequency modulation (frequency typically f_{mod} ~10 kHz). The sine wave used for this modulation is created in the lock-in amplifier and used to wobble the frequency (f_{RF}) of a synthesizer, which is the oscillator for the whole experiment. This signal of the modulated frequency (f_{RF}) is amplified and multiplied by a factor of 3 in an active multiplier chain (AMC), which creates a very stable radiation output level at the multiplied frequency ($3 \cdot f_{RF}$). Changing the multiplier and thus the multiplication factor allows the desired frequency range up to 2 THz to be covered with just this one instrument.

As a result of the modulation, the signal intensity changes over time in a sinusoidal fashion, alternating, e.g., between a lower and a higher signal level, assuming we are at the edge of a molecular line. In the lock-in amplifier, this periodic signal is demodulated such that just the difference between the high and low levels of signal is determined. The demodulation at the same frequency as the modulation (1f-demodulation) results in a signal that follows the first derivative (slope) of the line profile when the center frequency (synthesizer) is scanned slowly across the absorption line. If the line is observed on top of a gradual change in intensity, e.g., due to a standing wave in the instrument, the line would be shifted due to the way the difference signal is detected. In order to avoid such systematic errors, the modulated signal is demodulated with a signal of twice the frequency (2f-demodulation). Doubling the frequency also means that an intermediate signal level between the lower and the higher signal level is considered by the lock-in amplifier. In fact, it forms the difference of the differences between intermediate and low signal level versus that of the high and intermediate signal level. It thus determines the slope of the slope of the line profile that corresponds to the second derivative of the original absorption profile. These signals are intrinsically immune against drifts in the background and therefore result in the most reliable center frequencies, in which we are interested foremost in order to use these values for the model description of transition frequencies discussed in the previous section. Apart from the immunity against background drifts, the modulation detection yields very high signal-to-noise (S/N) ratios because the signal is only detected in a very narrow frequency window around f_{mod}, and therefore any noise contributions from other frequencies (in particular, low-frequency drifts, etc.) are perfectly rejected.

The result of such an absorption experiments is depicted in Figure 6.8, where one rotational line of a linear rotor, OCS, is shown. For this linear molecule, the $J = 7 \leftarrow 6$ transition is detected near 85.1391 GHz. Using the simple $2B(J + 1)$ formula for these transitions, B is determined as 6.08136 GHz. The CDMS value is 6.08149 GHz, deviating only by 2×10^{-4} just from this coarse reading. This demonstrates the power of rotational spectroscopy in identifying molecules from their fingerprint-like transitions. More can be learned from this measurement. The measured line has a line width on the order of 200–300 kHz. The experiments were conducted at low pressures, and therefore the line width should be dominated by the Doppler effect of all the absorbing molecules, which travel at different speeds with respect to the direction of the radiation exciting this transition. In fact, the velocity distribution of the thermal ensemble of molecules follows the Maxwell–Boltzmann distribution $f(v_z)$,

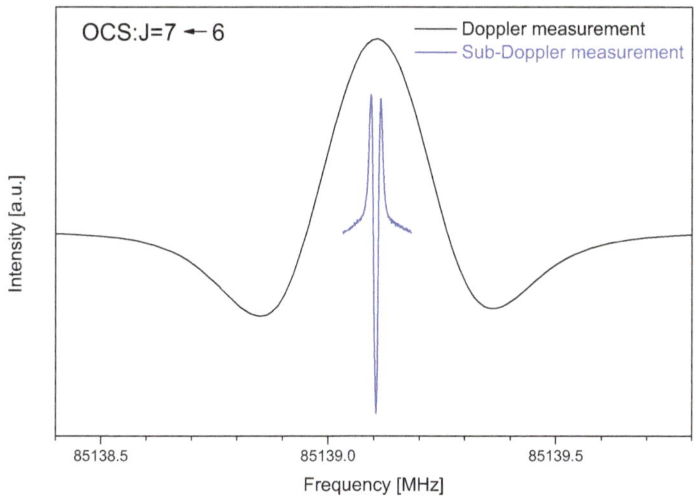

Figure 6.8. Rotational transition line of the $J = 7 \leftarrow 6$ transition of OCS near 85,139.1 GHz. The line shape looks like the second derivative of a bell-shape function. Its width is about 200–300 kHz and corresponds to the Doppler shift of molecules traveling at different speeds with regard to the direction of the exciting radiation in the cell depicted in Figure 6.7. The center frequency of this transition can be determined to an accuracy beyond 100 kHz. Therefore, the relative line position is determined to about 10^{-6}. As a second trace, the same line is recorded using saturation spectroscopy (Lamb-dip), which allows the Doppler width to be reduced by more than one order of magnitude. Therefore, transition frequencies of rotational lines can routinely be determined to a level better than 10^{-7}. (Figure courtesy of Oliver Zingsheim and Jakob Maßen.)

$$f(v_z) \sim e^{\frac{-mv_z^2}{2k_BT}} = e^{-\frac{v_z^2}{v_p^2}}, \tag{6.25}$$

Here, v_z is the velocity of the molecule along the direction of propagation of the radiation, which is given by the wave vector $\vec{k}$, and $v_p = \frac{2k_BT}{m}$ is the most probable velocity. Due to the Doppler effect, the transition frequency for absorption ω_a is thus shifted according to

$$\omega_a = \omega_0 + \vec{k} \cdot \vec{v} = \omega_0(1 + v_z/c). \tag{6.26}$$

Combining these two relations leads to the Doppler profile

$$I(\omega) = I_0 e^{-\frac{(\omega-\omega_0)^2c^2}{\omega_0^2 v_p^2}} \tag{6.27}$$

which is simply a bell-shaped function with the center frequency ω_0 and a full width at half maximum (FWHM) given by

$$\Delta\omega = 2\sqrt{ln2}\,\omega_0\frac{v_p}{c} = \frac{\omega_0}{c}\sqrt{8k_BT\ \ln 2/m}. \tag{6.28}$$

Due to the FM modulation technique, the actual line shape is not the Gaussian from the velocity distribution, but rather its second derivative. Of course, the line width will be modified by this process. However, the FWHM of the observed line in

Figure 6.8 can be estimated by Equation (6.28) to be 160 kHz, which is in the right order of magnitude. Closer inspection of Equation (6.28) shows that the line width sensitively depends on the mass of the molecule and the temperature of the environment. Thus, it is clear that line profiles like the one shown here can be used to analyze remote sensing data for the kinetic temperature of the ambient gas.

In Figure 6.8, a second trace shows a much narrower line of the same transition. This measurement was done with the same setup, but at even lower pressures in order to avoid any pressure-broadening effect, which occurs when the transition lifetime is shortened due to collisions with the other molecules and the corresponding effective wavelet of absorbed radiation is shortened in time. As a consequence, the frequency spectrum of such a transition becomes broader. In contrast to this broadening effect, the molecular line shown as a second trace is much more narrow. This narrow feature arises from a saturation effect in the absorption cell. On the way through the cell, many molecules are excited and therefore could not be excited by another beam passing the same way. Thus, the sample is bleached to some extent. This second beam is that which is reflected by the rooftop-shaped mirror. However, the reflected beam is probing different velocity components of the velocity distribution, therefore exciting molecules not bleached by the incoming beam. However, for molecules with $v_z = 0$, both beams probe the same ensemble because they are counter-propagating. Therefore, the absorption is reduced at the center frequency, which creates the so-called Lamb-Dip seen as the second trace in Figure 6.8. Apparently, the line width is much narrower than the Doppler width and the line center can be determined with even better precision. This Lamb-Dip technique is particularly useful when a Doppler-limited absorption feature is comprised of several overlapping lines with only small splitting, one example being hyperfine splitting when, e.g., the angular momentum of a nucleus interacts with molecular rotational angular momentum. Such features might be resolved or remain unresolved in astrophysical observations. Still, when finer details are available from the laboratory, as depicted here, a more reliable simulation of blended lines for astrophysical interpretations becomes possible by these laboratory means.

6.4 The Symmetric Rotor

6.4.1 Energy Terms and Spectra

Most molecules of interest have a more complex structure than diatomics like hydrogen (H_2) or carbon monoxide (CO). Molecules like water (H_2O) and ammonia (NH_3), as well as more complex species like methanol (CH_3OH) or dimethyl ether (CH_3OCH_3), are among the most prominent molecules found in space. Due to the specific bonds, each molecule has very characteristic molecular vibrations, which will be discussed later. Due to the extension in all directions of space, the moment of inertia is no longer a scalar as it appeared for the diatomics or linear molecules. Instead, the moment of inertia is a tensor, which means—from the perspective of classical mechanics—that the angular momentum vector, $\vec{J}$, no longer needs to point in the direction of angular rotation, $\vec{\omega}$. The relation is given by

$$\vec{J} = I\vec{\omega} \tag{6.29}$$

where I is the moment of inertia tensor. Any molecular rotor therefore has three principal axes with three moments of inertia associated. As a result, the simple Hamiltonian for the molecular rotation of a linear molecule given by Equation (6.8) has to be expanded for all components of the angular momentum

$$\hat{H} = \frac{\hat{J}_X^{\,2}}{2I_{XX}} + \frac{\hat{J}_Y^{\,2}}{2I_{YY}} + \frac{\hat{J}_Z^{\,2}}{2I_{ZZ}}. \tag{6.30}$$

The solution of this problem depends on the three principal values of the tensor, I_{XX}, I_{YY}, and I_{ZZ}. It is useful to sequence these three components in rising order, naming them $I_A \leqslant I_B \leqslant I_C$ for convention. In fact, it turns out that the case of the linear molecule is a special case of Equation (6.30) with $I_A = 0$, as all the mass of the heavy nuclei is concentrated on the molecular axis and $I_B = I_C$. As a result, Equation (6.30) collapses to Equation (6.8)

$$\hat{H} = \hat{T}(\theta, \phi) = \frac{\hat{J}^2}{2I} \tag{6.31}$$

with $\hat{J}^2 = \hat{J}_X^{\,2} + \hat{J}_Y^{\,2}$ and $I = I_B$.

Together with the case of the linear molecule, five cases of molecular rotation are distinguished based on the relation between $I_A \leqslant I_B \leqslant I_C$ and summarized along with examples in Table 6.1.

In this section, the symmetric rotor will be described in more detail. It will also form the basis for explaining the most complex, asymmetric rotor case. As it is found in many textbooks, the discussion starts with the quantum mechanical Hamiltonian operator Equation (6.30) for $I_A < I_B = I_C$ (prolate case)

$$\hat{H} = \frac{\hat{J}_Z^{\,2}}{2I_A} + \frac{\hat{J}_X^{\,2} + \hat{J}_Y^{\,2}}{2I_B} \tag{6.32}$$

Table 6.1. Classification of the Molecular Rotors by the Relation of their Moments of Inertia

Rotational Constants	Type of rotor	Examples
$I_A = 0;\ I_B = I_C$	Linear rotor	CO
		OCS
		N_2O
$I_A < I_B = I_C$	Prolate symmetric top	CH_3CN
$I_A = I_B < I_C$	Oblate symmetric top	BF_3
		H_3^+
		CH_3^+
$I_A = I_B = I_C$	Spherical top	CH_4, SF_6
$I_A < I_B < I_C$	Asymmetric top	H_2O
		CD_2H^+
		CH_3OCH_3

Here, the Hamiltonian is expressed in the frame (coordinate system) of the molecule where the molecular axis is the Z-axis with the associated angular momentum component $\hat{J}_Z$. Likewise, the two other components are given as $\hat{J}_X$ and $\hat{J}_Y$, respectively. The total angular momentum, and therefore also $\hat{J}^2 = \hat{J}_X{}^2 + \hat{J}_Y{}^2 + \hat{J}_Z{}^2$, are "good" quantum numbers and thus conserved in any reference frame, which here is the frame given by the principle axis of the molecule. Expanding the second part of the right-hand side of Equation (6.32) to contain $\hat{J}^2$, the term $\frac{\hat{J}_Z{}^2}{2I_B}$ is added and subtracted to obtain

$$\hat{H} = \frac{\hat{J}_X{}^2 + \hat{J}_Y{}^2 + \hat{J}_Z{}^2}{2I_B} + \hat{J}_Z{}^2\left(\frac{1}{2I_A} - \frac{1}{2I_B}\right) \tag{6.33}$$

$$\hat{H} = B\hat{J}^2 + (A - B)\hat{J}_Z{}^2 \tag{6.34}$$

with A and B representing the rotational constants $\frac{\hbar^2}{2I_A}$ and $\frac{\hbar^2}{2I_B}$, respectively. Those constants are the constants in the picture of a rigid rotor with chemical bonds at the equilibrium configuration, i.e., fixed bond length and angles. Using this Hamiltonian, the Schrödinger equation can be solved easily as it is derived in many textbooks, to yield the rotational energy

$$E_{\rm rot} = BJ \cdot (J + 1) + (A - B)K^2. \tag{6.35}$$

Here, J is the quantum number for the magnitude of the angular momentum vector as discussed before, and K is the quantum number for the projection of the $\hat{J}$ vector onto the molecular axis. This is very much in line with the discussion of the magnetic quantum number m_l, which is the projection quantum number of the angular momentum vector $\hat{L}$, except it is the projection onto the z-axis of the laboratory frame. As m_l can take $2l + 1$ values, K can take $2J + 1$ values ranging from $-J$ to $+J$. Apparently, the energy term $E_{\rm rot}$ does not depend on the sign of K, and thus these states are degenerate in energy. Figure 6.9 shows the energy level diagram of a prolate symmetric rotor as derived from Equation (6.35). For each $|K|$, there is a ladder in J where the energy increases quadratically with J, just like for the linear molecule. Each K-ladder is shifted to somewhat higher energies, as the pre-factor $(A - B)$ in Equation (6.35) is positive because $A > B$. For $J = 0$, only the $K = 0$ level is present, for $J = 1$ the $K = 0$ and ±1 level, etc. Selection rules for electric dipole allowed transitions are depicted in Figure 6.9 by vertical arrows following $\Delta J = \pm1$ and $\Delta K = 0$.

Based on these rules, the spectrum of the prolate symmetric top rotor and that of the linear molecule would be the same, with rotational lines every $2B(J + 1)$ as can be easily gleaned from Equation (6.35) by following the selection rules. This expectation should be compared to the actual spectrum of a prolate symmetric top molecule. Figure 6.10 shows an emission spectrum of methyl cyanide, CH_3CN, a prolate symmetric top molecule, with $A = 158{,}099.0$ MHz and $B = 9198.8992$ MHz according to the CDMS catalog (Müller et al. 2001, 2005; Endres et al. 2016). The molecule has a CCN linear backbone. The methyl (CH_3) subgroup at one end is a

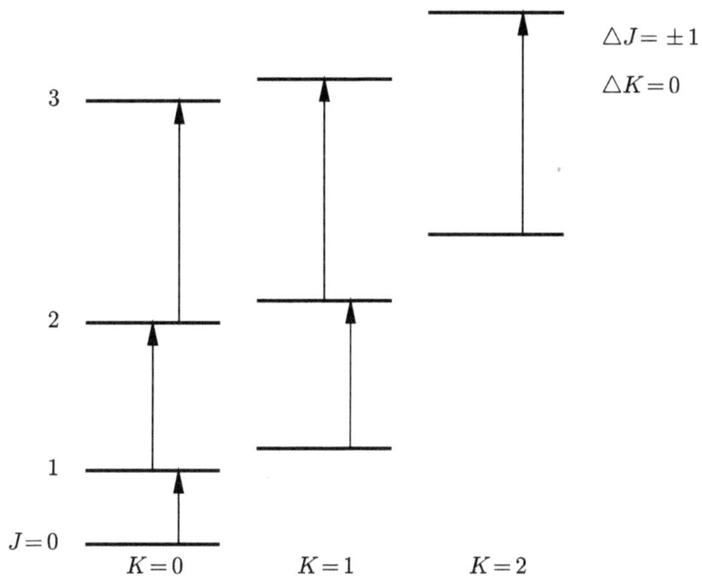

Figure 6.9. Energy-term diagram of a prolate symmetric top molecule. Arrows indicate transitions (here depicted in absorption), which are allowed only within each K-ladder as explained in the text.

chemical group that appears in many other molecule. It spreads in the plane perpendicular to the molecular Z-axis, which leads to the additional moment of inertia I_A. Due to the light hydrogen atoms on this group, $I_A \ll I_B = I_C$ and thus $A \gg B$. The spectrum of this typical prolate symmetric rotor molecule is shown near 331 GHz, where one would expect the transition for the $J = 18 \rightarrow 17$ to occur at 331.16 GHz using B. Instead of just one line, many lines are observed in the 2.2 GHz window shown. A closer look at the spectrum shows two series of lines where one progression ends near the expected position of the $J = 18 \rightarrow 17$ rotational transition. In fact, this one series at lower frequencies belongs to the $J = 18 \rightarrow 17$ transition, which exhibits different lines for different $|K|$ values, a progression that shows decreasing intensities at lower frequencies. This progression emerges as a result of centrifugal distortions for states of higher rotational excitation in J and K. As was already seen for the diatomic molecule, terms of higher order in $J(J + 1)$ appear in the energy formula due to the anharmonicity of the interaction potential. The associated centrifugal forces change the bond lengths, and thus the rotational energies are distorted. These effects are also at play for more complex molecules like the CH_3CN discussed here. In fact, distortions can be expected, not only for stretching vibrations, but in all vibrational coordinates of the molecule. As a result, additional correction terms in powers of (K^2) will modify Equation (6.35). However, all these terms cannot explain the K progression seen in the spectrum, as there is no term depending on K and J to create a shift of the different K states for a given J. Therefore, there are mixed distortion terms like $D_{JK} \cdot J(J + 1) \cdot K^2$ and higher-order terms that lead to the observed spectral richness. In fact, for the $J = 18 \rightarrow 17$ transition, a splitting of $2D_{JK}(J + 1) \cdot K^2$ can easily be derived. Inspecting Figure 6.10, the frequency difference between the $|K| = 7$ and 8 transition amounts to

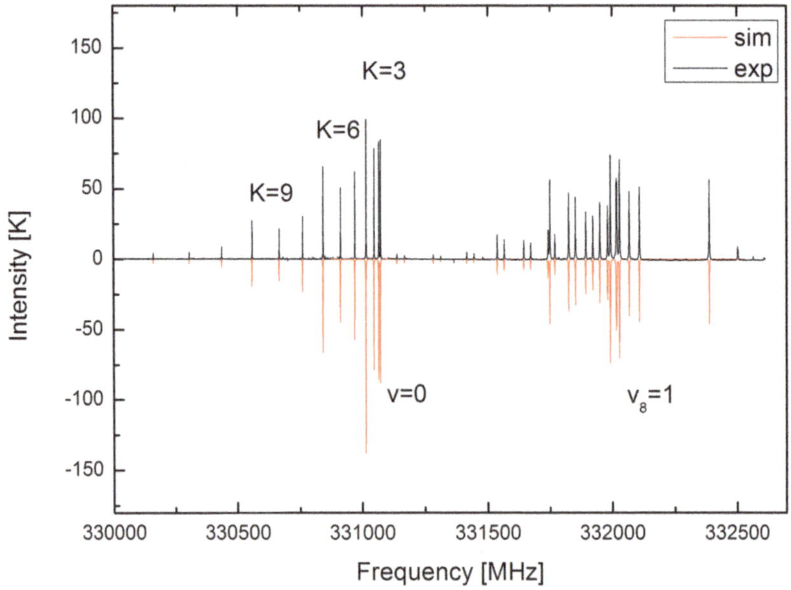

Figure 6.10. (Upper panel) Laboratory emission spectrum of the prolate symmetric rotor molecule methyl cyanide, CH_3CN, near 331.2 GHz where the $J = 18 \rightarrow 17$ transition is expected. Close to this frequency, one progression of lines for different K values extending to lower frequencies is observed. This progression shows a Boltzmann distribution with dropping intensities at lower frequencies, i.e., higher K values. At higher frequencies, rotational transitions for the vibrational excited molecule are observed with a similar progression but a more complex intensity pattern. (Lower panel) Simulated spectrum using the rotational constants and the centrifugal distortion constants for the vibrational ground state and the first excited vibrational state ($\nu_8 = 1$, C–C–N bending mode). Very good agreement is reached between the observed spectrum and the simulation obtained with the pgogher program using the molecular parameters from the CDMS catalog (Endres et al. 2016). (Figure reproduced from Wehres et al. 2018b.)

roughly 100 MHz. From this, the D_{JK} distortion constant would be determined to be 170 kHz. The actual fit uses a slightly larger value, $D_{JK} = 177.408$ kHz, which again shows the high accuracy with which molecular spectra can be predicted. The smallness of D_{JK} compared to A and B leads to the fact that all K states for a given J state are found in a very small portion of the spectrum, as shown here. Again, the different J states for CH_3CN are found every $2\ B = 18.4$ GHz.

6.4.2 Intensities and Remote Sensing

One additional, very useful aspect of spectroscopy is associated with the intensity dependence of the observed K states. Going from lower to higher K states, the intensity drops in a very systematic fashion. This observation is closely related to the (J, K) level population. According to the Boltzmann distribution, the population of the rotational states ($J = 18, K$) should roll off as $e^{-\frac{(A-B)K^2}{k_BT}}$, an exponential decay with rising K. This agrees very well with the observation, except that the K = 3n states show roughly twice the intensity, for which the K = 0 state is an exception. This general finding has fundamental consequences, as the intensity dependence of such a

K progression obviously can be used to derive the ambient temperature T used in the Boltzmann distribution. In order to derive the correct temperature, the proper transition dipole moments have to be used, but the main dependence is indeed the population distribution. Therefore, observations like the one shown from the laboratory emission experiment displayed in Figure 6.10 can be used for remote sensing of the physical quantity temperature. Of course, the intensity will also depend on the number of molecules in the field of view, and therefore this remote sensing can be also used to derive the optical depth of a sample, which is the amount (column) of molecules per m^2 observed. These derivations of physical quantities are done in many astronomical observations, and thus the temperatures of very distant places in the universe are often measured through the "simple" spectroscopy described here. The comparison between experiment and simulation by the pgopher program shows that most details of both spectra, including the complete intensity relation, agree very well. The odd behavior for the $K = 3n$ states is explained by the nuclear spin statistical weight, which will be explained later in combination with molecular symmetry.

As it turns out, the second progression seen in Figure 6.10 is associated with rotational transitions of the vibrationally excited CH_3CN ($\nu_8 = 1$, CCN bending mode), which happen to lie very close to the transitions of the molecule in its vibrational ground state. Both structures are almost identical with tiny differences associated with small changes in the vibrationally averaged structure, i.e., $\langle R \rangle$ values as first discussed for the diatomic molecule. The energy to excite, i.e., to populate this doubly degenerate vibrational mode, is on the order of 365 cm^{-1} ~ 1/20 eV. Comparing this to the thermal energy in the experiment, $k_B T \sim 1/40$ eV (room temperature, 300 K), reveals a substantial population of this vibration. In fact, the observed intensity of the associated lines is substantially smaller than those of the vibrational ground state but easily detectable in the experiments. The spectral features are more complex when compared to the simple K structure of the ground state. This has to do with the degenerate vibrational state $\nu_8 = 1$, which is associated with a vibrational angular momentum that will not be covered in this section but can be found in many more detailed textbooks on molecular spectroscopy. Again, the comparison of the experiment with simulation reveals almost perfect agreement, which means that these details can also be taken into account in a quantitative fashion.

6.4.3 Toward the Asymmetric Rotor

The next level of complication concerns the asymmetric rotor. Again, a rigorous mathematical description can be found in several textbooks. Here, we want to convey some fundamental ideas that allow the reader to understand the basic structure of the associated energy levels and to become ready for a rough interpretation of molecular spectra. The discussion starts with a semiclassical picture of the angular momentum vector that is depicted in Figure 6.11. The magnitude of this vector—and thus the quantum number J—is conserved in both reference frames. Furthermore, M, the quantum number for the projection of $\vec{J}$ onto the

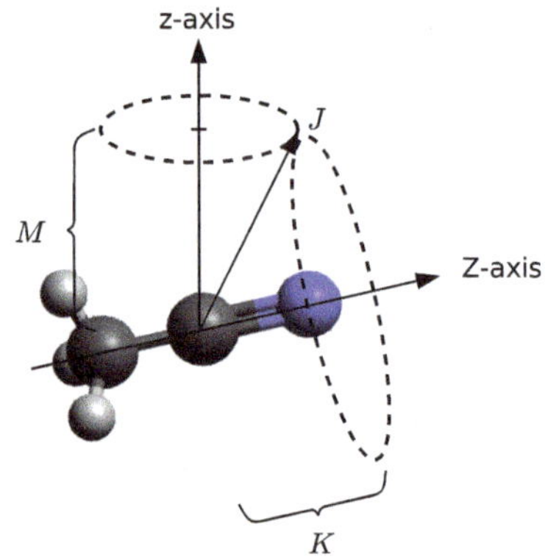

Figure 6.11. Semiclassical picture of the angular momentum vector $\vec{J}$ with its projection onto the z-axis of the laboratory frame. The associated quantum number is called M. The projection of $\vec{J}$ onto the Z-axis of the molecular frame is also shown. The associated quantum number is K as it is used for the symmetric rotor throughout this section.

laboratory frame z-axis, as well as K, the quantum number for the projection of $\vec{J}$ onto the molecular frame Z-axis, is conserved as used in this section. As a result, the rotational eigenstate of the symmetric rotor is presented by the set of quantum numbers (J, K, M) with the associated wave functions written in the Dirac notation $|J, K, M\rangle$. Heisenberg's uncertainty principle forbids the determination of a second component of $\hat{J}$ in any of these reference frames because only the magnitude and one component of the angular momentum can be measured accurately. Therefore, there is no more detailed wavefunctions than the $|J, K, M\rangle$ for describing the rotational problem, even for the more general asymmetric rotor. In particular, there is no further "good" quantum number associated. However, at this point, only the prolate symmetric rotor has been considered. The solution for the oblate symmetric top molecule is derived in analogy to the prolate top molecule discussed thus far. The energy eigenvalues are given by

$$E_{\text{rot}} = BJ \cdot (J + 1) + (C - B)K^2. \qquad (6.36)$$

Here, $I_A = I_B < I_C$, and thus only the rotational constants B and C appear in Equation (6.36). The quantum number K denotes the projection of $\vec{J}$ along the C-axis of the molecule. Therefore, K can be denoted as K_c. For the case of the prolate top molecule, K denotes the projection of $\vec{J}$ along the A-axis of the molecule. Therefore, K can be denoted as K_a. Thus, $|J, K_a, M\rangle$ describes an eigenstate of the prolate symmetric rotor, and $|J, K_c, M\rangle$ that of an oblate symmetric top rotor. Figure 6.12 depicts the energy levels of both symmetric rotors in one term diagram with the prolate symmetric top to the left and the oblate symmetric top to the right. For the prolate case, the energy levels increase quadratically with increasing J. The energies also increase with K_a^2 as described in this section. A similar picture appears for the oblate case. The energy levels increase as J^2. However, as $C < B$, the term $(C - B)K_c^2$ in Equation (6.36) leads to a decrease of energies with increasing K_c. In this framework, the symmetric rotor energies are very well-described, as was shown

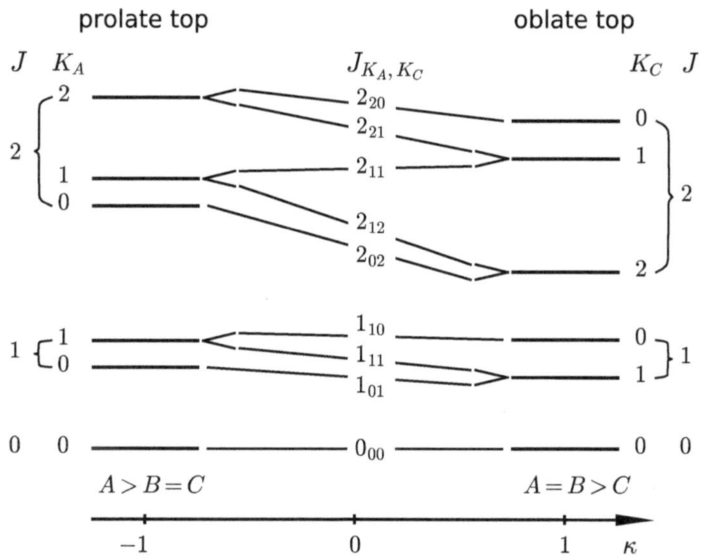

Figure 6.12. Generic energy-term diagram for the asymmetric rotor with quantum number labeling $J_{K_a,\,K_c}$. On the left-hand side, the levels are shown in the prolate symmetric top limit with $\kappa = -1$. On the right-hand side, the corresponding labeling for the oblate symmetric top limit ($\kappa = +1$) is given. The symmetry of the wave functions for the asymmetric top molecule are given by the symmetries of the contributing symmetric top wave functions. This symmetry does not change going from the prolate to the oblate limiting cases, and therefore these states are connected by the states labeled $J_{K_a,\,K_c}$.

for the CH_3CN molecule in this section. In the asymmetric rotor molecule, all moments of inertia are different. Following the thorough description of the prolate top molecule, one might consider cases where $I_A < I_B \approx I_C$ holds for an asymmetric rotor being close to the prolate symmetric rotor limit.

In fact, the asymmetry of the molecule leads to a splitting of the former degenerate $|K_a|$ levels into the level for $\pm K_a$. This splitting is shown schematically in Figure 6.12 where, e.g., the $J = |K_a| = 1$ level is split into two components. Instead of naming the states just by their quantum numbers $|J, K_a, M\rangle$, which according to the discussion on the properties of angular momenta is fully sufficient to describe the molecular state, a different labeling is used (as shown in Figure 6.12), which will be explained below. As with the picture of a slightly distorted prolate top molecule, this picture could arise from a slightly distorted oblate symmetric top description as shown on the right-hand side of Figure 6.12. Here, the former degenerate $|K_c|$ levels split into the level for $\pm K_c$. None of these descriptions is superior, and therefore it is a general convention to label the states by (J, K_a, K_c). As both K_a and K_c range from 0 to J, a new quantum number $\tau = K_a - K_c$ could be used to label all states with τ ranging from $-J$ to J. Instead, the (J, K_a, K_c) convention is used. The actual degree of asymmetry of a molecule is described by Ray's asymmetry parameter, defined as

$$\kappa = \frac{(B - A) + (B - C)}{(A - C)}, \tag{6.37}$$

which ranges from $\kappa = -1$ for the prolate top molecule to $\kappa = +1$ for the oblate case, as depicted in Figure 6.12. The calculation of the actual energies goes beyond the scope of this chapter and should be looked up in detail from respective text books. In fact, while complete analytic solutions are not available, very good approximations are built-in features of computer simulation programs like pgopher. Furthermore, the derivation of selection rules is rather cumbersome and should not be treated in this text.

It should be noted that the dipole moment in a complex molecule can point in any direction, in the general case. Therefore, its principle components μ_a, μ_b and μ_c are considered and respective selection rules for these cases are derived in some books. The asymmetry splitting and the various selection rules make the spectra of asymmetric rotors extremely rich and more difficult to interpret. Often, however, symmetries in the molecular structure lead to the cancellation of some μ components, which helps a lot. For this reason, e.g., water, H_2O, obviously has only one non-vanishing component μ_b. In a later section, another case, the CH_3^+ molecule[9] and its deuterated versions,[10] is treated in more detail. Here, it could be seen how the energy-term diagram of the asymmetric rotor arises from that of the prolate and oblate symmetric top molecule.

6.5 Laboratory Rotational Emission Spectroscopy

6.5.1 Experimental Setup

A schematic overview of a new laboratory heterodyne emission spectrometer is shown in Figure 6.13 (Wehres et al. 2018b). It is the first of its kind, and so far it has been used to record the spectrum of methylcyanide, CH_3CN, in the range from 270 GHz to 390 GHz. In Figure 6.10, just a very small part of the spectrum is displayed to discuss the spectral features of a symmetric rotor molecule. Another instrument of this kind operating in the 100 GHz regime is described in another publication (Wehres et al. 2018a). With this emission spectrometer, the spectra are obtained without the use of any excitation source. The molecules just emit their radiation against a cold background. Practically, this instrument is the same as a receiver from one of the many radio telescopes where the warm molecules are observed against the cold sky. In the laboratory experiment, the molecules are housed in a Pyrex glass cell of about 2 m length and about 10 cm in diameter. The molecules are evaporated from a glass flask into the glass cell under pressures of about 10^{-2} mbar. The pressures allow for recording the spectrum under local thermodynamic equilibrium (LTE) conditions. The emission of the room-temperature molecules is then detected in front of a cold surface, which acts like a blackbody at a certain temperature. Here, a Dewar, in which microwave absorber foam is immersed in liquid nitrogen (77 K), is used as a calibrated blackbody radiator. Replacing the Dewar with a room-temperature absorber yields much higher emission signal from that surface than from the cold background. Such

[9] A symmetric rotor molecule.

[10] Some of which are asymmetric rotor molecules.

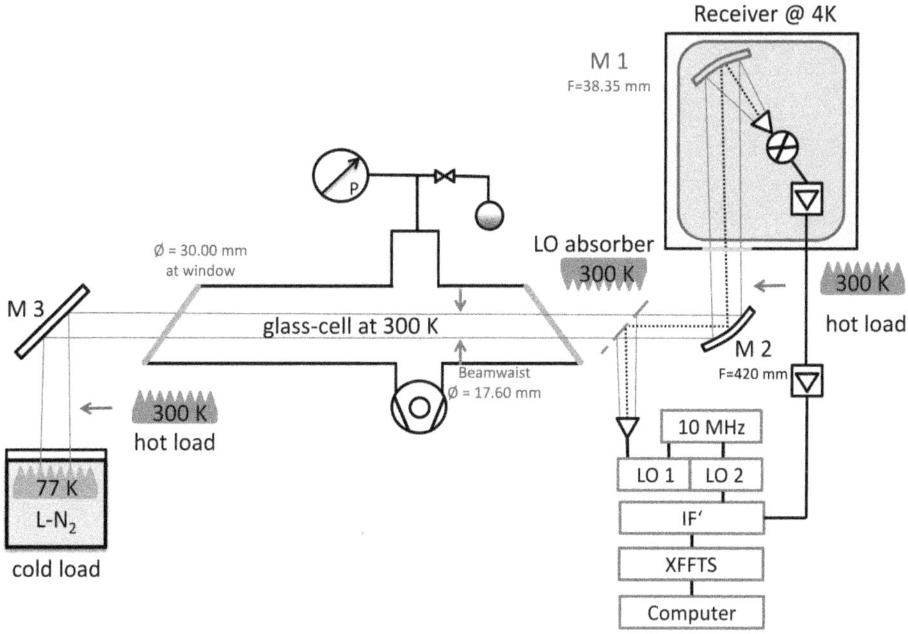

Figure 6.13. Schematic of the SURFER emission spectrometer. The setup consists of a receiver frontend housed in a Dewar that is kept at 4 K (upper right), an intermediate frequency (IF) processor (lower right), a gas cell (center), and various hot and cold loads for calibration purposes. Light from the emitting molecules is directed toward the horn antenna of the receiver via two elliptical mirrors (M2 and M1). Under these conditions, a beam-filling microwave absorber foam immersed in liquid nitrogen (l-N_2 Dewar) is used as a cold load background radiation source (left part). Alternatively, a room-temperature absorber (hot load) is used as a hot intensity calibrator. A dual pumping stage allows for a constant pressure of about 10^{-2} mbar during measurements. The molecular sample, here methyl cyanide (CH_3CN), is supplied from a reservoir via a needle valve to the vacuum chamber (glass cell). (Credit: Wehres et al. 2018b.)

control measurements are used to calibrate the emission spectrum of the molecules. Again, it is the same procedure as used at radio telescopes.

The sensitivity of such an instrument is determined by the signal detected even without receiving signal from a cold or hot load. This signal, which does not show any spectral features but appears as a blackbody of a certain temperature, is called the noise temperature. It is obtained when comparing the room temperature (T_{hot} = 297 K) signal to the cold (T_{cold} = 77 K) signal. The difference should correspond to a $T_{hot} - T_{cold}$ = 220 K signal, but each individual signal will carry the contribution associated with the noise signal of the instrument. The difference signal of the gas-filled cell in front of the cold background and the cold background alone is then just the signal from the molecules, as shown in Figure 6.10. Calibration of this signal to the known $T_{hot} - T_{cold}$ signal leads to an intensity in units of temperature, which is the traditional way to report astrophysical spectra and therefore also the laboratory emission spectrum shown in Figure 6.10. The noise level of this measurement, taken in 300 s, is on the order of 50 mK, a value expected for a system noise temperature of some 100 K for the complete system. This value for our laboratory instrument is comparable to what is obtained for radio telescope receivers.

The emission signal at radio frequencies (RF) are handled in the so-called frontend (or receiver housing), which is shown on the right-hand side of Figure 6.13. In this receiver, the RF signal is collected with a horn antenna and fed into a Superconducting/Insulating/Superconducting (SIS) mixer, where it is superimposed with the signal of a local oscillator (LO), which is operated at about the same RF frequency as the desired radio signal from the molecules. Combining the signals on the mixer, which has a non-linear Current–Voltage (*I*–*V*) characteristic, leads to signals at the sum and difference frequencies of the fixed LO frequency and the frequency of the emitting molecules. Only the difference frequency, called the resulting intermediate frequency (IF), is then further amplified: first by a cryogenic HEMT (High Electron Mobility Transistor) amplifier, and then further in the IF processor, which is shown schematically in Figure 6.13. The purpose of this signal amplification is to match the signal intensity and frequency bandwidth of the molecular emission signal to the input requirements of another spectrometer, called the backend, where the complex emission signal can be recorded simultaneously over a sizable frequency range. Here, the backend is an eXtended Fast Fourier Transform Spectrometer (XFFTS), which acts like a fast oscilloscope recording the signal and transforming the result into a frequency spectrum.

Again, this process of receiving the radio-frequency (RF) signal of the molecules, converting it to a signal at lower IF frequencies, and recording it with the help of a spectrometer is the principle of any radio telescope. Thanks to the rapid advancement of fast electronics, this technology has now also become available for laboratory experiments, which can record spectra of molecules to be expected in astronomical observations. A 300 s integration of methyl cyanide is shown in Figure 6.10 and discussed in detail in the previous section. It will be interesting to see how such a laboratory instrument can be used to record spectra of unstable molecules or ions, as well as how the receiver can be incorporated into a chirped-pulse emission spectrometer, which will be introduced in another section below. In any case, the broad band and fast timing response are new features for future approaches of laboratory spectroscopy.

6.6 Molecular Symmetry—Group Theory in a Nutshell

6.6.1 Introduction, Separation of Variables

Many molecules are rather symmetric in their structure, e.g., H_2 or the methyl CH_3 rotor group of CH_3CN discussed in detail in the previous section. These symmetries have several important consequences. For example, the intensity alternation of the K structure of CH_3CN for values of $K = 3n$ is associated with the symmetry of the respective wave functions, $|J, K, M\rangle$. Many more examples could be given. Due to its importance, this topic is the subject of several classical textbooks. Like in other sections of this chapter, it is not the idea to present a comprehensive mathematical description, but rather to illustrate the importance of the topic in a nutshell.

Before entering the topic in some detail, the discussion starts with the Schrödinger equation for the molecule of interest. The simplicity of the equation $\hat{H}\Psi = E\Psi$ should not lead us to underestimate the challenge to solve it. However, symmetry is a great helper in this respect, as shall be seen below. At first, however, we are dealing with another aspect of simplification. As molecules in general consist of many electrons and atomic nuclei, the problem becomes complex right from the start. Already in the discussion of the diatomic molecule, the principle of separation of variables was introduced to simplify the treatment. First, the Born–Oppenheimer approximation is used to separate the fast electronic motion from that of the heavy nuclei. Formally, the complete Hamiltonian, which depends on the coordinates of the electrons r_i and those of the nuclei R_α, is split into two parts

$$\hat{H}(r_i, R_\alpha) = \hat{H}(r_i, \{R_\alpha\}) + \hat{H}(R_\alpha). \tag{6.38}$$

The first term on the right-hand side contains the electronic problem with the nuclear coordinate $\{R_\alpha\}$ as a parameter. The second part contains the nuclear motion, depending on R_α. Second, as discussed for the diatomic molecule, the rovibrational problem is split further into a vibrational and a rotational one. In fact, this is a very common concept in quantum mechanics, to start the problem from a set of Hamiltonians, depending only on subsets of coordinates. For the internal motion of the nuclei of the diatomic molecule, the Hamiltonian is split into two parts

$$\hat{H}_{\text{ro-vib}}(R, \theta, \phi) = \hat{H}_{\text{vib}}(R) + \hat{H}_{\text{rot}}(\theta, \phi). \tag{6.39}$$

In this approach, a separation of the variables (R,θ,ϕ) is involved. This separation of variables allows us to also express the total wave function of the problem as a product of (independent) wave functions

$$\Psi_{n,\,J,\,M}(R, \theta, \phi) = R_n(R) \cdot Y_{JM}(\theta, \phi) \tag{6.40}$$

here for the vibrational and rotational motion of the diatomic. We have seen that this is only an approximation, as the internuclear distance is not fixed at $R = R_e$, but the non-rigidity of the anharmonic rotor leads to corrections in the energy terms. Inclusion of these corrections represent, however, an extremely good description of the energies of the internal motion of the molecule, as seen in previous sections of this chapter.

The same approach is used for more complex molecules, separating the rotational and the vibrational problem. In fact, the interactions between different degrees of freedom are mostly treated as (small) perturbations that lead to correction terms in the respective energy terms. In any case, the resulting wave functions, like the ones for the rotational and vibrational motion of the diatomic molecule given in Equation (6.40), present a complete basis set for the problem, and the proper wave functions can be expressed as linear combinations of these elementary wave functions.

The concept of separation of variables is also used for other degrees of freedom in molecules. One example concerns the interaction of the nuclear spin with the motion of the molecule. Typically, this interaction is rather weak and therefore the product ansatz for the respective wave functions is a very good approximation. As a result of this, the total wave function of a molecule is written as

$$\Psi_{tot} = \psi_{el} \cdot \psi_{ro-vib} \cdot \chi_{nuc} \tag{6.41}$$

where ψ_{el} describes the electronic wave function of the molecule, ψ_{ro-vib} describes the motional part of the nuclei, and χ_{nuc} the nuclear spin part. Although very often one is only concerned with one aspect of the internal motion within a molecule, like the rotation that dominated the discussion in the previous sections, the presence of the other parts of the wave function can play an important role due to symmetry considerations, as will be seen in the following.

6.6.2 Pauli Principle

Among all the different postulates in quantum mechanics, the Pauli exclusion principle does not have a counterpart in classical mechanics, and therefore it should receive special attention. According to the Pauli exclusion principle, the total wave function, as given, e.g., in Equation (6.41), has to be antisymmetric with respect to the exchange of any two identical Fermions (particles with half integer spin like electrons or protons). It has to be symmetric for the exchange of any two identical Bosons (particles with integer spin like the deuteron or other nuclei). This principle has fundamental consequences for the properties of molecules.

As a very simple case, molecular hydrogen is discussed. The total wave function may be written as in Equation (6.41). In order to evaluate the symmetry of the total wave function, Ψ_{tot}, which is considered for the Pauli principle, each individual part of the wave function has to be inspected individually. The electronic ground state of molecular hydrogen is totally symmetric with respect to the exchange of the two electrons or the two protons. Therefore, the product of the two other wave functions $\psi_{ro-vib} \cdot \chi_{nuc}$ shall be antisymmetric with regard to exchange of the two protons. It is therefore customary to inspect them term-by-term.

First, the nuclear spin function χ_{nuc} of the two protons shall be considered. Each spin can point up[11], $|\uparrow\rangle$), or down, $|\downarrow\rangle$, resulting in four different possibilities to combine those elementary states. Again, these spins are treated as independent and the net wave function of the two spins may be written as products, like $|\uparrow\uparrow\rangle = |\uparrow\rangle|\uparrow\rangle$. In nature, only symmetric or antisymmetric linear combinations of these elementary states are realized, leading to three symmetric linear combinations

$$\begin{aligned}
\chi_1 &= |\uparrow\rangle|\uparrow\rangle \\
\chi_2 &= \frac{\sqrt{2}}{2}(|\uparrow\rangle|\downarrow\rangle + |\downarrow\rangle|\uparrow\rangle) \\
\chi_3 &= |\downarrow\rangle|\downarrow\rangle
\end{aligned}$$

and one antisymmetric linear combination

[11] $|\uparrow\rangle$ referring to the $m_I = +1/2$ state and $|\downarrow\rangle$ referring to the $m_I = -1/2$ state.

$$\chi_4 = \frac{\sqrt{2}}{2}(|\uparrow\rangle|\downarrow\rangle - |\downarrow\rangle|\uparrow\rangle).$$

The first three functions represent the total spin state $I = 1$, with $m_I = 1, 0, -1$, respectively, for χ_1 through χ_3, and the fourth state with the total spin $I = 0$.

The rovibrational state of molecular hydrogen is further broken down into the $\psi_{ro-vib} = R_{vib}Y_{rot}$, as discussed above in detail. The vibrational ground state is totally symmetric with regard to the considered proton exchange, and thus leaves the symmetry of the product state unchanged. Finally, the symmetry of the rotational state $Y_{rot} = Y_{JM}$ has to be considered. The symmetry of the spherical harmonics Y_{JM} eigenfunctions changes with the prefactor $(-1)^J$, and thus the even (odd) numbered J states are symmetric (antisymmetric). In total, for the electronic and vibrational ground state of the hydrogen molecule, the symmetry of the total wave function is determined by the symmetry of the rotational and the nuclear spin state. Combinations obeying the Pauli exclusion principle are $\chi_i \cdot Y_{JM}$, with either $i = 1, 2, 3$ together with odd J or $i = 4$ with even J. Thus, all rotational states J of molecular hydrogen appear in nature. However, for each even J state, there is only one nuclear spin state χ_4, whereas there are three nuclear spin states for each odd J state. As there is no spontaneous transition between the symmetric and antisymmetric nuclear spin states, hydrogen comes in two independent nuclear spin configurations: ortho H_2 for χ_1, χ_2, χ_3, and para H_2 for χ_4. Para H_2 exhibits only even-numbered J states and ortho H_2 only odd-numbered ones. In cold environments, such as in interstellar space, the rotational states of hydrogen relax to the lowest rotational states. However, the $J = 1$ state can not relax to the $J = 0$ state—and thus, at low temperatures, the rotational population of molecular hydrogen generally will not be thermal,[12] due to this quantum mechanical symmetry constraint. For molecules like CO_2, some rotational levels are even missing.

The interaction of the nuclear spins with other motions in the molecule are so small that the energy terms for the different nuclear spin states lead to a threefold degeneracy of states associated with the ortho nuclear spin states. In contrast, there is only one elementary para nuclear spin state. As a result of these considerations, intensities of rotational transitions would show a 3:1 ratio when comparing transitions of ortho and para hydrogen. This alternation for the odd and even states of H_2 is, of course, not seen in ordinary rotational spectroscopy because H_2 does not have a permanent dipole moment and therefore lacks a regular rotational spectrum. Only when considering electrical quadrupole transitions, as already observed in hot interstellar media, does the intensity alternation appear.

The same discussion leads to the intensity alternation observed for CH_3CN for the states with $K = 3n$, which are twice as much populated as the neighboring K levels. However, it is a bit more cumbersome to derive the relations for three identical nuclei in the CH_3 unit of CH_3CN, where $2^3 = 8$ elementary states are realized.

[12] Unless the presence of a para magnetic material helps the nuclear spins to flip.

$$\chi_1 = |\uparrow\rangle|\uparrow\rangle|\uparrow\rangle$$
$$\chi_2 = |\uparrow\rangle|\uparrow\rangle|\downarrow\rangle, \chi_3 = |\uparrow\rangle|\downarrow\rangle|\uparrow\rangle, \chi_4 = |\downarrow\rangle|\uparrow\rangle|\uparrow\rangle$$
$$\chi_5 = |\uparrow\rangle|\downarrow\rangle|\downarrow\rangle, \chi_6 = |\downarrow\rangle|\uparrow\rangle|\downarrow\rangle, \chi_7 = |\downarrow\rangle|\downarrow\rangle|\uparrow\rangle$$
$$\chi_8 = |\downarrow\rangle|\downarrow\rangle|\downarrow\rangle.$$

Here these states are grouped w.r.t. the magnetic quantum number 3/2, 1/2, −1/2 and −3/2. Obviously, χ_1 and χ_8 are totally symmetric states. For χ_2 through χ_4, with the magnetic quantum number +1/2, one symmetric linear combination and two degenerate linear combinations are formed, as one finds when exchanging for example the two proton spins 1 and 2, represented by the so-called (12) exchange (symmetry) operation when applied to χ_3 or χ_4, which converts χ_3 into χ_4 and vice versa. The same is true for the three functions χ_5 through χ_7. With these two-dimensional representations, it is not straightforward to apply group theory in the pedestrian way used to demonstrate the 3-to-1 ratio in ortho and para states of H_2. Instead, the proper mathematical toolbox allows the symmetric rotor states to be combined with the associated nuclear spin states discussed here, which finally leads to the 2:1 intensity alternation for methyl cyanide. The reader is referred to the literature on group theory for this evaluation, or to respective tables that list the statistical weights for this and even more complicated cases.

In any case, group theory is a very powerful mathematical tool to determine the symmetry of wave functions and other properties of a molecule, in general, by use of the (irreducible) representations of the associated molecular symmetry group. From the discussion so far, it is obvious that these properties are needed to properly interpret molecular spectra. However, it is far beyond the scope of this chapter to derive the elements of molecular group theory. The reader is referred to the extremely rich and detailed literature, e.g., Bishop (1993), and in particular to the textbooks of Bunker & Jensen (1998, 2005). In this chapter, another problem regarding molecular symmetry—namely, finding the normal modes of molecular vibration—is used to illustrate the use of molecular symmetry, group theory, and in particular, character tables.

6.6.3 Normal Modes of Molecular Vibration

For the diatomic molecule, there is only one vibrational motion, i.e., the stretching vibration in the coordinate R, as seen in the first sections of this chapter. For any other molecule, it is not so obvious which or how many are the vibrational motions of a molecule. For example, for a linear molecule like the OCS presented above or the N_2O molecule discussed below, there are four normal mode vibrations, which will be treated in this section.

One basic concept of molecular physics is the separation of the translational, rotational, and vibrational degrees of freedom. This allows us to at least evaluate the number of (independent) molecular vibrational coordinates. Each nucleus in a molecule adds three degrees of freedom, leading to a total of $3N$ degrees of freedom if the molecule consists of N nuclei. Of these degrees of freedom, three degrees belong to the overall translation of the molecule. For a linear molecule, there are two directions perpendicular to the molecular axis about which the molecule can rotate. Non-linear molecules have instead three such axes. Therefore, the number of vibrational modes are $3N - 5$ for the linear molecule and $3N - 6$ for any other molecule. Thus, OCS and N_2O with $N = 3$ have four vibrational coordinates.

Due to the separation of translation, rotation, and vibration, no vibrational mode should have any contribution to rotation or translation. It is obvious that, for a diatomic ($3N - 5 = 1$), this leads to a relative motion of the two nuclei outward and inward in phase. In this process, the bond is stretched or shortened, respectively. These vibrations are consequently also called stretch vibrations in larger molecules. The sizes of the displacements of the respective nuclei are given by the ratio of their masses, in order that the center of mass of the molecule will not change.

Considering next a linear triatomic molecule like N_2O, with one bond between the two N atoms and one between the central N and the O atom, it is clear that there should be two stretch vibrations characteristic for the NN bond and the NO bond. Following the counting rules above, the remaining vibration can only be the NNO bending. This bending vibration is doubly degenerate because there are two independent planes in which N_2O might be bent. Thus, all four vibrations are identified. However, this exercise rapidly becomes difficult to manage in this bootstrap fashion. In particular, in order to satisfy the separation of the translation, rotation, and vibration, symmetrized displacements of the participating atoms have to be constructed, like for the diatomic where the displacements are scaled by the relative masses. For larger molecules, these proper vibrational motions within a molecule are called normal modes. Routes constructing these normal modes are laid out in textbooks, e.g., Wilson et al. (1980). It is not the purpose of this section to repeat these constructions. Instead, normal modes will be inspected for different types of molecules, characterizing typical molecular vibrations like the ones mentioned already. On the one hand, this is important when interpreting vibrational spectra of some molecules, which will be done later in this chapter. On the other hand, the symmetrized displacements in molecules can be used as simple examples for analyzing so-called representations of molecular symmetries. This will give reason to introduce some basic concepts of molecular symmetry groups.

Figure 6.14 shows examples of various triatomic molecules with their respective normal mode vibrations. Considering first the case of the linear N_2O, see Figure 6.14(a), we find the three fundamentally different modes that were previously introduced above. From left to right, they are the symmetric stretch, the bending, and the asymmetric stretch, which are all different in energy (ν_1 through ν_3)[13]. In fact, the second mode for N_2O is twofold degenerate, as the bending vibration might not only occur in plane as drawn in Figure 6.14 but also out of the plane, which is a linear independent motion. Thus, following the $3N - 5$ rule and drawing the simple pictures like in Figure 6.14(a) is sufficient to obtain a complete picture of the vibrational modes for this molecule.

This is as simple for other triatomic molecules, as depicted in Figure 6.14(a)–(c). For the omnipresent water molecule, H_2O, see Figure 6.14(b), there are $3N - 6 = 3$ independent modes of vibration. Again, we observe symmetric stretching, bending,

[13] Each of the vibrations denoted as ν_i correspond to a fundamental vibrational frequency, in the sense of the diatomic molecule with the vibrational frequency ω.

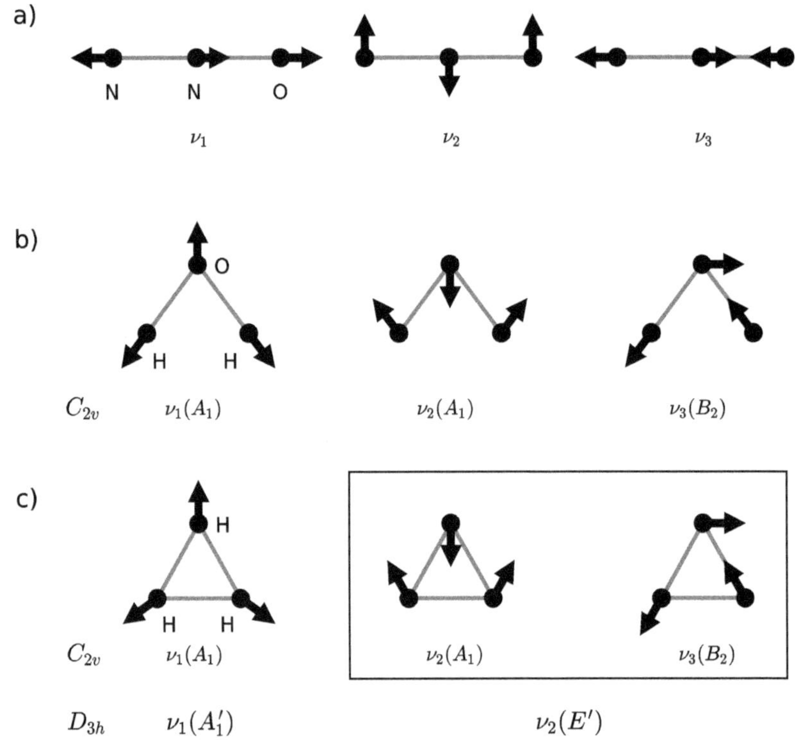

Figure 6.14. Normal modes of some triatomic molecules of different geometry: (a) N_2O, a linear molecule; (b) H_2O, a bend molecule; and (c) H_3^+, a triangular molecule. Note the similarity in the displacement vectors of the normal modes of the different molecules.

and asymmetric stretching. As a third example, the astrophysically relevant H_3^+ molecule, see Figure 6.14(c), is considered. It forms an equilateral triangle with all three protons exhibiting the same bonds to the two neighbors. Still, the three vibrational motions are the same as for the water molecule, as can be seen comparing the different rows in Figure 6.14(a)–(c). However, as it turns out, the bending and the asymmetric stretching vibrations for H_3^+ are degenerate in energy. Therefore, these two configurations form a degenerate, i.e., two-dimensional representation.[14] This finding is associated with the symmetry of the molecule and the respective bond strengths, i.e., force constants.

If one replaces one proton by a deuteron, forming the astronomically detected H_2D^+, the triangular symmetry of the molecule is broken. Despite the fact that the chemical bonds are still equally strong between the three constituents, the degeneracy of the two vibrational modes, ν_2 and ν_3, is lifted. Concerning molecular symmetry, the case of H_2D^+ therefore compares to the situation depicted for water rather than to that of H_3^+. From these simple examples, it can be seen that the normal modes of vibration are strongly related to molecular symmetry. As it turns out, H_3^+, which has

[14] Note that the displacement vectors for each atom are orthogonal when comparing the second and third pictures in Figure 6.14(c).

an equilateral triangular shape, belongs to the D_{3h} molecular symmetry group and is much more symmetric than the bend H_2D^+ or H_2O molecules belonging to the C_{2v} group. In fact, the higher symmetry that arises from the inspection of Figure 6.14(b) and (c) will become apparent when the molecular symmetry groups are introduced.

Later in this chapter, the four-atomic molecule CH_3^+ and its isotopically labeled siblings, CH_2D^+, CD_2H^+, and CD_3^+ are considered. These molecules are thought to play an important role in astrochemistry, which is the subject of another chapter in this book. Thus, these ions appear to be present in the interstellar medium but can only be found when their spectra are known. Recording their spectra in the laboratory is a challenge, and the topic of a later section. There also, the vibrational spectra of these molecules are investigated, and therefore it is interesting to inspect their vibrational modes. In Figure 6.15, the normal modes for these molecules are depicted in a representation that can be understood as an extension of the previous triatomic systems discussed so far.

Compared to H_3^+, just one more carbon atom is added to form CH_3^+. Following the $3N - 6$ rule, this leads to three more vibrational modes. Thus, six normal modes are expected. These modes are shown in Figure 6.15. The displacement vectors of the three atoms in the lower H–C–H triangle are the same as those in the three rows depicted in Figure 6.14(a)–(c). Again, the latter two are degenerate in energy, as in the case of H_3^+. A similar picture arises when inspecting the modes ν_4, ν_3, and ν_6, which are an out-of-plane bending, and two H–C–H bends. Again, the latter two are degenerate in energy for CH_3^+ and CD_3^+, which also belong to the molecular symmetry group D_{3h}, while this degeneracy is lifted for the mixed isotopologs, CH_2D^+ and CD_2H^+, which belong to the symmetry group C_{2v}. As a consequence of these symmetry considerations and the resulting degeneracies or lifting thereof, it can be expected to observe four normal modes for CH_3^+ and CD_3^+ and six modes for CH_2D^+ and CD_2H^+, which will be discussed below through inspection of the experimental vibrational spectra of these molecules.

In order to make a link between the inspection of normal mode vibrations and a more rigorous mathematical description used to derive the normal modes without explicit evaluation of the displacement vectors, the molecular symmetry groups shall now be invoked below.

6.6.4 Molecular Symmetry Representations

A molecule like H_2O exhibits some evident symmetries, as do the vibrational modes depicted in Figure 6.14(b). The two protons are indistinguishable Fermions, and thus the exchange of the two should not change the energy of the system, i.e., the Hamiltonian should remain unchanged when swapping the coordinates for the two. In mathematical language, there are symmetry operations, $\hat{O}$, that leave the Hamiltonian, $\hat{H}$, unchanged. Therefore, $\hat{H}$ and $\hat{O}$ commute, $[\hat{H}, \hat{O}] = 0$, and thus $\hat{H}$ and $\hat{O}$ share the same eigenfunctions. With regard to the picture of normal modes, $\hat{O}$ will not change the energy of the molecule. Therefore, it is interesting to consider those symmetry operations that leave the normal mode unchanged versus those that change the displacement vectors. First of all, it is interesting to find all symmetry

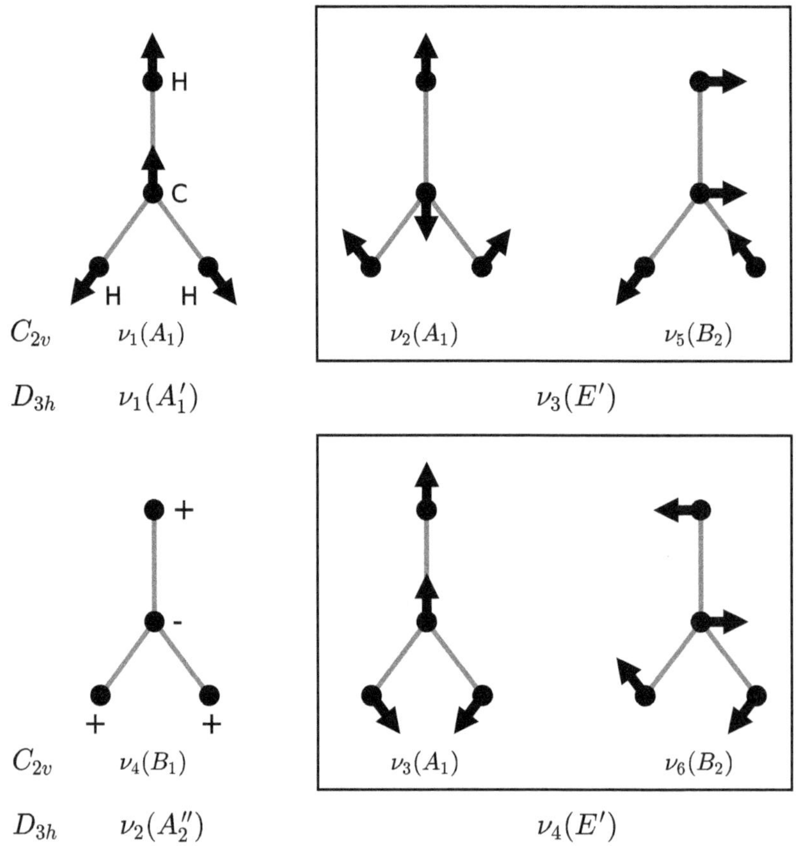

Figure 6.15. Normal modes of the four-atomic molecule CH_3^+. It exhibits three more normal modes than the triatomic case. The six modes belong to two one-dimensional modes and two two-dimensional, i.e., degenerate ones. This degeneracy is lifted for the mixed isotopologs CH_2D^+ and CD_2H^+ as described in the text. The experimentally observed modes are shown in Figure 6.18 and listed in Table 6.4 and are discussed there in more detail.

operations of a molecule. This set of operations forms the mathematical object of a group with the well-known group axioms:

The identity operation, $\hat{E}$ is always a member of the group.

If the operations $\hat{P}$ and $\hat{Q}$ are members of the group, the product $\hat{R} = \hat{P}\hat{Q}$ is also a group member.

As long as the elements are not interchanged, the order of application does not matter: $(\hat{P}\hat{Q})\hat{R} = \hat{P}(\hat{Q}\hat{R})$.

There is always an inverse element to $\hat{R}$ called $\hat{R}^{-1}$, such that $\hat{R}^{-1}\hat{R} = \hat{E}$.

Following the description of the point groups,[15] the water molecule exhibits four symmetry operations. First, a rotation about the molecular symmetry axis (z-axis)

[15] Which are isomorphic to the more appropriate Complete Nuclear Permutation Inversion (CNPI) groups; see, e.g., Bunker & Jensen (1998, 2005). However, we consider the point groups here, as the symmetry operations are nicely related to the displacement vectors of the normal mode vibrations.

by 180° called $\hat{C}_2$. In general, a $\hat{C}_n$ rotation is a rotation by $2\pi/n$ about a symmetry axis in the molecule. Water exhibits also two mirror planes, one is the plane of the molecule called σ_v'', and the other is perpendicular to this plane, called σ_v'. The label v for vertical indicates that the molecular axis lies in this mirror plane. Together with the identity operation, $\hat{E}$, which is a group element of any of these groups, water exhibits the symmetry group C_{2v} with the group elements $C_{2v} = \{\hat{E}, \hat{C}_2, \hat{\sigma}_v', \hat{\sigma}_v''\}$.

As it turns out, the group has as many so-called irreducible representations as there are classes in the group. In the case of water, all four elements of the group represent a class of their own, and thus there are four irreducible representations, $\Gamma_1 - \Gamma_4$, called A_1, A_2, B_1, and B_2 by convention. While the action of the group operations (elements) are rather clear, the representations appear somewhat mysterious at first. They are, of course, mathematical terms and it takes many more proper definitions, etc., to become acquainted with these terms and to take advantage of this mathematical tool. Instead, we look at the normal mode vibrations of water in order to inspect their symmetry properties and thus to identify their associated representations. As pointed out before, there are $3N - 6 = 3$ vibrational modes, i.e., vibrational coordinates, to be expected for the water molecule. The vibrational motions depicted in Figure 6.14(b) show three sets of displacement vectors representing motions of all nuclei. Let us now apply the symmetry operations to these displacement vectors to see how they change. These changes distinguish the normal modes and associate them with the irreducible representations of the molecular symmetry group.

Before doing so, we inspect the character table of the C_{2v} group, another helpful tool in group theory, which is listed in Table 6.2. The character table is organized as a matrix with the action of the group elements (symmetry operations) as columns and the irreducible representations $\Gamma_1 - \Gamma_4$ as rows. The elements of this table are the characters, i.e., numbers $\chi_i[\hat{O}]$, which contain information on the behavior of each symmetry species (irreducible representation, row i) with respect to each symmetry operation (column, $\hat{O}$). Here, these characters are just the values ± 1, indicating whether or not a symmetry operation leaves a property of the molecule—in our case, the displacement vectors of the normal mode vibration—unchanged ($\chi = +1$) or not ($\chi = -1$). A quick inspection of the table shows that the characters of any irreducible

Table 6.2. Character Table of the Molecular Symmetry Group C_{2v}

C_{2v}	$\hat{E}$	$\hat{C}_2(z)$	$\hat{\sigma}_v'(x, z)$	$\hat{\sigma}_v''(y, z)$
A_1	+1	+1	+1	+1
A_2	+1	+1	−1	−1
B_1	+1	−1	+1	−1
B_2	+1	−1	−1	+1

Note. This is relevant for identifying symmetry species for the water molecule, but also for the CH_2D^+ molecule discussed below.

representation (row of the matrix) are different from those of any other irreducible representation. This is the reason why inspection of the action of the symmetry operations on the normal mode vibrations will associate them with one and only one of the irreducible representations shown in the character table.

Applying the identity operation to these vectors of the normal mode vibrations depicted in Figure 6.14(b) leaves everything as it is. Therefore, the character $\chi = +1$ indicates that all vectors are the same for all of the four irreducible representations. This is why the first column in Table 6.2 contains +1.[16] Considering the action of the $\hat{C}_2$ operation, the result is different for the three normal modes. The displacement vectors for ν_1 and ν_2 are either staying where they are or they are swapped among the two protons, resulting in an unchanged set of displacement vectors. The resulting character is therefore $\chi = +1$. Ergo, the two vibrations ν_1 and ν_2 transform as the A_1 or A_2 representations and not as the B_1 or B_2 representations, for which the result is $\chi = -1$ upon the action of $\hat{C}_2$. For ν_3, the picture is quite different and rotating the molecule by 180° inverts the direction of each displacement vector, resulting in a character $\chi = -1$, and thus associating this mode with the B_1 or B_2 representations. Further inspection of the action of the reflections $\hat{\sigma}_v'$ and $\hat{\sigma}_v''$ reveals that, for the ν_1 and ν_2 modes, the displacement vectors are again left unchanged ($\chi = +1$). In view of all symmetry operations, ν_1 and ν_2 are said to be fully symmetric. In fact, the characters for these modes are the same as for the A_1 representation, and thus we say ν_1 and ν_2 are both transforming as the A_1 representation.

Different results are obtained for ν_3, as the reflection in the plane of the molecule leaves everything unchanged ($\chi = +1$) and that about the mirror plane perpendicular leads to displacement vectors with changed signs ($\chi = -1$). Comparing all transformations for ν_3 with those of the $\Gamma_1 - \Gamma_4$ shows that ν_3 transforms as $\Gamma_4 = B_2$. Thus, each normal mode for the water molecule corresponds to one so-called irreducible representation of the molecular symmetry group. Thus, the normal modes should be taken as symmetrized linear combinations of the displacement vectors. Of course, this was the way they were constructed, as symmetrized linear combinations.[17] Thus, it could be considered a nice mathematical trick that the normal modes can be classified in these irreducible representations. However, this property turns out as an invaluable tool, as shall be seen below.

In order to make use of the properties of the character table, we inspect the rows and columns of the character table in Table 6.2. Consider the characters of a row as elements of a vector, $\chi_i[\hat{O}]$, for one i and all $\hat{O}$. Multiplication of the vector of the first row with that of any other row vector

[16] To be more accurate, the symmetry operation has to be applied to the sum of all displacement coordinates; the x, y, and z components of the displacement vectors of all nuclei in the molecule. Only those numbers constitute a one-dimensional representation as expected for a (1D) normal mode vibration.

[17] There are elegant ways to find the irreducible representations of normal modes of a molecule without any physical intuition, which we used here to illustrate the simplicity of molecular symmetry. This is when molecular symmetry unfolds its magic in solving seemingly hard problems.

$$\sum_{j=1}^{4} \chi_1[\hat{O}_j] \cdot \chi_i[\hat{O}_j] = 0 \tag{6.42}$$

shows that the first row vector is orthogonal to each other row vector. In fact, all row vectors are orthogonal to each other. The same behavior is found for the vectors defined by the columns. The product of any of these vectors with itself reveals the order of the group, $g = 4$. This finding is expressed in the great orthogonality theorem, which can be written for the characters of any point group as

$$\sum_{j=1}^{4} \chi_i[\hat{O}_j] \cdot \chi_k[\hat{O}_j] = g \cdot \delta_{ik}. \tag{6.43}$$

This mathematical description should be read as follows. Consider two irreducible representations (rows) of a group. Sum the products of the characters for each symmetry operation, $\hat{O}$, i.e., multiply the row vectors. All irreducible representations are orthogonal. For the irreducible representations A_1 and B_2 for the normal modes in C_{2v}, this means that they cannot be broken down into a simpler set of displacement vectors. The three independent modes, $2A_1 \oplus B_2$,[18] are the three (independent and symmetrized) modes expected for the water molecule. They transform as one of the irreducible representations, and therefore it is customary to name them after their symmetry classification. For water, the A species remain unchanged under the $\hat{C}_2$ operation. The index 1 or 2 distinguishes between the outcome of the reflection operations.

It is not the denomination of these symmetry species but their orthogonality that is of utmost importance and help. For water, there are only three one-dimensional normal mode vibrations and they are easy to find by constructing the displacement vectors as shown in Figure 6.14(b). However, for more complex molecules, just adding one more atom causes the number of normal modes to increase rapidly (see Figure 6.15 for examples) and a classification becomes rather cumbersome. A more formal application of group theory to these situations instead reveals the irreducible representations for the normal mode vibrations in a rather straightforward way. This is when group theory becomes very valuable.

Another very important application for group theory is related to selection rules. Earlier in this chapter, the transition dipole moment $\langle \Psi_f | \mu | \Psi_f \rangle$ was considered. The corresponding integral determines whether a transition can be observed or not. Thus, the selection rules emerge from this calculation. Instead of evaluating the integral, only the product of the three functions $\Psi_f \mu \Psi_f$ needs to considered. Depending on the oddness or evenness of this triple product along each direction in space, the integral will be zero or non-zero, respectively. Due to the symmetry of the molecule, the eigenfunctions of $\hat{H}$ and $\hat{O}$ are the same, and thus Ψ_i, Ψ_f, and μ

[18] The symbol $\oplus$ is the direct sum, which means that these are not the sum of the numbers, but in a vectorial sense, the three representations span a three dimensional space—the one for the three vibrational degrees of freedom.

must transform as representations of the molecular symmetry group. Using the orthogonality of the irreducible representations of a symmetry group, it becomes easy to evaluate the product of two such functions in terms of the resulting representation. Let us consider the character table for the C_{2v} group shown in Table 6.2 as an example. When multiplying two functions of the triple product transforming as A_1 and A_2, the resulting product representation can be found by multiplying the characters of the two representations for each symmetry operation (column). Simple inspection of the character table reveals that $A_1 \cdot A_2 = A_2$ or $B_1 \cdot B_2 = A_2$, etc. Once the representation of the product has been found, the triple product can thus be evaluated from multiplying those double product representations by that of the third term, i.e., another irreducible representation. In the end, only the totally symmetric representation A_1 for the triple product will lead to a non vanishing integral. Thus, group theory is essential when it comes to finding the allowed transitions in a molecule. Things get very involved when considering all the different degrees of freedoms in a molecule. However, using computer aid boils this down to a very manageable problem. In programs like pgopher and others, the aid of group theory is built-in.

Therefore, the first step in the analysis of molecular spectra is to determine the molecular symmetry group, which is equivalent to finding all symmetry elements of the respective group. Nice systematic flow diagrams are given in the literature (Bernath 1995) to find the respective molecular symmetry group. Figure 6.14(c) shows the normal modes of the highly symmetric H_3^+ molecule. As stated before, this molecule belongs to the D_{3h} group. The character table is shown in Table 6.3. Many more symmetry operations are feasible for this molecule than for one in the C_{2v} group discussed before. In particular, two rotations $\hat{C}_3$ and $\hat{C}_3^{\,2}$, which are rotations by 120° and 240°, respectively, about the molecular axis are additional symmetry operations. All the different symmetry operations fall into six classes, which leads to a total of six irreducible representations with names different from those of the C_{2v} group. Among the irreducible representations of D_{3h}, there are two-dimensional ones, named as E' and E''. In the picture of the normal modes, the two vibrational frequencies for the bending vibration and the asymmetric stretch for H_3^+ as given in Figure 6.14(c) are degenerate and belong to the E' symmetry species. In practice, this

Table 6.3. Character Table of the Molecular Symmetry Group D_{3h}

D_{3h}	$\hat{E}$	$2\hat{C}_3(z)$	$3\hat{C}'_2(z)$	$\hat{\sigma}_h(x, y)$	$2\hat{S}_3$	$3\hat{\sigma}_v$
A'_1	+1	+1	+1	+1	+1	+1
A'_2	+1	+1	−1	+1	+1	−1
E'	+2	−1	0	+2	−1	0
A'_1	+1	+1	+1	−1	−1	−1
A'_2	+1	+1	−1	−1	−1	+1
E'	+2	−1	0	−2	+1	0

Note. This is relevant for identifying symmetry species for molecules like H_3^+ and CH_3^+, discussed below.

means that the number of vibrational bands should be reduced in molecules with higher symmetry.

It goes beyond the scope of this chapter to inspect the symmetry of all these motions, but it is instructive to see what happens if one of the protons is replaced by a deuteron. It has already been found by simple inspection of Figure 6.14(c) and (b) that the D_{3h} symmetry is broken, and in this case the lower C_{2v} group is appropriate. This transformation also occurs in the character table because many of the symmetry operations have to be removed from the table. Complete columns will be missing, and the only operations left are: $\hat{E}$, one of the three $\hat{C}_2$ operations, one of the $\hat{\sigma}_v$ reflections, and the $\hat{\sigma}_h$ reflection—which turns into a $\hat{\sigma}_v$ operation as the molecular symmetry axis changes from being perpendicular to the plane to in-plane. As a result, the number of symmetry operations drops to four, and consequently the number of irreducible representations also shrinks to four, as discussed before. The result of the $\hat{C}_3$ operation in the group D_{3h} distinguishes between the A and E irreducible representations, as seen from the different characters of these species for the $\hat{C}_3$ operation. As this operation is no longer a symmetry operation, the E species drop from the table and one is left with the character table of the C_{2v} group. This implies that the degeneracy of the E' vibrational mode in H_3^+ is lifted for the case of H_2D^+ or D_2H^+. This is exactly what is observed. Later in this chapter, the case of the four-atomic CH_3^+ molecule (D_{3h} group) with its deuterated siblings CH_2D^+, CD_2H^+ (C_{2v} group), and CD_3^+ (D_{3h} group) will be considered for illustration. There, the number of observed vibrational modes will be discussed and studied in greater detail. To begin the discussion on molecular vibrations for polyatomic molecules, IR spectra of the simpler N_2O molecule will be discussed next.

6.7 Vibrational Spectroscopy

6.7.1 N_2O, a Linear Molecule

A triatomic linear molecule such as N_2O features $3N - 5 = 4$ fundamental vibrational modes, as discussed in the previous section. It exhibits two stretching modes (ν_1 and ν_3), both of which are non-degenerate (see Figure 6.14(a)), and a doubly degenerate bending mode ν_2, i.e., the molecule is free to bend in two orthogonal planes and the energy associated with any of these two vibrational motions is the same. The fundamental vibrational modes of N_2O are observed at 1285 cm^{-1} (ν_1), 589 cm^{-1} (ν_2), and 2224 cm^{-1} (ν_3; see, e.g., Toth 1991). Figure 6.16 shows the vibrational spectrum of N_2O obtained with a commercial (teaching-grade) FTIR spectrometer in our laboratory, obtained at a resolution of 0.5 cm^{-1} in the wavenumber regime from 1000 to 4000 cm^{-1}. This spectrum shows a total of twelve vibrational bands (see Bryant et al. 2008). In the following, the notation $n_1n_2^{\ell}n_3$ will be used to specify the degree of vibrational excitation, with n_i being the vibrational quantum number of vibrational state i ($n_i = 0, 1, 2, 3, \ldots$; $i = 1, 2, 3$). Upon excitation of the degenerate ν_2 mode, the bending vibrational motions in the two orthogonal directions can have different phases. Depending on this phase relation, the bending vibration is associated with an angular momentum, as if the bent molecule rotates about the molecular axis. Therefore, ℓ, as denoted in the $n_1n_2^{\ell}n_3$

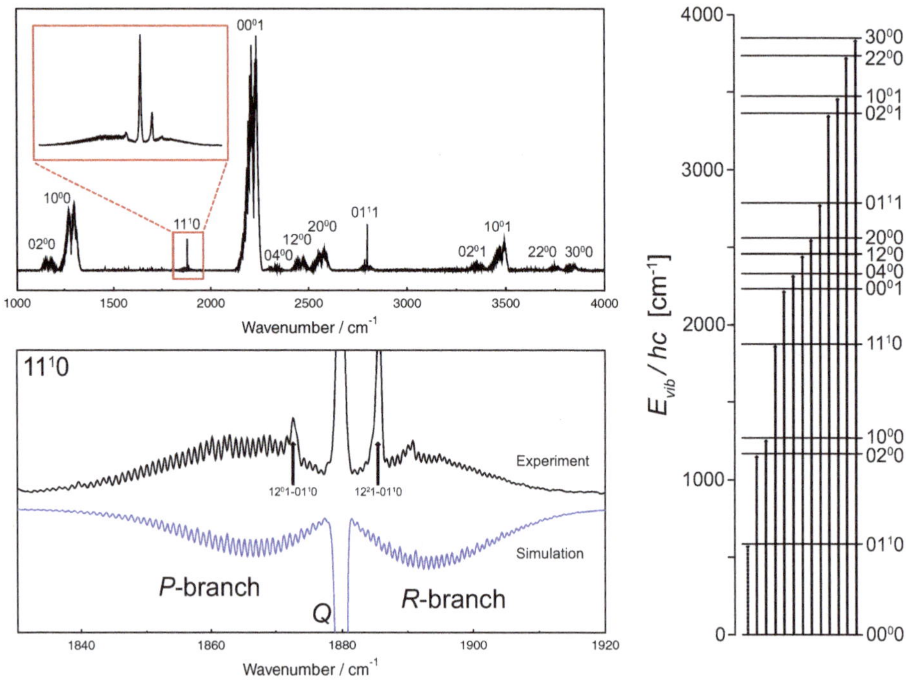

Figure 6.16. Vibrational survey spectrum of N_2O from 1000 to 4000cm^{-1}, observed at a resolution of 0.5 cm^{-1} (top left). All vibrational transitions visible in the survey originate from the ground vibrational state (00^{00}) and are indicated in the energy level diagram (right). The bending fundamental ν_2 (i.e., the transition $01^10 - 000$, indicated by the leftmost arrow) corresponds to a wavenumber of 589 cm^{-1} and has not been covered here. The bottom-left spectrum shows a detailed view of the 11^{10} band centered at 1880cm^{-1} with resolved rotational structure in the *P*-branch, partially resolved rotational structure in the *R*-branch, and an intense and unresolved central *Q*-branch. The simulation of the 11^{10} band was performed using the Pgopher program (Western 2017), based on the data reported in Amiot & Guelachvili (1976). In addition, weaker *Q*-branches are observed in the experimental 11^{10} spectrum (indicated by arrows) from a transition not originating from the ground, but rather an excited vibrational state (01^{10}), causing a *hot band*; see text.

labeling of the vibrational excitation of the molecules, is a quantum number accounting for angular momentum $\ell\hbar$ about the figure axis when $n_2 \geqslant 1$ (and can take the values $\ell = n_2, n_2 - 2, n_2 - 4, \ldots, -n_2$, e.g., if $n_2 = 1$ then $\ell = \pm 1$).

As indicated in the corresponding vibrational energy level diagram shown in Figure 6.16, all these vibrational transitions originate from the ground vibrational state 00^00. The energetically lowest vibrational fundamental 01^10 (the ν_2 bending mode) is outside the range covered experimentally, but the two strongest features observed are the two stretching vibrational fundamentals 10^00 and 00^01. In addition, several more transitions are observed in which the molecules ends up in a state $n_i > 1$ of one particular mode (also designated as *overtone*) or in a *combination mode* when at least two different vibrational modes of the molecule are excited at the same time. The vibrational modes are not just observed as the transition between the two vibrational states, but also the rotational states of the molecule changes. Therefore, the vibrational transitions come as more extended bands, which are composed of the individual rovibrational transitions. When this substructure is not resolved—or only partly resolved, as in the present case—the bands appear as contours with particular substructure.

The vibrational transitions observed in the experimental spectrum show such additional structure from the $\Delta J = -1$ and $+1$ rotational selection rules, which still hold for the rovibrational transitions observed here. This leads to the occurrence of *P*- and *R*-branches, respectively, observed for all vibrational transitions in Figure 6.16. In other words, upon excitation, the molecule undergoes a transition from a rotational level J'' of the ground vibrational state 00^00 into a rotational level $J' = J'' - 1$ or $J'' + 1$ of an excited vibrational state $n_1n_2n_3$.

Participation of the bending state leads to a slightly more complicated picture: For a first excited bending mode in a rotating molecule, $\ell = \pm 1$, i.e., the rotational levels of the vibrationally excited state split into two components of different parity. As a consequence, for vibrational bending modes (or combination modes involving one or more quanta of the bending mode) a $\Delta J = 0$ selection rule also holds, leading to very compact *Q*-branch features. In Figure 6.16, this is clearly apparent through the presence of intense and sharp *Q* branches of the 11^10 and 01^11 combination modes at 1880 and 2798cm^{-1} in the top survey scan. This $\Delta J = 0, \pm 1$ extension of the earlier fundamental selection rules can be understood from the combination of angular momenta of the different degrees of freedom (vibration and rotation), and as for many aspects of quantum mechanics, similar behaviors are found for electronic transitions in atoms with more than one electron and thus more than one orbital angular moment. Therefore, the reader might be reminded of this analogy. The interesting subject of couplings of angular momenta in molecules will not be covered in this chapter.

Close inspection of the 11^10 combination mode in the bottom left spectrum in Figure 6.16 highlights all characteristics of the reasoning above. The rotationally resolved *P*-branch covers the region from 1835 to 1880, the dominant (unresolved) *Q*-branch is centered at 1880cm^{-1}, and the partially resolved *R*-branch follows from 1880 to some 1910 cm^{-1}. As can be estimated from the appearance of the *P*- and *R*-branches, the spacing between different rotational levels is not strictly uniform, a consequence of the fact that the rotational constants in the lower (here, the ground vibrational state) and the vibrationally excited state are not identical but slightly different. For N_2O, the ground-state rotational constant B_{00^00} is 12,561.6 MHz, whereas B_{11^10} is 12,528.9 MHz. Exactly this trend became apparent already for the diatomic molecule as $\langle R \rangle$, which was close to R_e for the vibrational ground state but became larger for higher vibrational (and rotational) excitation as depicted in Figure 6.4. Thus, the moment of inertia increases and the rotational constant decreases as observed for the vibrational excitation of N_2O. It is gratifying to see to what extent the simple (semiclassical) physics picture of the diatomic molecule holds for more complex molecules.

The numbers for the rotational constants B_{00^00} and B_{11^10} may be obtained from a least-squares analysis of rotationally resolved spectra, such as the one for the 11^10 band in Figure 6.16, using the method of *combination differences*: transition wavenumbers of transitions in the *P*- and *R*-branches sharing the same upper level may be used to evaluate the lower (here, ground-state) rotational constant from the relation $\tilde{\nu}_{R-\text{branch}}(J) - \tilde{\nu}_{P-\text{branch}}(J+2) = 2B''(2J+3)$. Likewise, the relation $\tilde{\nu}_{R-\text{branch}}(J) - \tilde{\nu}_{P-\text{branch}}(J) = 2B'(2J+1)$ holding for transitions with common lower states can be used to determine B' of the upper (vibrationally excited) state.

As can be seen from a comparison of the experimental and simulated 11^10 bands in Figure 6.16, the experimental spectrum also shows contributions from other vibrational transitions such as the *hot bands* $12^20 - 01^10$ and $12^00 - 01^10$ that do not originate in the ground vibrational state, but rather from the first excited bending mode. Rovibrational transitions of many molecules are recorded in much higher resolution, such that the vibrational transition frequencies and the rotational constants can be determined with much higher precision. Here, the main interest was to demonstrate that vibrational spectra of even a triatomic molecule already deliver great spectral richness, and many more than just the principal modes of the molecule, i.e., the normal modes, can be observed.

In the next subsection, the subject of infrared spectroscopy at various levels of spectral resolution will be covered in greater detail. First, vibrational bands of the CH_3^+ molecule shall be identified by covering a wide spectral range in the infrared region. Second, the rovibrational transitions are recorded with unprecedented accuracy using very narrowband laser systems. Third, rotational transitions of an isotopic variant of this molecule will be found based on the very same idea of combination differences, here shown for the N_2O molecule.

6.7.2 Vibrational and Rovibrational Spectroscopy CH_3^+—A Symmetric Top Molecule

In this subsection, the vibrational, rovibrational, and (when applicable) rotational fingerprints of symmetric and asymmetric top rotors shall be illustrated with the example of the CH_3^+ ion and its isotopologs CH_2D^+, CD_2H^+, and CD_3^+. These kinds of ions have traditionally been investigated via absorption experiments in which the light (laser or mm-wave) traverses a discharge tube containing methane, and in which these ionic species are generated in trace amounts. In this section, however, we will mainly focus on the spectroscopy of these species in cold ion traps, and therefore a short introduction of the methodology is given here.

Such ion trap experiments offer the advantage of operating with a mass-selected and cooled ion sample that can be detected (ion counting) with very high efficiency. It has to be stressed here that typical ion counts in such trapping experiments are on the order of 10^5, so that one has to resort to a different spectroscopic method, in which one takes the ion counts as the spectroscopic signal and not the change in the detected light as in absorption spectroscopy. This so-called action spectroscopy scheme will be explained below. A 4 K ion trapping machine specially designed for (sub)mm-wave spectroscopy is depicted in Figure 6.17. The central part of such a machine is the ion trap, which is attached to a coldhead that can cool down the experiment, i.e., the trap, to as low as 4 K. Ions generated in the ion source are mass-filtered and pulsed into this trap. Helium buffer gas injected into the trap before the arrival of the ion pulse ensures that the ions cool down close to the nominal trap temperature. During the trapping time of typically 1 s, the ions are irradiated with

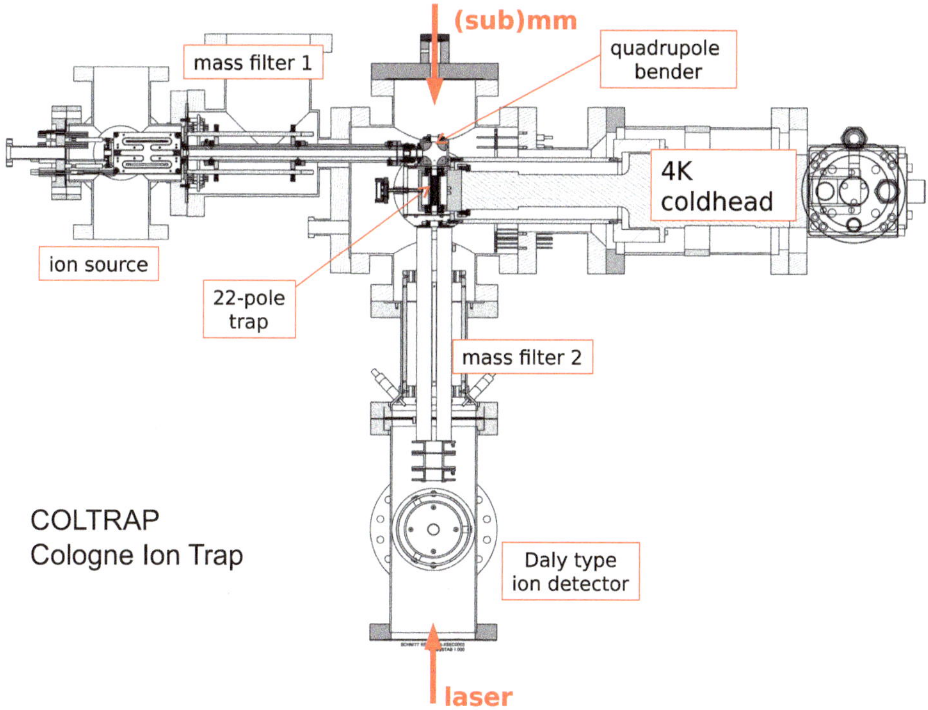

Figure 6.17. Setup of the Cologne-built 4 K ion trap machine COLTRAP (Asvany et al. 2014). It comprises an ion source (top left), a first quadrupole mass selector (mass filter 1), an electrostatic quadrupole bender, the 22-pole ion trap mounted on a 4 K coldhead (top right), the second quadrupole mass selector (mass filter 2), and finally a Daly-type ion detector (bottom). While windows on both end flanges (close to quadrupole bender and detector) can be used for laser access, the flange closer to the 22-pole trap is particularly suited for introducing (sub)mm-wave radiation.

IR light (for vibrational or rovibrational spectroscopy) or with (sub)mm-wave radiation (for rotational spectroscopy). In the applied action spectroscopic schemes, the resonant absorption of the radiation by the ions first leads to an excitation of the molecular ion of interest as shown in Equation (6.12), and then to a chemical reaction. As a result, the mass composition of the trapped ion ensemble changes, which is finally detected by the high-efficiency ion counter, as first demonstrated in a scheme termed laser induced reactions (Schlemmer et al. 1999). For taking spectra, these 1 s trapping cycles are repeated while the frequency of the radiation source is stepped.

Now we turn our attention to the molecule of interest. The CH_3^+ ion is a planar molecule where the three protons are arranged in a triangle about the central carbon atom as indicated in Figure 6.19. Substitution of the protons with deuterium, of course, does not change its geometry. It has $N = 4$ atoms and thus has $3N - 6 = 6$ fundamental vibrational modes. The normal modes of this molecule are depicted in Figure 6.15. With three covalent bonds, it will have three stretching motions, and therefore three modes are left for the low-frequency bending motions, as shown in the upper and lower panel of Figure 6.15, respectively. Depending on the isotopologue, the symmetry group is either D_{3h} for CH_3^+ and CD_3^+, and the molecule is a symmetric rotor; or C_{2v} for CH_2D^+ and CD_2H^+, which are asymmetric rotor molecules like water. As a consequence of the respective molecular symmetry, as was already pointed out before, some of the vibrational modes of the respective molecules can be IR inactive or degenerate, as summarized in Table 6.4 and illustrated in Figure 6.18, where the different observed vibrational modes are denominated according to Table 6.4 and Figure 6.15. Like for the N_2O example recorded with a broadband FTIR instrument the low-resolution spectra displayed in Figure 6.18 illustrate the vibrational bands of CH_3^+ and its isotopologues, which will be further discussed below.

The spectroscopy of CH_3^+ has been pioneered by Takeshi Oka and coworkers, and the simple simulation shown for CH_3^+ in Figure 6.18 is based on their high-resolution

Table 6.4. Numbering of Modes for CH_3^+ and CH_2D^+ (Similar for CD_3^+ and CD_2H^+)

D_{3h}	CH_3^+		C_{2v}	CH_2D^+	
A_1'	ν_1	Symm. stretch	A_1	ν_1	Sym. C–H stretch
A_2''	ν_2	Out-of-plane bend	A_1	ν_2	C–D stretch
E'	ν_3	Asym. stretch	A_1	ν_3	In-plane C–H bend
E'	ν_4	In-plane bend	B_1	ν_4	Out-of-plane bend
			B_2	ν_5	Asym. C–H stretch
			B_2	ν_6	In-plane C–D bend

Notes. As mentioned in the text, the CH_3^+ isotopologues all have three stretching modes and three bending modes. The modes are numbered first according to the irreducible representations of the respective symmetry group (D_{3h} or C_{2v}) and then with ascending frequency (see also Figure 6.18). For CH_3^+, the mode with A_1' representation, the symmetric breathing mode ν_1, is IR-inactive. The modes with representation E' (the asymmetric C—H stretch ν_3 and the in-plane bend ν_4) are doubly degenerate. For CH_2D^+, all modes are IR active. The displacement vectors of these modes are illustrated in Figure 6.15.

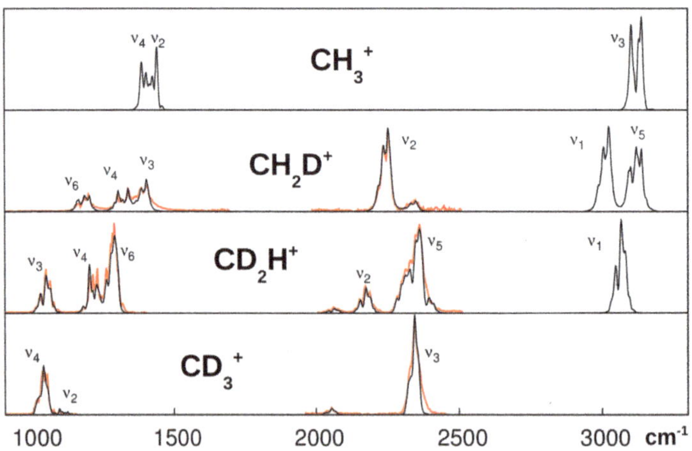

Figure 6.18. Measured and simulated vibrational spectra of isotopologues of CH_3^+ (Asvany et al. 2018). The measured spectra (red) have been recorded in low resolution with the free-electron laser FELIX (Oepts et al. 1995), and the spectroscopic simulations (black) were done with PGOPHER (Western 2017). The mode numbers of the IR active fundamental vibrations are given. The numbering scheme is explained in Table 6.4. All features in the 3 μm (3000 cm^{-1}) region have been measured previously in high resolution (Crofton et al. 1988; Jagod et al. 1992; Gärtner et al. 2013; Jusko et al. 2017a). As an example, the ν_1 mode of CD_2H^+ is shown in Figure 6.19.

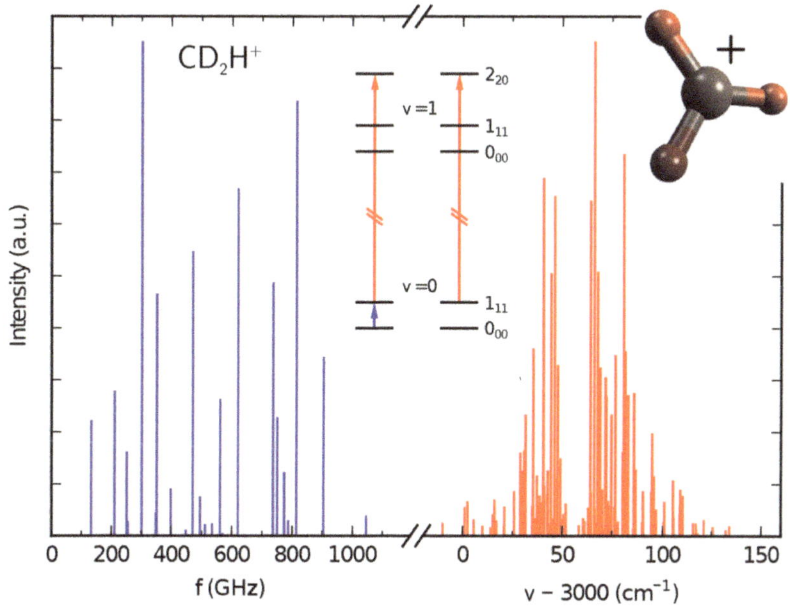

Figure 6.19. Stick spectra summarizing the 108 rovibrational (red, right) and 25 pure rotational (blue, left) lines recorded for CD_2H^+ in an ion trap experiment (Jusko et al. 2017a). Please observe the different frequency units (cm^{-1} vs. GHz) on the abscissae. The rovibrational lines belong to the ν_1 mode whose envelope is depicted in Figure 6.18. The insets illustrate the applied spectroscopic methods: The rovibrational lines have been measured in a direct manner (right), while the rotational lines were detected with a double-resonance scheme (left). (Credit: Jusko et al. 2017a.)

work (Crofton et al. 1988; Jagod et al. 1992), which revealed spectroscopic parameters such as the rotational constants, distortion constants, etc., with high accuracy. For CH_3^+, the symmetric top with D_{3h} symmetry, the totally symmetric stretching vibration ν_1 (the breathing motion) is IR-inactive and will not be detected in IR absorption experiments. This is to say that not only the dipole moment of CH_3^+ vanishes due to the coincidence of the center of mass and the center of charge of the molecule, and therefore it does not exhibit a pure rotational spectrum. However, even the excitation of a totally symmetric vibration does not change the dipole moment, and therefore the vibrational excitation cannot be invoked by a (single) photon. Therefore, the only detectable vibrational mode in the 3 μm region is the asymmetric C–H stretch ν_3. This mode has the representation E' and is doubly degenerate, as is the in-plane bend ν_4. This latter vibration and the ν_2 mode are bends and are typically found at much lower frequencies, with the ν_2 mode only detected recently (de Miranda et al. 2010). These normal mode vibrations are already depicted in detail in Figure 6.15, where the modes are denominated for the D_{3h} and the C_{2v} cases.

On substitution of one or two H atoms with deuterium, obtaining CH_2D^+ or CD_2H^+, the ion turns into an asymmetric top belonging to the C_{2v} symmetry group. At this stage, all modes are non-degenerate and IR-active; see Table 6.4. This allows one to be reassured that there are indeed three stretching motions and three bending motions; see Figure 6.18. The stretches involving C–H motions are located around 3050 cm^{-1}, and the stretches with the heavier D atom are redshifted to about 2300 cm^{-1}. It can also be clearly seen that the number of H and D atoms in CH_2D^+ and CD_2H^+ corresponds to the number of the respective C–H and C–D stretches. Similarly, it can be discerned that the bends containing the D atoms form the lowest-frequency bending modes, which are ν_6 for CH_2D^+ and ν_3 for CD_2H^+. Finally, for the perdeuterated CD_3^+, its symmetry and spectroscopy is again quite similar to CH_3^+, with the vibrational fingerprints merely redshifted.

Before continuing the discussion of the spectra in more detail, it is interesting to inspect the observed redshift upon deuteration, because this redshift can be understood from the derivation of the vibrational frequency for the diatomic molecule in the beginning of this chapter. The vibrational transition frequency in the diatomic molecule was just given by $\omega = \sqrt{k/\mu}$, with k being the force constant of the chemical bond and $\mu = \frac{m_A \cdot m_B}{m_A + m_B}$ the reduced mass of the AB diatomic molecule. When applying this relation to the C–H (C–D) stretching vibration of CH_3^+ and CD_3^+, we realize that the force constant k is the same, as the chemical bond does not change. However, due to the mass ratio of carbon and hydrogen (deuterium), μ is close to 1 u for hydrogen and 2 u for deuterium. As a result, the vibrational frequency is expected to change by a factor of $\sqrt{2}$. Thus, a 3050 cm^{-1} C–H stretch should shift toward 2200 cm^{-1}, which is remarkably close to the experimental finding. Thus, the simple physics pictures derived for the diatomic molecule are again applicable to much more complex molecules.

The overview spectra shown in Figure 6.18 (red trace) have been measured with the free-electron laser FELIX (Oepts et al. 1995) and the action spectroscopy method described above. This special laser allows us to cover this very impressive spectral range, and thus allows for the interesting overview spectra that show all the allowed bands. However, due to fundamental experimental boundaries, the spectral resolution of this type of laser only permits us to record the contours of the vibrational bands, but not single rovibrational lines. High-resolution infrared spectroscopy is technically best realized in the 3 μm (3000 cm^{-1}) region, and all displayed C–H stretches of CH_3^+, CH_2D^+ and CD_2H^+ have been investigated in rovibrational detail (Crofton et al. 1988; Jagod et al. 1992; Gärtner et al. 2013; Jusko et al. 2017a). An example for such a measurement is shown in Figure 6.19, depicting the ν_1 stretch of CD_2H^+. As it was explained briefly for N_2O, the rovibrational spectrum exhibits a P-branch and an R-branch for $\Delta J = -1$ and $+1$ rotational transitions, respectively. As a result the lines of the P-branch correspond to energies lower than the energy for the vibrational transition and the R-branch extends to higher frequencies, starting from the pure vibrational energy difference, which is called the band origin. Thus, the P- and the R-branch for the ν_1 stretch of CD_2H^+ are clearly visible, with the band origin at 3056.177571 cm^{-1} (Jusko et al. 2017a).

In this case, high resolution could be obtained by applying an optical parametric oscillator (OPO) as the light source and by cooling the investigated ions in the cryogenic ion trap depicted in Figure 6.17. For CD_2H^+, the ν_1 mode is the C–H stretch (with A_1 representation; see also Figure 6.15). Its transition dipole moment thus points in the direction of the C–H bond, which is the molecular b-axis (for CD_2H^+, the b-axis points along C–H and the c-axis is perpendicular to the molecular plane). Therefore, the allowed rovibrational transitions are of b-type, meaning that they obey the selection rules $\Delta J = 0, \pm 1$, $\Delta K_a = \pm 1$, $\Delta K_c = \pm 1$ for their rotational quantum numbers J_{K_a, K_c}. With the high power of the applied OPO light source and the sensitivity of the trap instrument, even a few intrinsically weak transitions with $\Delta K_a = \pm 3$ and/or $\Delta K_c = \pm 3$ could be detected in that work (Jusko et al. 2017a). In Figure 6.19, the two branches, consisting mainly of transitions with $\Delta J = -1$ (P branch) and $\Delta J = +1$ (R branch), can be clearly distinguished. Because CD_2H^+ is an asymmetric rotor molecule, the rovibrational spectrum does not show a simple rotational structure, as was visible for the N_2O case where a $2B$ progression (in particular, for the P-branch depicted in Figure 6.16) is visible. The simulated spectrum also show this clear pattern. The rotational structure for CD_2H^+ shown in Figure 6.19 appears much more irregular, and only the P- and R-branches are clearly visible. Nevertheless, this complex spectrum can be analyzed in great detail with the help of programs like Pgopher. Furthermore, the concept of combination differences, which isolates energy levels of the upper or lower vibrational manifold, is applicable, and the rotational energy level diagram of a polyatomic asymmetric rotor molecule can be determined with high accuracy on the basis of its infrared transitions. This accuracy is orders of magnitude higher than in the low-resolution study for the N_2O case based on the high spectral purity of the OPO system compared to the low-resolution FTIR spectrometer.

As a consequence, the infrared spectroscopy of molecular species in rovibrational resolution is typically the first step to obtain information of the ground-state energy-term diagram of that molecule. The rovibrational spectrum contains information about the level structure of the ground state and of the vibrationally excited state. Thus, in the case of a well-behaved molecule without any perturbations, a model rotational Hamiltonian can be applied to fit the data and to simultaneously obtain the rotational parameters for both states with high confidence. This is the case for CH_2D^+, for example, for which the rotational ground state has been determined by Oka and coworkers by IR spectroscopy (Jagod et al. 1992; Gärtner et al. 2013). In contrast, the ν_1 vibrational band of CD_2H^+ shown in Figure 6.19 is slightly perturbed by another band, the most probable candidate being the combination band $\nu_2 + \nu_3$ due to symmetry reasons (ν_2 and ν_3 have A_1 symmetry (see Table 6.4), and therefore also $\nu_2 + \nu_3$, and can thus interact with ν_1) and due to its energetic proximity (Jusko et al. 2017a). Nevertheless, in such a case, the ground state can be determined well by the use of the so-called ground-state combination differences (GSCDs). A GSCD is the difference of two rovibrational lines that share a common upper (vibrationally excited) rovibrational state. By this definition, the GSCD is the difference between two levels in the ground state, as exemplified by the red arrows in Figure 6.20, which displays the energy-term diagram of the vibrational ground state of CD_2H^+ The advantage of taking such differences is that the perturbation in the vibrationally excited state is canceled out and the non-perturbed ground state can be completely recovered by a GSCD fit. With 108 rovibrational lines measured for the

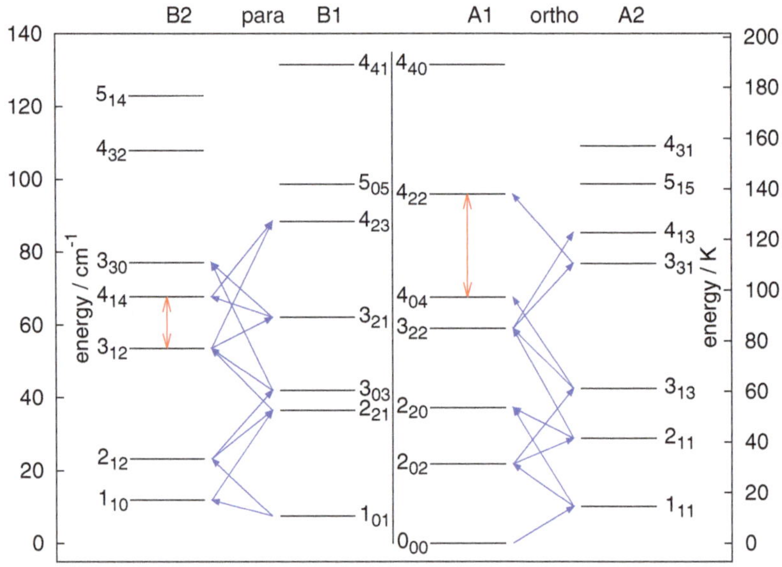

Figure 6.20. Illustration of the rotational levels of CD_2H^+ and the 25 rotational transitions measured in an ion trap experiment (Jusko et al. 2017a). The asymmetric rotor levels are labeled as $J_{K_a,\ K_c}$. The color coding is the same as in Figure 6.19. The measured rotational lines are drawn in blue, whereas two examples for GSCDs (derived from the IR data) are drawn in red. As can be seen, the GSCDs connect different sets of rotational levels as the rotational transitions. (Credit: Jusko et al. 2017a.)

ν_1 band of CD_2H^+, 142 GSCDs could be obtained (Jusko et al. 2017a). Due to the complete coverage of the low-J transitions, many of these GSCDs are coincident, i.e., they appear more than once. For example, the difference between the ground-state levels 4_{22} and 4_{04} (red arrow in Figure 6.20) is about 28.321 cm^{-1}. It could be formed in six different ways, one example being the difference of the rovibrational transitions $5_{15} \leftarrow 4_{22}$ and $5_{15} \leftarrow 4_{04}$ at 3058.276383 and 3086.597407 cm^{-1}, respectively. In a well-calibrated experiment, these different values should coincide within their experimental uncertainties, and any deviation from this indicates potential spectroscopic errors. Such coincidences thus help to check the quality and the assignment of the data, and can help to identify, e.g., misassignments or lines from other vibrational bands.

6.7.3 Rotational Spectroscopy of an Asymmetric Top

Once the ground state is understood by rovibrational spectroscopy, one can try to execute pure rotational spectroscopy in cases where the molecule has a permanent dipole moment. For the CH_3^+ isotopic family, CH_3^+ and CD_3^+ are symmetric tops without a dipole moment—and thus without a rotational spectrum—but the mixed isotopologs CH_2D^+ and CD_2H^+ are accessible. As CH_2D^+ is assumed to be responsible for high-temperature ($T > 20$ K) deuteration processes in interstellar space, its radioastronomical detection in space—and therefore detection in the laboratory—has been of particular interest in the last decade (Roueff et al. 2013).

The first pure rotational lines of CH_2D^+ were presented by Amano (2010), and in the meantime, a complete set of high-resolution transitions is available from ion trap experiments for CH_2D^+ (Töpfer et al. 2016) as well as for CD_2H^+. For the latter molecular ion, the measurement scheme shall be explained with the help of Figure 6.19. As direct rotational spectroscopy in ion traps can be challenging, a double-resonance scheme has been applied for CD_2H^+. As shown in the inset of Figure 6.19, a rovibrational transition (red arrow) has been used to create a spectroscopic signal, just as was the case for the rovibrational spectrum shown before. By scanning the mm-wave frequency of a connected pure rotational transition (blue arrow), the signal for the rovibrational transition can be modulated and the rotational transition thus detected with a high signal-to-noise ratio. In this way, 25 transitions of CD_2H^+ were detected in the range up to 1.1 THz, shown as blue sticks in Figure 6.19. Due to the accurate predictions of the reliable rovibrational data, no tedious searches were necessary, and the targeted lines were found quite quickly. Furthermore, the high S/N of the double-resonance method and the low-temperature operation of the trap experiment allowed the transitions to be measured with a relative uncertainty down to 10^{-9} (about 1 kHz for rotational transitions). The 25 rotational transitions are also depicted as blue arrows in the ground-state level scheme in Figure 6.20. These transitions are again b-type transitions ($\Delta K_a = \pm 1, (\pm 3)$, $\Delta K_c = \pm 1, (\pm 3)$) as the permanent dipole moment points along the C–H bond (the molecular *b*-axis). It can also be seen that the rotational transitions connect symmetry species of the C_{2v} group with different parity ($A_1 \leftrightarrow A_2$, $B_1 \leftrightarrow B_2$), but do not change the nuclear spin (ortho, para). In contrast, the GSCDs connect rotational levels of the same parity (red arrows). Again, however, comparisons can be made to check the obtained spectroscopic data. To give an example, the GSCD between the levels 4_{14} and 3_{12} (red vertical arrow in Figure 6.20) can be formed in four different ways by the available IR lines, but also in two different ways by the rotational data (one difference and one sum, see blue diagonal lines in the figure). In this manner, a very stringent test of all rotational and rovibrational transitions can be obtained, and consequently so can high-resolution parameters for use in molecular physics and astronomy.

In summary, an energy-term diagram as shown in Figure 6.20 is the most detailed information yet obtained for the energy pattern of the rotational states of a molecule. Here, we find it for a polyatomic asymmetric rotor. For the astrophysicist, this is very interesting information, but what they appreciate most is the precision with which this information is available. From just the rovibrational transitions, the energy terms were known to an accuracy of sometimes even below 1 MHz. This is rather impressive, taking into account that the ~3000 cm^{-1} for those transition frequencies amount to 9×10^7 MHz, thus reaching a relative precision of a few times 10^{-8}. For the detection of astrophysical molecules, this might still not be satisfactory, as some lines are observed with a precision of only a few 100 kHz at frequencies up to 1 THz. Therefore, recording individual rotational transitions, as demonstrated here, improves this accuracy by two to three orders of magnitude, which is good enough to unambiguously identify molecular transitions in space.

All this precision is not visible in the energy-term diagram of Figure 6.20. Instead, for a neophyte in spectroscopy, the energy level diagram is still as confusing as the irregular rovibrational and rotational spectra shown in Figure 6.19. However, a closer inspection of the term diagram reveals the close agreement with the generic-level diagram for an asymmetric rotor shown in Figure 6.12. There, the energy-term diagram was derived from the symmetric rotor model in the two limiting cases of the prolate and oblate top molecule. In fact, the states J_{K_a} or J_{K_c} are split as the $K = \pm 1$ degeneracy is lifted by the asymmetry of the molecule because the three moments of inertia are different from each other. It is obvious that this asymmetry splitting kicks in gradually as the asymmetry parameter $\kappa = \frac{(B-A)+(B-C)}{A-C}$ departs from the limiting prolate ($\kappa = -1$) or oblate ($\kappa = +1$) cases, respectively. However, the question is what do all the cases have in common, which therefore should also be found for the special case of the CD_2H^+ molecule discussed here.

The rotational constants are A = 217,431.5125 MHz, B = 140,618.0669 MHz, and C = 84,406.6900 MHz which leads to $\kappa = -0.15$ (Jusko et al. 2017a), which is very close to the most intermediate $\kappa = 0$ case. Therefore, CD_2H^+ is a clear intermediate case of a very asymmetric rotor. Inspecting the energy-term diagram of Figure 6.20 in view of this finding first reveals that the ordering of the energy states labeled as $J_{K_a,\,K_c}$ is as expected from Figure 6.12. Most conveniently, this is checked for the $J = 1$ levels with increasing energy: 1_{01}, 1_{11}, 1_{10}. A closer view of all $2J + 1 = 9$ levels for $J = 4$ shows some interesting energy patterns. First, the energy levels are spread over a wide range of energies from 4_{04} (~70 cm^{-1}) to 4_{40} (~135 cm^{-1}). This wide spread is associated with the large asymmetry of the molecule. In the view of the prolate symmetric top molecule, the two states 4_{04} and 4_{40} would correspond to the prolate symmetric rotor levels $4_{K_a=0}$ and $4_{K_a=4}$, which would be split by $(A - B) \cdot K_a^2$; for $K_a = 0$ and 4, this amounts to an energy difference of ~40 cm^{-1}. Starting the discussion from the oblate limit, these states would correspond to the levels $4_{K_c=4}$ and $4_{K_c=0}$, which are split by $(C - B) \cdot K_c^2$; for $K_c = 4$ and 0, this amounts to an energy difference of ~30 cm^{-1}. Interestingly, the sum of these energy differences agree with the amount of splitting observed for CD_2H^+ for these two states, 4_{04} and 4_{40}. Other than the large overall spread of energies, some states are very close in energy; for example, the 4_{04} and 4_{14} states and the 4_{41} and 4_{40} states.[19] Their small splitting arises from the low K_a or K_c quantum numbers, respectively, which plug in as $(A - B) \cdot K_a^2$ (~2.5 cm^{-1}) and $(C - B) \cdot K_c^2$ (~−1.9 cm^{-1}), respectively, and therefore the splitting is hardly visible in Figure 6.20. The energy differences of these pairs of transitions are even smaller than those estimates suggest. This is caused by the fact that the 4_{04} and 4_{14} states belong to different K_a values but the same K_c. Because of the $\kappa \sim 0$ intermediate situation, the two states are not separated as for the prolate limit, but the two states for the same K_c get closer together. In contrast to these small splittings, states that emerge from one K_c level but are not equal to $J = K_c$ are split further, e.g., the 4_{13} and 4_{23} states, which show a

[19] This small asymmetry splitting will occur again for the case of methacroline, a molecule with the extra challenge of a large amplitude torsional motion.

substantial splitting as expected for this intermediate case. The same picture arises when discussing levels for the same K_a value. Splittings for $J = K_a$ or $J = K_c$ states are smallest. In the semiclassical physics picture of the molecules we consider in this chapter, this finding corresponds to a molecule that rotates (close) to the *a*-axis or *c*-axis. As for the classical rotor, under these conditions the influence of the asymmetry appears to be much smaller than for intermediate *K* values.

In conclusion, a closer look at the energy diagram of an asymmetric rotor reveals a lot of details on the specific molecule. However, there are some very general trends that have to be observed for any case. In that respect, the level diagram shown in Figure 6.20 is gratifying to astrophysicists and molecular physicists alike, just from different perspectives.

6.8 Large Amplitude Motion: Tunneling and Internal Rotation

Up to this point, it was quite impressive how well even simple models like the distorted rigid rotor and the harmonic oscillator, as well as the normal mode picture in several dimensions, work for predicting molecular spectra. As it turns out, the corrections introduced to the simple models by perturbation theory improve the model predictions so well that the spectra for many molecules can be described by a rather limited set of model parameters. This raises curiosity as to the limits of this approximation. As long as the amplitudes of the vibrational motions of a molecule are rather small compared to its dimension, the interaction between vibrations or vibration and rotation is rather small, and consequently the corrections to the eigenenergies are small. However, for large-amplitude motions (LAM), the situation can be quite different and the models discussed so far can fail terribly. Obvious examples for such LAMs are motions in weakly bound molecules, such as van der Waals bonded systems like H_2–CO, or hydrogen bonded systems like water complexes. It is beyond the scope of this chapter to treat such systems, and because no aggregate like this has been found in space so far, it is certainly not of immediate concern in the astrophysical context. However, the description of the spectroscopy of loosely bound clusters is an interesting and challenging topic of molecular physics. Aside from that, even strongly bound molecules like the astrophysical relevant protonated methane, CH_5^+, also have many LAMs, and therefore it may still be considered a challenge for astrophysics to predict its lowest energy transitions in order to find this and other key molecules in space.

6.8.1 Tunneling: Chirped-pulse Fourier Transform Spectroscopy

Even for ordinary molecules like the omnipresent methanol, CH_3OH, or ammonia, NH_3, LAMs are of concern and thus should be treated to some extend. Here, we start the discussion with the tunneling motions in molecules, for which ammonia is the classical example. Figure 6.21 shows a potential energy diagram for ammonia where the energy is plotted versus the N–H distance representative for the N–H stretch vibration. The nitrogen atom can tunnel through the potential barrier that is present when N penetrates the plane of the three Hs. This tunneling can be considered as a LAM because the amplitude of the stretching vibration is no longer

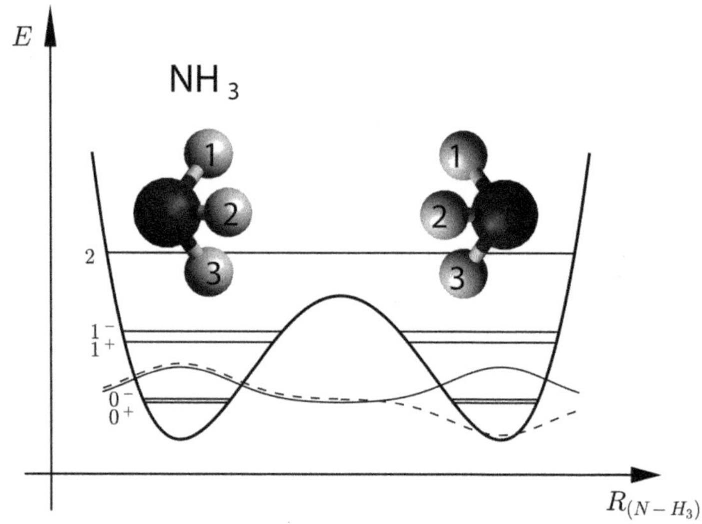

Figure 6.21. Double well potential for the umbrella or inversion motion in NH_3. The vibrational levels for $n = 0$ and 1 are split because the phases of the wave functions for the left and right potential wells are different, as indicated by the + and − superscript to the vibrational level that indicates the symmetry (parity) of the state with respect to inversion of all coordinates of the molecule. Transitions between these tunneling states can be observed at wavelengths of about one cm, as shown in Figure 6.22.

very small compared to the N–H equilibrium distance. As a result of the tunneling motion, the nitrogen atom has a probability to be on each side of the barrier. For the vibrational ground state ($n = 0$), the wave function is a Gaussian in each potential well, very much like the ground state wave function in any harmonic oscillator.

However, there is a symmetric and an antisymmetric way to superimpose the wave functions of the right and the left part of the potential well. These linear combinations are depicted for the vibrational ground state ($n = 0$) in Figure 6.21. The energy state of the symmetric linear combination, 0^+, is lower than that of the asymmetric linear combination, 0^-. This splitting can be rationalized in the following way. The symmetric wave function creates a larger probability to find the N nucleus in the classically forbidden region of the potential compared to that of the antisymmetric wave function, the probability function of which, $\sim\langle\Psi||\Psi\rangle$, has a zero-crossing in the center. Based on this observation, ammonia in the 0^+ state experiences an extended range of N–H distances, and therefore the zero-point vibrational energy is reduced compared to the 0^-, which is restricted to a smaller range of N–H distances. This observation—and thus, the ordering of the n^+/n^- states —is found for all vibrational states. Moreover, the splitting increases for higher n states. This finding is related to the increased tunneling probability, which depends exponentially on the distance to the barrier height, $V_B - E$.

The tunneling splitting can be observed directly, as transitions between the rovibrational states $|n, J, K, M, \pm\rangle \leftrightarrow |n, J, K, M, \mp\rangle$ of ammonia are dipole allowed due to the parity change ($\pm\leftrightarrow\mp$). Figure 6.22 shows a 1.2 GHz stretch of such a tunneling spectrum. Many transitions fall into this particular region, which is

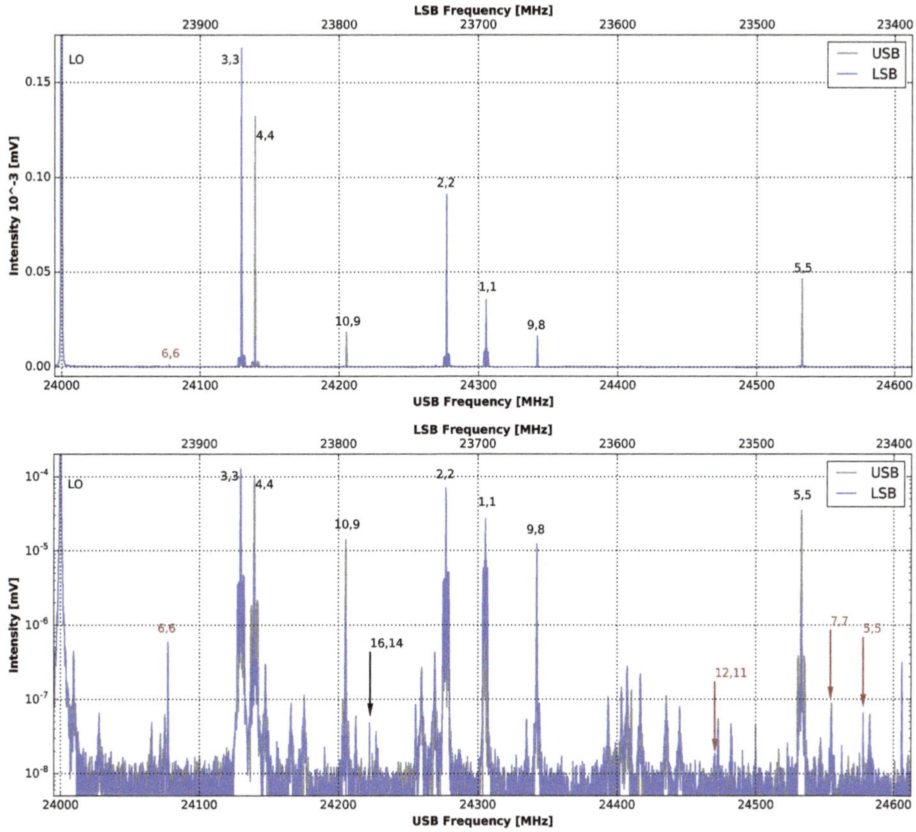

Figure 6.22. Chirped-pulse Fourier transform microwave spectrum of ammonia, NH_3. The recorded spectrum contains both the upper- and lower-side bands observed in a heterodyne detection scheme shown in Figure 6.23. Upper panel: several tunneling transitions $|n, J, K, M, \pm\rangle \leftrightarrow |n, J, K, M, \mp\rangle$ of the vibrational ground state ($n = 0$) all fall into the 1.2 GHz bandwidth and are recorded at once. They are labeled by their J, K quantum numbers. Lower panel: the spectrum displayed with logarithmic intensities shows the large S/N of such an experiment. Many more transitions become apparent.

close to 24 GHz, the splitting between the symmetric and the antisymmetric states as depicted in Figure 6.21. The individual transitions are shifted for the different rotational states, J_K, in this oblate symmetric molecule.

These transitions of an oblate symmetric rotor molecule are easily observable in absorption using a setup similar to that shown in Figure 6.7. In fact, such an instrument is used in several teaching classes. However, here the spectrum has been recorded using the method of chirped-pulse Fourier transform microwave spectroscopy. One such instrument is shown schematically in Figure 6.23. Here, all molecules in the path of the beam are excited coherently by a short pulse of strong radiation. A very cold sample of molecules is created by a free jet expansion of ammonia from a pulsed valve, as depicted in the center of Figure 6.23. The free induction decay (FID) emission signal is recorded using a fast digitizer. Due to the broadband excitation and detection, many lines are observed at once. The spectrum

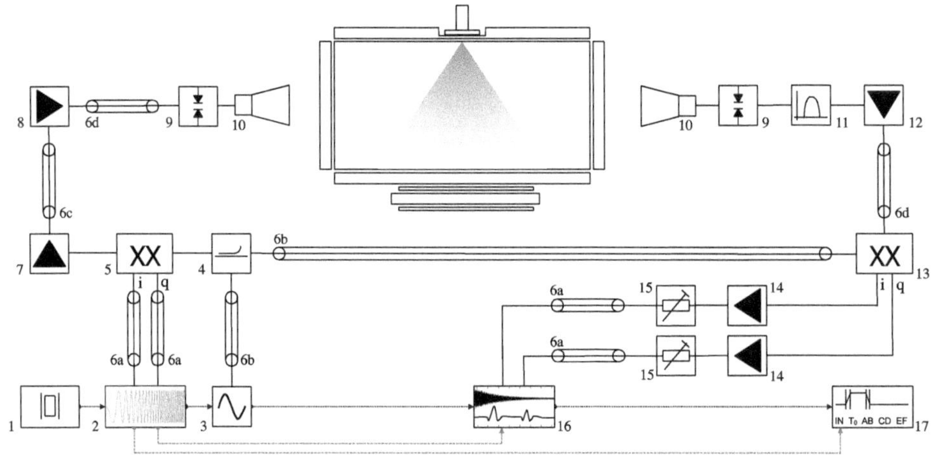

Figure 6.23. Schematic diagram of the chirped-pulse Fourier transform microwave spectrometer used to record the spectrum shown in Figure 6.22. Molecules are admitted to the vacuum chamber (top center) by a pulse free jet. They are excited by a short microwave pulse coming from the antenna shown on the left. The coherent emission signal, called the FID is collected by the antenna on the right. The excitation signal is created in an arbitrary waveform generator (AWG, 2), which is multiplied (xx,5) with the signal from a synthesizer (3). The FID signal is recorded with a fast digital oscilloscope (DSO, 16). All signals are referenced to a 10 MHz reference clock (1). The rest of the circuitry is used for synchronization and optimized detection.

shown in Figure 6.22 has been recorded in a waveguide cell rather than from the cold jet, in order to record many more rotation-tunneling transitions for the demonstration purpose here. The tunneling splittings of all observed rotation-tunneling transitions fall into a very small range near 24 GHz. However, the small differences for the different $|n = 0, J, K, M\rangle$ states show that the splitting depends not only on the vibrational state but also on the rotational state. In effect, there must be a coupling of both degrees of freedom in order to generate the small shifts in the tunneling spectrum. The tunneling spectrum may be predicted using the pgopher program. In the perturbation section for the symmetric rotor molecules of the web description of the program (Western 2017), the additional operators responsible for the shift in the splitting are given. Here, the two $0^{\pm}$ sub-states are introduced as different vibrational states, and then operators $\hat{J}_{\pm}^{m}$ with $m = 6$ and 12 are introduced. As a result, different distortion constants for the symmetric and antisymmetric wave functions are determined and fit to the spectrum. As will be seen below, tunneling also occurs in molecules other than NH_3.

Both panels in Figure 6.22 show the same spectrum. In the lower panel, the intensity is plotted logarithmically in order to distinguish signals in the spectrum from the noise floor. Here, even very weak transitions or transitions involving low populations are recorded with high S/N. This demonstrates the great sensitivity of chirped-pulse Fourier transform microwave spectroscopy, which makes this technique very popular. Thus, today it is used in many laboratories, with increasing ranges of frequencies from a few GHz, where high-power amplifiers (up to kW) guarantee the coherent excitation of even large molecules with small dipole moments and rather unfavorable partition functions, up into the sub-millimeter range, where more powerful amplifiers (mW) would be desirable for a similar performance. The short-time spectroscopy possible with these new techniques opens up new options for experiments, like effective double-resonance spectroscopy, which is instrumental in the assignment of complex spectra. Due to the high sensitivity, spectra of radicals and ions are becoming detectable. Last but not least, the time information contained in the FID signal can be used to infer rate coefficients for inelastic collisions, a property as important as Einstein coefficients for optical transitions when it comes to the interpretation of astrophysical spectra. In fact, collisional excitation and radiative transfers are competing, and which of the two processes dominates is dependent on the number density. The quantity that describes the break-even situation is called the critical density. A more thorough discussion of the excitation of astrophysical molecules is given in the chapter by Paul Dagdigian.

6.8.2 Internal Rotation: CH_3 Torsion

Aside from the inversion tunneling of ammonia, the most prominent large amplitude motion (LAM) appearing in many molecules is the tunneling motion associated with internal rotation, the study of which is an active field in microwave spectroscopy. Many standard molecules carry, for example, a methyl, CH_3, side group. Therefore, it is instructive to consider such a prototypical case. This subgroup is attached to the rest of the molecule. In the minimum energy configuration, the methyl group has a fixed orientation with respect to the other part of the molecule. As an example of such a molecule, the methacrolein molecule, C_4H_6O, will be discussed. Its equilibrium structure is shown in Figure 6.24. It is an asymmetric rotor with an asymmetry parameter of $\kappa = -0.49$, thus making it a near-prolate asymmetric top molecule. The methyl side group shown at the top of Figure 6.24 has a fixed orientation with respect to the planar structure of the rest of the C_3H_3O molecule.

The internal rotation of the CH_3 group about the bond axis leads to a periodic potential with three equivalent minima, each time when the methyl group turns into the same equilibrium orientation. This potential is shown in Figure 6.25 as a function of the torsional angle α. The situation is very similar to the tunneling

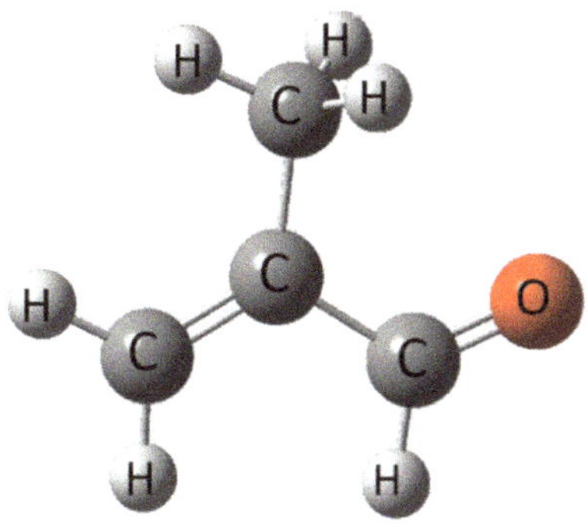

Figure 6.24. Equilibrium structure of the methacrolein molecule, a near-prolate asymmetric top molecule with internal rotation of the CH_3 subgroup.

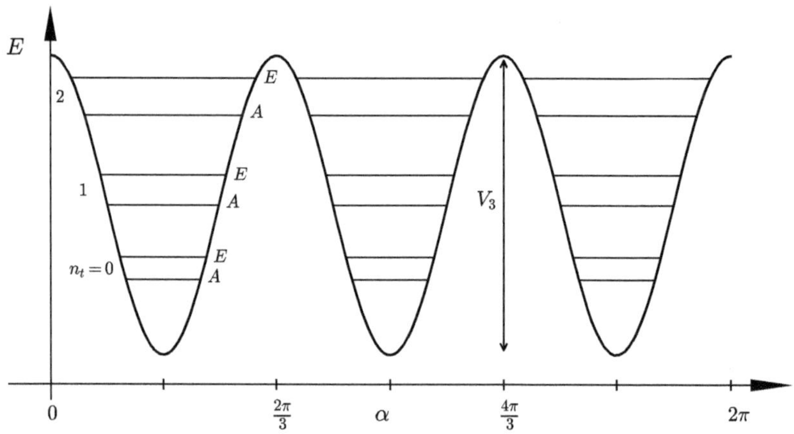

Figure 6.25. Potential energy curve for the torsional motion of a methyl group, CH_3, along the torsional coordinate α. Tunneling the threefold symmetric potential leads to a splitting of the states corresponding to a doubly degenerate E state and a non-degenerate A state. This splitting increases for torsionally excited states ($n_t = 1, 2...$).

motion in NH_3, but this time the potential has three equivalent minima and not two. This potential can be parameterized as

$$V(\alpha) = \sum_k a_k \cos kN\alpha \tag{6.44}$$

which may be rewritten to incorporate the potential energy barrier height

$$V(\alpha) = \frac{V_3}{2}(1 - \cos 3\alpha) + \frac{V_6}{2}(1 - \cos 6\alpha) + \dots \tag{6.45}$$

Here, the first term is the leading term due to the threefold (C_3) symmetry of the problem. The Schrödinger equation for the torsional problem may thus be written as

$$-F\frac{d^2}{d\alpha^2}\Phi(\alpha) + V(\alpha)\Phi(\alpha) = E\Phi(\alpha) \tag{6.46}$$

where $-F\frac{d^2}{d\alpha^2}$ is the operator for the kinetic energy. Here, E and $\Phi(\alpha)$ are the eigenvalue and eigenfunction of the internal rotor and $F = \frac{\hbar^2}{2rI_\alpha}$ is the internal rotational constant with $r = 1 - \sum_g \lambda_g^2 \frac{I_\alpha}{I_g}$. The λ_g with $g = X, Y, Z$ are the direction cosines of the rotational axis (i) of the torsional top in the coordinates of the principle axis system of the molecule, $\lambda_g = \cos\theta(i, g)$. It is apparent that the solution to Equation (6.46) will depend on the relative orientation of the methyl group with respect to the rest, which is why these angular functions have to appear in the Schrödinger equation.

For the case of a free rotor ($V_3 = 0$), Equation (6.46) only contains the kinetic energy part with the solution $E_m = Fm^2$, where $m = 0, \pm1, \pm2, \dots$. For the infinitely high barrier, the harmonic oscillator solution is $E_{n_t} = h\nu_t(n_t + \frac{1}{2})$ with $n_t = 0, 1, 2, \dots$. The vibrational frequency $\nu_t \sim (FV_3)^{\frac{1}{2}}$ depends on both the barrier height and the torsional rotational constant. In this case, ν_t is triply degenerate because the torsional oscillation happens in one of the three equivalent potential wells.

For a finite barrier height, however, quantum-mechanical tunneling leads to the separation of the energy levels. The eigenfunctions must be periodic, $\Phi(\alpha + 2\pi) = \Phi(\alpha)$. Therefore, without a detailed solution of the general Schrödinger equation, Equation (6.46), the eigenfunction can be written as a superposition of periodic functions

$$\Phi(\alpha)_{n_t\sigma} = \sum_{k=-\infty}^{\infty} A_k^{(n_t)} e^{i(3k+\sigma)\alpha} \tag{6.47}$$

with coefficients $A_k^{(n_t)}$ for any given torsional state n_t. Here, k and σ are integer values to impose the periodicity of the solution. The phase factor $e^{i\sigma\alpha}$ distinguishes three different (symmetry) cases, $\sigma = 0$ describes the non-degenerate A state, and the case $\sigma = \pm 1$ a doubly degenerate E state. As the inversion tunneling states of ammonia with different wave functions led to different eigenenergies, here, the states with $\sigma = 0$ and $\sigma = \pm 1$ do as well. Thus, the energies are different for $\sigma = 0$ and $\sigma = \pm 1$. The latter two states are obviously degenerate because $e^{i\pm\alpha}$ are just phase factors with no effect on the eigenenergy. It is exactly this energy structure of the torsional states that becomes apparent in Figure 6.25. First, the levels $n_t = 0, 1, 2$ describe the torsional vibrational states. Subsequently, each of these levels is split into an A and an E state following the description above.

In the sense of the quantum mechanical treatment throughout this chapter, the rotational and torsional problem could be separated, as the Hamiltonian could be separated in the respective coordinates, $\hat{H} = \hat{H}_{\text{rot}} + \hat{H}_t$. Thus, for each torsional vibrational state, one would observe a set of rotational states, or vice versa. In fact, the latter splitting is depicted in Figure 6.26. On the left of this figure, energies for the $|J, K\rangle$ states of a symmetric top molecule are given. The asymmetry splitting leads to a separation of each $K = K_a$ level[20] into two levels with $K_a + K_c = J$ and $K_a + K_c = J + 1$, as also seen in Figure 6.12. These states are further split into the A and E states by the torsional tunneling splitting of the hindered rotor. As depicted in this figure, the selection rules for torsional states are $\Delta\sigma = 0$, which results because the dipole moment is independent of the internal rotation angle. Transitions from one internal sub-level to another are thus not allowed. The transition energies for the A and E states as depicted in Figure 6.12 would not differ. As a consequence, both rotational spectra would overlap, and A and E states could not be distinguished from rotational spectra. Only from transitions when the torsional state n_t changes could this be inferred, because the splitting for A and E states is different for the ground and excited states. Nevertheless, an A/E splitting may also be observable for the torsional ground state only. Therefore, there must be a substantial coupling between the two degrees of freedom (rotation and torsional vibration).

[20] The discussion is equivalent for K_a and K_c. Here, $K = K_a$ is chosen, as this is the case closer to the methacrolein molecule discussed later.

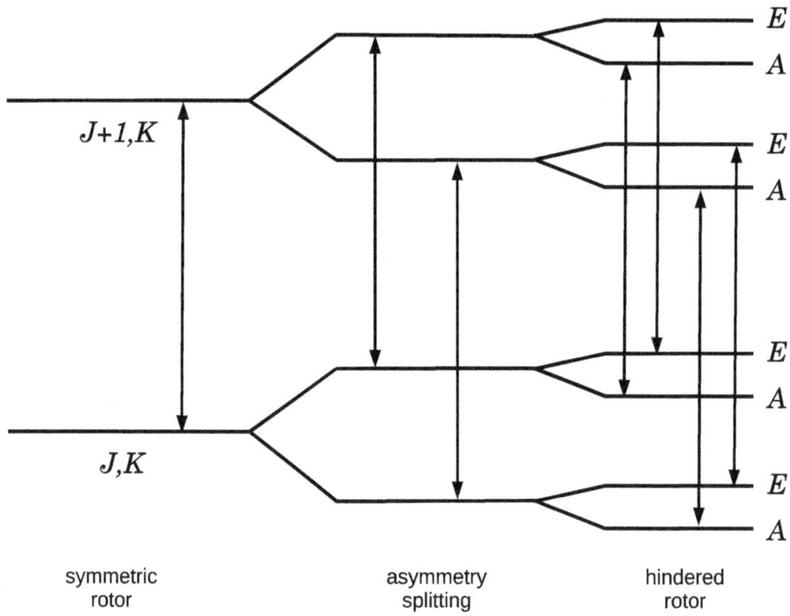

Figure 6.26. Energy-term diagram of an asymmetric rotor with internal rotation of a methyl, CH_3, group. Reading from left to right, one transition for a symmetric rotor splits into two due to the asymmetry splitting. Each of these transitions is further split into one A and one E transition. Depending on the amount of distortion due to the asymmetry or the methyl torsion, up to four lines can be observed.

6.8.3 Rotation Torsion Coupling

Because of the large amplitude motion in α, the molecule can no longer be considered as a rigid molecule. In practice, there is a considerable coupling of torsional motion with the end-over-end rotational motion of the molecule. Following the description in the principle axis of the molecule, including the torsional motion along a different axis of (internal) rotation, the complete Hamiltonian might be written as $\hat{H} = \hat{H}_{\text{rot}} + \hat{H}_t + \hat{H}_{\text{rot},\,t}$. So far, solutions for the first two separated parts have been considered and the multitude of states will be expressed in terms of the quantum numbers $|J, K, M\rangle$ for the rotational problem ($\hat{H}_{\text{rot}}$) and $|n_t, \sigma\rangle$ for the torsional problem ($\hat{H}_t$). Coupling terms in $\hat{H}_{\text{rot},\,t}$ obviously contain the angular momentum operators for the end-over-end rotation, $\hat{J}_g$, and the internal rotation, $\hat{J}_\alpha$, $\hat{H}_{\text{rot},\,t} \sim F\hat{J}_g\hat{J}_\alpha$, which leads to further shifts of the respective states depicted in Figure 6.12. In principle, the treatment is rather involved and it is beyond the scope of this chapter. However, for simplicity, we assume that we are concerned with a near-prolate, (slightly) asymmetric top molecule with a methyl rotation axis that (almost) coincides with the molecular a-axis. Analysis reveals eigenenergies for such a case that can be written as (Gordy & Cook 1984)

$$E_{\text{rot},\,n_t,\,\sigma} = E_{\text{rot}} + F\sum_k W^{(k)}_{n_t,\,\sigma}(\hat{J}_Z\hat{J}_\alpha)^k \tag{6.48}$$

where E_{rot} are the energies of the unperturbed asymmetric rotor and $W^{(k)}_{n_t,\sigma}$ are terms calculated in kth-order perturbation theory. Only a very few of these terms need to be considered to model spectra of many asymmetric rotors with internal rotation. As it turns out, the solution might be rewritten as a modified rotational constant $A_{n_t,A}$ or $A_{n_t,E}$, where the W-terms appear as a correction in the following form

$$A_{n_t,A} = A_{n_t,0} + F\rho_X^2 W^{(2)}_{n_t,A}. \tag{6.49}$$

Here, $A_{n_t,0} = \frac{\hbar^2}{2I_X}$ is the rotational constant of the unperturbed asymmetric rotor, and $\rho_X = \lambda_X \frac{I_\alpha}{I_X}$ with λ_X is the direction cosine as discussed before. Thus, depending on the value of $W^{(2)}_{n_t,A}$ and $W^{(2)}_{n_t,E}$, the rotational constant A changes—and therefore, the rotational spectra for the A and E torsional states differ and may appear as different transitions, depending on the experimental resolution. Therefore, depending on these influences, A and E transitions might be split or blended, which has to be inspected carefully in the recorded spectra.

6.8.4 Example: Methacrolein

In order to evaluate the simplified treatment of a molecule with internal rotation, a spectrum of methacrolein, C_4H_6O (see Figure 6.24), is shown in Figure 6.27 (Zakharenko et al. 2016). Only a 300 MHz section of the very rich and seemingly confusing spectrum is shown. The quantum numbers J_{K_a,K_c} indicate the rotational levels in the convention used so far for the semirigid asymmetric rotor. In the spectrum shown, at around 192.590 GHz, some a-type transitions with $\Delta K_a = 0$ and $\Delta K_c = 1$ for the torsional ground state molecule are identified. These transitions are rather intense because the dipole moment along the molecular a-axis is considerable: $\mu_a = 2.67$ Debye (compare $\mu_b = 0.84$ Debye). The E and A transitions ($32_{0,32} \leftarrow 31_{0,31}$ and $32_{1,32} \leftarrow 31_{1,31}$) are blended and appear as one peak with the characteristic second derivative line shape of an FM absorption spectrum. Please note that these transitions are emerging from $J = K_c$ levels. Their asymmetry splitting is rather small and not resolved. Therefore, the asymmetry splitting and the E/A splitting are unresolved.

In the same stretch of the spectrum, the lines for which J is decreased by 1 while K_a is increased by 1 and K_c is reduced by 1 appear at around 192.360 GHz. Here, the E and A transitions ($31_{1,30} \leftarrow 30_{1,29}$ and $31_{2,31} \leftarrow 30_{2,29}$) are separated by a few MHz. It shows the increase in the splitting with rising K_a. This would be expected if the rotational A constant is affected by the rotational–torsional coupling, i.e., $A_{n_t,A}$ and $A_{n_t,E}$ are different. In fact, in the prolate symmetric rotor limit, the energies would scale as $(A - B)K_a^2$, and thus the differences would increase with increasing K_a. Again, the asymmetry splitting is still blended for this set of transitions emerging from the $30_{1,29}$ and $30_{2,29}$ rotational states. Of course, many more coupling terms and interactions discussed in the previous section on the rotation–torsional coupling are included in present day simulations of such spectra, but the difference between

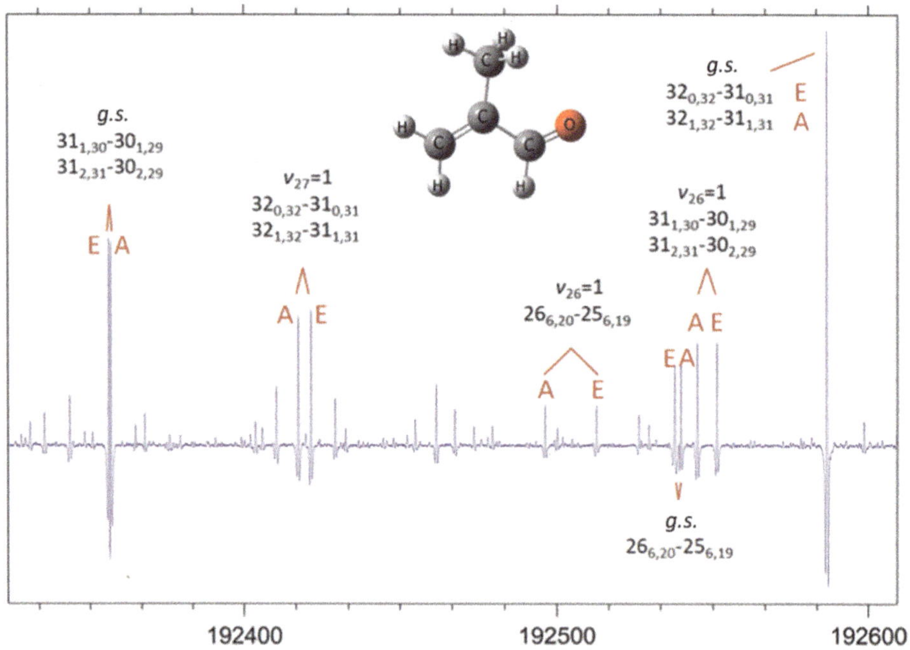

Figure 6.27. Small portion of the rotational spectrum of methacrolein near 192.5 GHz. Splittings due to the tunneling motion of the torsion of the methyl side group are blended or only partly resolved for the ground-state molecules. These *E*/*A* splittings vary with the level of vibrational excitation (ν_{27} CH_3 torsional mode, ν_{26} CCC out-of-plane bending mode). Details of the vibrational rotational coupling are explained in the text. (Inset) The CH_3 side group is shown at the top of the molecular structure. (Credit: Zakharenko et al. 2016.)

blended and separated lines is easily observed in this sample spectrum of methacrolein.

The spectrum shown in Figure 6.27 does not only show the transitions of vibrational ground-state methacrolein. In fact, C_4H_6O has 11 nuclei—and thus, 27 vibrational degrees of freedom. Among those, the ν_{27} mode is the CH_3 torsional motion with the lowest vibrational frequency (130.8 cm^{-1}). This is the motion for which we are also concerned with tunneling. The next lowest in energy vibration is ν_{26} (169.8 cm^{-1}), which is the C–C–C out-of-plane bending mode that does not include the carbon atom of the methyl group. To visualize these motions, see Figure 6.24. Both these modes are excited in a considerable fraction of a methacrolein sample held at room temperature. Therefore, rotational transitions of these vibrational excited molecules are also observed in the spectrum displayed in Figure 6.27.

As it turns out, the *E* and *A* splitting is much larger for the vibrationally excited methacrolein. Around 192.560 GHz, the same rotational transitions ($31_{1,\,30} \leftarrow 30_{1,29}$ and $31_{2,\,31} \leftarrow 30_{2,\,29}$ for $\nu_{26} = 1$) as for the ground state are observed and the *E*/*A* splitting increases roughly by a factor of ten. A splitting of a similar size is observed for the ($32_{0,\,32} \leftarrow 31_{0,\,31}$ and $32_{1,\,32} \leftarrow 31_{1,\,31}$) transitions of the torsionally excited molecule ($\nu_{27} = 1$). This is expected, as the E/A splitting of the energy levels increases as shown in Figure 6.25 for the torsionally excited species.

The E/A splitting becomes larger when K_a is further excited. As an example, the splitting for the $26_{6,\,20} \leftarrow 25_{6,\,19}$ transition of the vibrational ground state is already sizable (~192.550 GHz) and amounts to about 20 MHz for the $\nu_{26} = 1$ excited state (~192.500 GHz). Here, the asymmetry splitting has separated lines corresponding to K sets, which are degenerate in the symmetric rotor limit and practically also in the $J = K_a$ or $J = K_c$ limit.

In summary, tunneling splittings and the manifold of vibrational excited states make the spectra of even moderately sized molecules like methacrolein, C_4H_6O, extremely rich and cumbersome to analyze in detail. The spectrum shown in this subsection belongs to the lower-energy conformer of methacrolein shown in Figure 6.24. Here, the three carbon atoms (without the one carbon in the methyl group) and the oxygen atom are in trans configuration, which gives this conformer its name, trans-methacrolein. A ~14 kJ/mole higher-energy conformer arises when the H–C=O end group of the molecule is rotated by 180° about the connecting C–C bond, thus putting the CCCO in a cis configuration. Again, the spectral complexity rises because these conformers also exist in a thermal environment. For many relevant molecules, even more conformers have to be considered or the number of internal rotations is larger than one like that in the astrophysical very abundant dimethyl ether (DME), CH_3OCH_3, with two equivalent methyl groups.

By now, the reader can already appreciate the additional complications arising when considering isotopologues of these molecules. Replacing a ^{13}C for the most abundant ^{12}C in a molecule like DME might appear to be a minor complication, as the spectra of the two isotopologs should be very similar because the rotational constants only change very little. In fact, the spectra differ quite significantly, as the symmetry of the molecule is affected by replacing just one carbon atom. Replacing both carbon atoms simplifies the spectroscopy significantly, but the spectrum might not be as relevant for the interpretation of astrophysical observations as the relative abundance of ^{13}C in relation to ^{12}C will probably not lead to intense spectra of doubly substituted DME, as one example. The situation becomes even more complicated when a proton is replaced by a deuteron and thus a CH_3 rotor, as another example, with a threefold symmetric potential (as discussed in this section in some detail) is replaced by a CH_2D or CD_2H group. Of course, spectra of these molecules are needed for astrophysical observations, but the spectroscopy becomes even more involved than shown in this chapter. Nevertheless, now it is time to just look at a few example observational spectra where the spectroscopy tools introduced in this chapter have helped to identify molecules in space.

6.9 Astrophysical Spectra

In this last section, three selected examples of molecular spectra from space are shown, where recent work related to astrophysics illustrates part of the content of this chapter.

6.9.1 Propanal: a Complex Molecule Seen with ALMA

Methanol, CH_3OH, was the first complex organic molecule found in space, in 1970. Astrochemists consider molecules having at least six atoms to be complex. Finding such molecules in space came as a bit of a surprise, as it was believed that polyatomic molecules would be difficult to form in the mostly low-temperature environment of molecular clouds. Furthermore, the harsh conditions of the interstellar radiation field and cosmic rays, which are mostly high-energy protons from supernova events, were thought to destroy those molecules even if they could form somehow. However, since then, more and more complex molecules have been found in space. Today, around 200 such species are known, with many additional isotopologues. Often, molecules come in groups of naturally growing complexity. Methanol, the simplest of all alcohols, is found in large abundance, not only in gas phase but also through broad ice features in infrared observations, showing that methanol is the second-most abundant ice component after water. Inserting CH_2 groups into this molecule leads to ethanol, which has been found in space, as well as more complex alcohols. A similar trend is seen for the methyl cyanide molecule, CH_3CN, for which a laboratory spectrum was previously shown as an example of a symmetric rotor. Including CH_2 groups, like in the case of methanol, ethyl cyanide and propyl cyanide are also formed—both of which have been found in space in considerable abundances.

Moderately large organic molecules can display isomerization, i.e., molecules with the same sum formula (same number of C, H, etc. atoms) but a different molecular structure. In addition, they may occur as different conformers, i.e., with the same chemical bonds but having sub-units that display distinguishable orientations, e.g., a rotation around a CC single bond by $\sim\pm120°$. This we have seen already for the methacrolein molecule, which comes in the two forms named cis and trans. For quite some time, propyl cyanide has been observed in space in a chain-like structure (n-propyl cyanide) (Müller et al. 2016). Recently, however, a different isomer, iso-propyl cyanide—a version of the molecule where a carbon group branches off from the carbon chain—has been detected for the first time (Belloche et al. 2014). This is a rather important step in astrochemistry, as many more complex molecules—in particular, building blocks of biological molecules like amino acids—exhibit a branched rather than a chain-like structure. These findings raise the interesting question of whether precursor molecules of life—in particular, amino acids like glycine, NH_2CH_2COOH—can be found in space, leading to an ongoing hunt for ever-more-complex molecules. Perhaps the closest shot to glycine is aminoacetonitrile (Belloche et al. 2008), NH_2CH_2CN, which already shares a substantial fraction of its components with this molecule. Other interesting aspects are concerned with the question about homo chirality of molecules. Recently, propylene oxide, CH_3CHOCH_2, has been detected as a first example of a chiral molecule in space (McGuire et al. 2016).

Another interesting molecule recently studied in the THz frequency range in greater detail in the Cologne laboratories is propanal, CH_3CH_2CHO. This molecule comes in two different conformers. Figure 6.28 shows the calculated potential

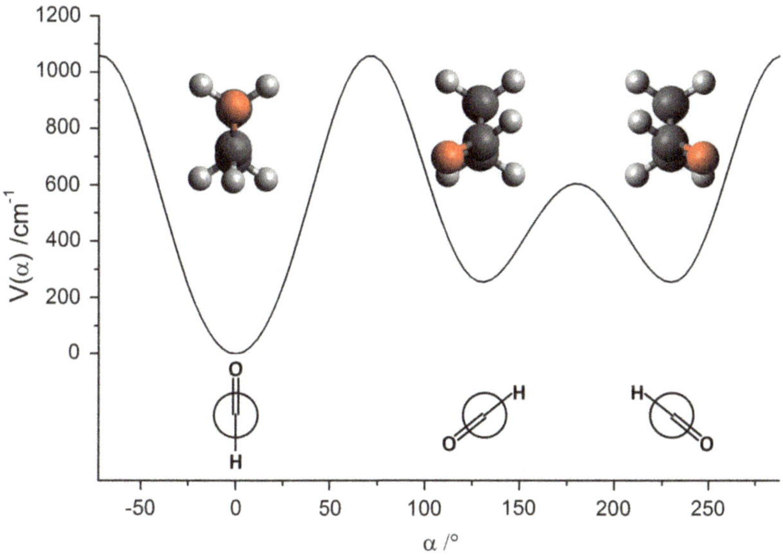

Figure 6.28. Potential energy curve of propanal, CH_3CH_2CHO. The most stable syn-configuration is achieved when the H–C=O aldehyde group is in plane with the C–C–C backbone of the molecule. The gauche configuration is twofold degenerate and the aldehyde group is rotated. Tunneling of the methyl, CH_3, group is present in both configurations. (Credit: Zingsheim et al. 2017.)

energy surface for propanal as a function of α, which is the relative orientation of the aldehyde, H–C=O, group with respect to the C–C–C skeleton of the molecule. The situation is similar to—but more complicated than—that for methacrolein, which has been discussed before. The most stable structure is the so-called syn-configuration, where the heavy elements CCCO all are in one plane ($\alpha = 0°$ in Figure 6.28). In the gauche conformer, the aldehyde, HCO, group is rotated by $\sim\pm128°$. Due to the symmetry of this molecule, these two orientations are equivalent. The low barrier between both conformations leads to tunneling and creates a symmetric and anti-symmetric tunneling state. For both of these conformers, there is an *E*/*A* splitting of the CH_3 torsional motion, similar to what has been found for methacrolein above.

It turns out that the spectrum of this one molecule is already extremely dense, and thanks to the high sensitivity of the absorption instruments it has been found in the laboratory spectrum that there are lines almost everywhere in the millimeter and sub-mm wavelength regime (Zingsheim et al. 2017). In order to analyze such spectra, it is therefore instrumental to consider the line intensities as well. Figure 6.29 shows only a 85 MHz part of the rotational spectrum near 210 GHz. The top line is the experimental spectrum. The lowest three panels show the contributions for the syn-conformer, i.e., from the vibrational ground state, the $\nu_{24} = 1$ excited aldehyde torsion and the $\nu_{23} = 1$ excited methyl torsion. The tunneling splittings are partly blended and partly resolved (vibrational excited states), as already seen in more detail for methacrolein. In fact, for propanal, the asymmetry splitting together with the *E*/*A* splitting leads to four transitions, as shown schematically in Figure 6.26.

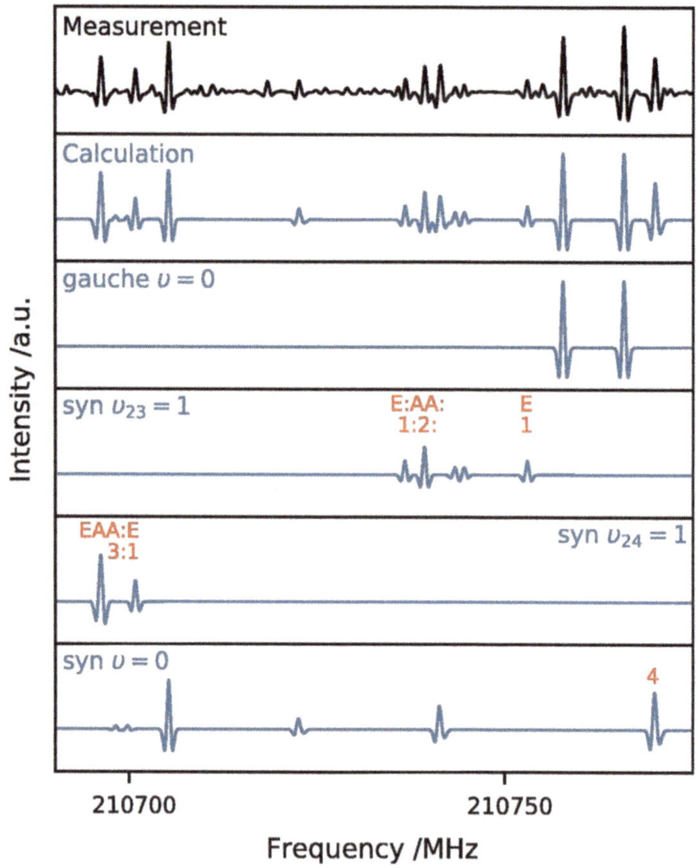

Figure 6.29. An ~85 MHz stretch of the rotational spectrum of propanal. The different panels show calculated contributions from the two syn- and gauche-conformers for different levels of torsional excitation. The second spectrum from the top is the composite of all spectral simulations that shall be compared to the experiment (top panel). (Credit: Zingsheim et al. 2017.)

Here, these four lines are only partly resolved for the syn-conformer, leading to the respective intensity ratios indicated in the respective panel of Figure 6.29. The fourth panel from the bottom displays the contribution to the spectrum from the ground-state gauche-conformer, which shows a well-separated doublet due to the fact that the barrier of the aldehyde torsion is so low—and therefore, in the ground state, there is a substantial splitting already. Altogether, the composite hypothetical spectrum (calculation) is in very good agreement with experiment (measurement), even though it shows many more features that belong to less-intense lines from even more excited vibrational states, etc., details that go beyond the aims of this introductory chapter.

The first detection of propanal was toward the galactic center, Sgr B2(N), and very recently in the Protostellar Interferometric Line Survey (PILS) of the low-mass protostellar binary source IRAS 16293-2422. The PILS study is based on observations with the Atacama Large Millimeter/submillimeter Array (ALMA) that allow detection of regions of less intense transitions than in previous studies. While the first studies found propanal in colder environments, the most recent detection by the PILS campaign found it in a warmer environment around 125 K, close to the protostar, where molecules desorb from grains and thus the column density is high. Luckily, the lines toward one component in the astrophysical object IRAS 16293-2422 are very narrow, so the rotational lines of propanal (and other molecules) can be critically investigated. Figure 6.30 shows small portions of the ALMA-PILS radio astronomical spectrum in order to identify lines associated with propanal. Spectral discrepancies between predictions prior to the recent THz laboratory investigations and the observations become apparent. They are seen when comparing the histogram-style observation with the dashed-line predictions. Based on the current molecular parameters, the new predictions (solid red lines) are now in much better agreement with the observations. Therefore, it can be concluded that the observed lines definitively belong to propanal. Moreover, more reliable temperatures and other relevant physical parameters can be extracted from the spectral analysis of the PILS spectra. This is but one example where highly accurate measurements lead to predictions that can unambiguously identify and quantify molecular species from radio astronomical observations. Many molecules, their isotopologues, vibrational excited states, torsions, and tunneling states are investigated this way by a number of laboratories worldwide. They guarantee that the billion-dollar investment of the telescopes turns into quantifiable science results. This chapter shows that a tremendous knowledge of molecular spectroscopy is needed for this purpose, but spectroscopy is the most accurate and detailed way to achieve this goal.

6.9.2 Protonated Hydrogen, H_3^+, Molecules as a Chemical Clock

Molecular ions attract the attention of astrochemists because many ion–molecule reactions are barrierless and therefore one way to form more complex molecules even in very cold astrophysical environments like dark molecular clouds ($T \sim 10K$). Among those ions, protonated hydrogen, H_3^+, has a special role. It is the most abundant ion because it is only composed of hydrogen, which is by far the most abundant element in space. H_3^+ is formed very rapidly in collisions of H_2^+ with H_2. Due to the high abundance of H_2, n(H_2) $\sim 10^4$ cm^{-3}, the timescale for this process is on the order of days to months under molecular cloud conditions. This is extremely short compared to other chemical reactions and to the timescale ($\sim 10^4$ yr) on which the chemical composition changes, according to chemical models presented in other chapters of this book, e.g., in the chapter entitled *The Molecular View Point of Interstellar Observations* by Evelyne Roueff. Therefore, this reaction can be thought of as happening instantly and the chemistry of H_3^+ is more determined by the formation of H_2^+, which is given by the abundance of cosmic rays (CR), which are mostly fast protons from highly energetic events like supernovae. The abundance of H_3^+ is thus a balance of this formation rate and its destruction rate.

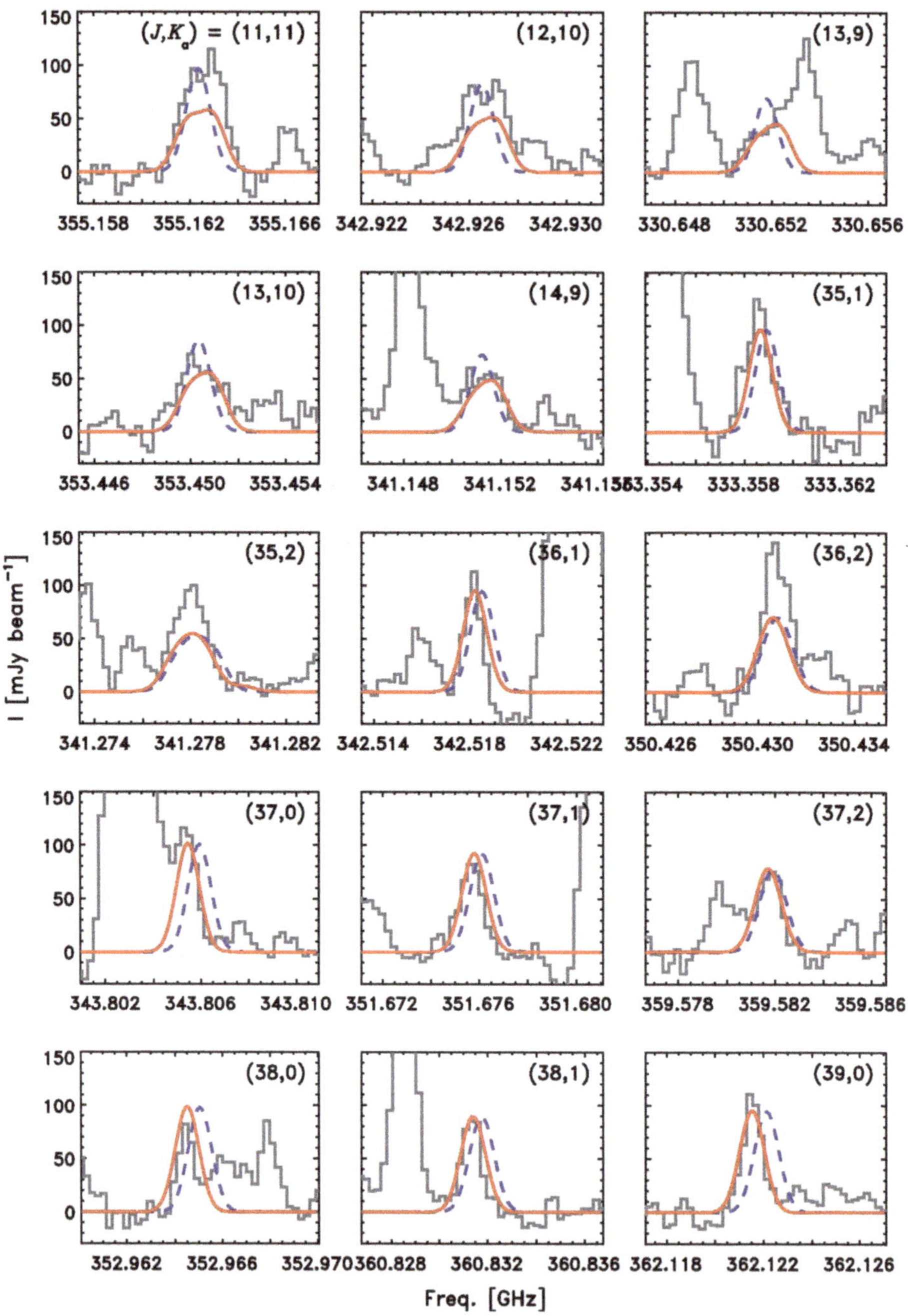

Figure 6.30. Fifteen pieces of spectra from the ALMA-PILS observational campaign, including propanal transitions. The observations are pointed toward the low-mass protostellar binary IRAS 16293-2422 in the vicinity of newly born low-mass stars, where ice from grain surfaces is evaporated leading to a large variety of complex molecules, among them propanal. Line predictions based on literature values are depicted by dashed lines and are compared to the observed spectral stretches (histogram). Lines predicted based on the most recent laboratory studies in the millimeter and sub-mm range (solid curve) show a clear detection for all lines associated with propanal. (Credit: Zingsheim et al. 2017.)

Both of these rates are of great interest for astrophysicists because it turns out that the CR ionization rate depends strongly on the penetration depth of the CR into a molecular cloud, which itself depends on the CR energy. Thus, the abundance of H_3^+ is intimately related to these quantities, and one could also learn a lot about cosmic rays and their energy distribution from H_3^+ observations. Astrophysical observation of the very interesting H_3^+ molecule is hampered because the symmetric triangular-shaped molecule (D_{3h} molecular symmetry group) does not have a permanent electrical dipole moment and therefore no classical rotational spectrum. In fact, H_3^+ was detected in 1996 for the first time in infrared absorption observations where the interstellar medium was observed against an IR-luminous star (Geballe & Oka 1996). Today, these IR observations of H_3^+ have been turned into a tool for astrophysicists because IR absorption features of this molecule have been found in many different environments that apparently have different ionization rates (Indriolo & McCall 2012). However, different destruction mechanism are also at work. Dissociative recombination is one important process where a cation meets an electron to neutralize.[21] As a result of the formation and destruction processes at work, the H_3^+ abundance is potentially a measure for the degree of ionization in an astrophysical environment.

Also, it should be noted that the proton in H_3^+ is only weakly bound to H_2, and therefore H_3^+ is a general proton donor to other molecules, such as the abundant CO or N_2 molecules. Of course, this can only happen when the environment is not so cold that these species are condensed on grains. Therefore, the presence of HCO^+ and N_2H^+ makes for a very valuable diagnostic for the coldest astrophysical environments. To summarize, H_3^+ is one of the key molecules that tell an astrophysical story about the physical conditions and the chemical evolution of the observed region, and thus their observation is of great interest. This importance arises from the fact that not too many processes are involved in the determination of its abundance, and therefore there is a chance to make a link between the abundances of a small set of molecules and a few chemical reaction rates. For complex molecules like propanal or others, the chemical networks in which they are but one product are too intricate to determine individual rates from observations. With H_3^+, the situation is quite different and much more favorable.

The process called isotopic fractionation comes into play when it becomes energetically favorable to substitute an abundant isotope with a less abundant one; see, e.g., the most recent references (Roueff et al. 2015; Bovino et al. 2017; Vastel et al. 2017) on this topic. For H_3^+, this process is well-known for collisions with HD

$$H_3^+ + HD \rightleftharpoons H_2D^+ + H_2. \tag{6.50}$$

The forward reaction is exothermic by 232 K. Under cold cloud conditions ($T \sim 10$ K), this leads to a strong enhancement of H_2D^+ with respect to H_3^+. While the statistical

[21] However, the bond energy of the electron to the cation is available; remember that these values are in the eV range where 1 eV corresponds to a temperature of about 11,600 K. Therefore, the molecule probably breaks up, often into more than just two heavy fragments. This is of concern when trying to understand which neutral molecules arise from dissociative recombination.

ratio of placing a deuteron in H_3^+ is on the order of 10^{-4}, based on the D/H ratio in space, the H_2D^+/H_3^+ ratio might reach values larger than 0.1. In fact, even higher degrees of deuteration leading to D_2H^+ or even D_3^+ are postulated in very cold environments, where it becomes more likely to meet HD than CO, which would be frozen out, while H_2 and its isotopologs are still in the gas phase. This behavior also makes H_3^+ and its isotopic variants a subject of astrophysical interest. Due to the asymmetry of the variants with mixed isotopes, these molecules possess small but significant dipole moments and thus exhibit a rotational spectrum. Recent laboratory studies from our group revealed the transition frequencies for those molecules, such that these species could be sought in observations (Asvany et al. 2008, Jusko et al. 2016, 2017b). As these molecules are light species, their moments of inertia are small and the subsequent rotational constants are rather large. Therefore, even the lowest-energy rotational transitions for H_2D^+ and D_2H^+ lie in the THz regime—at 1.37 and 1.48 THz, respectively. These frequencies are currently only accessible by the Stratospheric Observatory for Far-Infrared Astronomy (SOFIA) with the THz receiver instrument GREAT.

Figure 6.31 shows the observation (histogram) of the two lowest rotational transitions of H_2D^+ and the two lowest rotational transitions of D_2H^+ with SOFIA above 1 THz and with the Atacama Pathfinder EXperiment (APEX), which is a prototype ALMA single-dish telescope also located at the Atacama desert in Chile (Brünken et al. 2014; Harju et al. 2017). The source of this observation was the low-mass protostar region IRAS16293 where propanal was found close to the protostars. Here, light from one of the protostars is used to detect both THz transitions of H_2D^+ and D_2H^+ in absorption. This is necessary because the population of the rotational excited state is too low to lead to a classical rotational emission spectrum. In contrast, the lower-frequency transitions of H_2D^+ and D_2H^+ are observed (primarily) in emission. The spectrum is overlaid with simulations (solid lines) for the four lines. All four lines are fit fairly well. Therefore, it can be concluded that the rates for the processes described in Equation (6.50) and subsequent further deuteration steps are well-characterized, leading to a quantitative description of the astrochemical processes involved. It should be said that, despite the simplicity of the reaction processes described in Equation (6.50), pure experimental rate coefficients encompassing state-to-state processes are not available to date. Therefore, a proper comparison of the laboratory in space and the laboratory on Earth is not possible.

One subtlety of these seemingly simple reactions with just hydrogen and its isotopic variants lies in the nature of quantum mechanics, i.e., the Pauli exclusion principle, which was previously discussed for the states of H_2 in the section about symmetry. There, we found that H_2 comes in two nuclear spin isomers, one with the total nuclear spin $I = 1$, comprised of three elementary nuclear spin states, called ortho-H_2, and one with total nuclear spin $I = 0$, with just one elementary nuclear spin state, called para-H_2. Due to the Pauli principle, the odd (even) J quantum numbers are associated with ortho-H_2 (para-H_2). The same situation is found for the ortho and para configurations of H_2D^+ and D_2H^+, as already indicated by the ortho and para assignments in the observations; see Figure 6.31. In fact, the ortho and para populations of H_2 are considered to be statistical, i.e., in a ratio of 3:1, as given

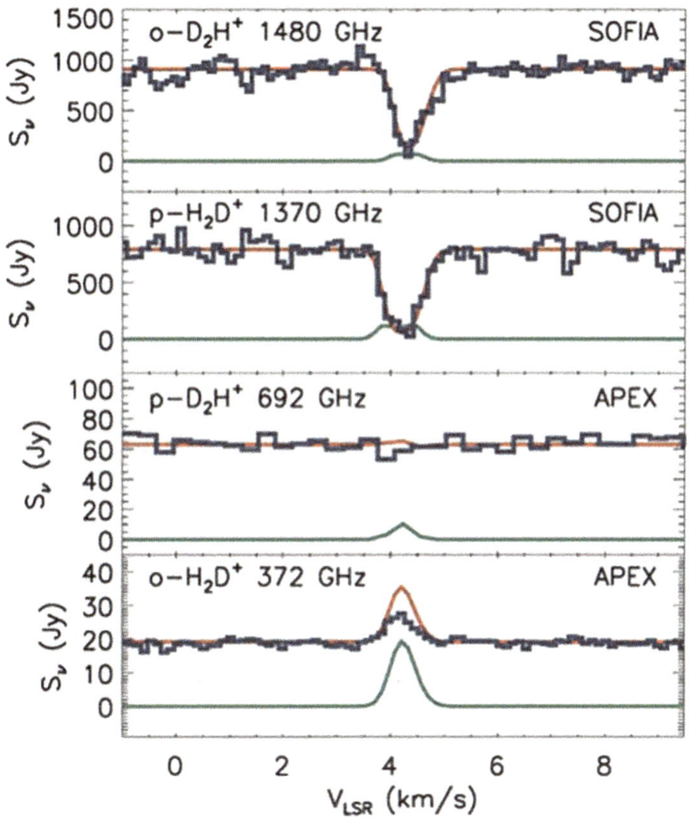

Figure 6.31. Observed and modeled ortho-D_2H^+, para-H_2D^+, para-D_2H^+, and ortho-H_2D^+ spectra toward IRAS 16293. The best-fit model (solid curves overlaid with observations; histogram) are from calculations using rate coefficients for a hydrogen chemistry model as described in the text. Analyses of the combined D_2H^+ and H_2D^+ data result in an age estimate for the cloud core of ~0.5 million years. Thus, ortho-to-para abundance ratios can be used as a chemical clock for astrophysical environments. (Credit: Harju et al. 2017.)

by the number of possible nuclear spin states upon the formation of H_2 on interstellar grains when just two hydrogen atoms meet. Based on this 3:1 ratio, the rotational population of H_2 is highly non-thermal, as many more molecules are in $J = 1$ than expected for a thermal J distribution. Likewise, the ortho-to-para ratio of H_2D^+ and D_2H^+ will be non-thermal but given by the outcome of reactions like Equation (6.50). As a result, detailed reaction rates including the different nuclear spin isomers need to be considered. In fact, so-called state-to-state rate coefficients still have to be measured in the laboratory. Some details on the ortho/para influence are available from previous laboratory work, e.g., Hugo et al. (2009), but the state-to-state rate coefficients are missing from the laboratory and are only available within reaction models, which so far mostly ignore the reaction dynamics (Aoiz et al. 2007).

Despite the fact that the details of these reactions are still hidden, it is very clear that collision processes will, over long periods in the interstellar medium, lead to a thermalized ensemble. Without going into further details, in reactions like

$$\text{para}-H_2D^{+}+\text{ortho}-H_2 \rightleftharpoons \text{ortho}-H_2D^{+}+\text{para}-H_2, \qquad (6.51)$$

a deuteron, D^+, might just be transferred from one H_2D^+ to H_2. Due to the differences in energies of the four partners,[22] rotational energy, e.g., stored in the $J = 1$ rotational state of ortho-H_2, is transferred to translation (heating) for exothermic collisions as in Equation (6.51). As a consequence, the ensemble of H_2D^+ and H_2 would cool internally. Of course, other processes plus the chemical reactions as described in Equation (6.50) have to be taken into account, to be quantitative, but the final result is a change of the ortho-to-para ratio of H_2, H_2D^+, D_2H^+, and surely that of the original H_3^+. The timescale of this process is only given by the gas-phase rate coefficients for reaction (e.g., Equation (6.50)) and inelastic processes (e.g., Equation (6.51)) and the abundance of hydrogen, deuterium, and H_3^+. Thus, the observed ortho-to-para ratios imprinted in the observations for H_2D^+ and D_2H^+ in Figure 6.31 are indicative for the time elapsed since the non-thermal hydrogen was formed, or say, that the cloud was formed. This time has been determined, from these first ortho/para observations of H_2D^+ and D_2H^+, to be on the order of 0.5 million years, which is much longer than the so-called free fall time where cloud material collapses into a dense region where a new star is born (Brünken et al. 2014; Harju et al. 2017). Therefore, molecular physics, in particular the detection of the ortho and para forms of this seemingly simple triatomic molecule, can be used as a chemical clock to date the age of the molecular cloud where these tracers of the coldest places in the universe are found. Thus, molecules and their spectra are the ultimate tools to evaluate the chemical evolution of the interstellar medium from which our Sun and the surrounding planets emerged.

6.9.3 C_{60}^+, One Carrier of the DIBs

The final example concerns another new detection of a key molecule in space and another ion, C_{60}^+. The DIBs were discussed previously in the introduction to this chapter. These many absorption features seen toward reddened stars have long been known, but not a single molecule could be identified as a carrier. However, the situation changed significantly in recent years because spectroscopy in ion traps, as described in some detail here, has improved the sensitivity—and most importantly, the selectivity—of ion spectroscopy, to the extent that now spectra of well-characterized and temperature-controlled molecular ions have become available. Another 22-pole ion trap has been built by John Maier's group in Basel to study the electronic spectra of molecular ions, which are potential carriers of the DIBs (Chakrabarty et al. 2013).

[22] Two on the reactant side and two on the product side, e.g., for the reaction Equation (6.51), but also for all other possible ortho and para combinations.

Figure 6.32 shows a laboratory spectrum of C_{60}^+ in the range of one set of the most intense DIBs at 9577 and 9632 Å (Campbell et al. 2015). The spectrum shown has been recorded using a messenger technique that has been in use since the late 1980s when it was introduced by Y. T. Lee; see Lisy (2006) for an overview. In the case discussed here, a He atom is attached to the parent molecule when it is cooled below 8 K. Upon the absorption of a near-infrared photon (around 960 nm), the He atom is released and the mass of the molecular ion changes by 4 u to 720 u, which is the mass of the bare C_{60}^+. It is instrumental to be able to attach this He atom to C_{60}^+ because only He is bound to C_{60}^+ so loosely that the absorption feature does not shift. This has been nicely demonstrated, as another He has been attached to C_{60}^+ and the unimolecular dissociation spectrum of the C_{60}^+-He_n aggregates look the same in position and width for $n = 1$ and 2. The red vertical lines in Figure 6.32 indicate the positions of the DIBs. This figure shows that the experimental spectrum agrees perfect with the DIB positions. Furthermore, it has already been shown that the width of the experimental features and the DIB features also agree very well. Therefore, it can be concluded that C_{60}^+ is indeed the first molecular carrier of two of the most prominent DIBs.

This could be considered a breakthrough for the question about the origin of the DIB features. Much smaller molecules were expected to be responsible for these features, as the spectral variations based on potentially different temperatures hint at varying rotational envelopes of molecules with only a few heavy atoms. Now with the detection of C_{60}^+ the story is very open again. The traditional scenario for molecule formation starts from atoms and yields molecules as complex as propanal, etc. Aside from this *bottom up* chemistry network, a *top down* scenario is also discussed, in which fragments of heavier structures, such as large pieces of graphitic

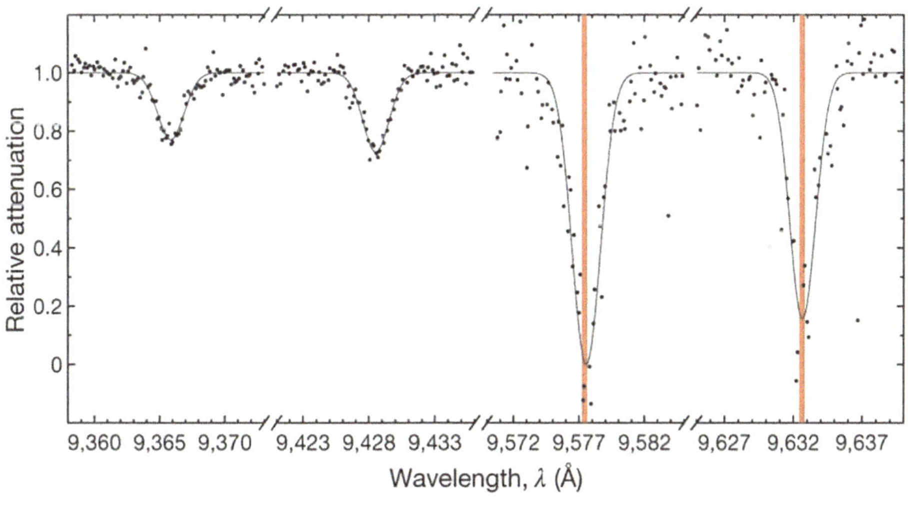

Figure 6.32. Laboratory spectrum of C_{60}^+ in the range of the most prominent DIB features. The laboratory absorption features coincide perfectly with astrophysical observations as indicated by the red vertical bars. (Credit: Campbell et al. 2015.)

clusters (very small grains), emerge as a result of a destruction process. See Huang & Oka (2015) for a current discussion of these ideas. In this context, C_{60}^+ is certainly a very large molecule, especially considering its high symmetry, which leads to narrow and intense spectral features due to high levels of degeneracies, as seen in Figure 6.32. The current findings certainly call for specific laboratory experiments where other pure carbon cluster ions and neutrals, as well as many other species, are investigated.

The detection of just one molecule as a carrier for the DIBs puts new questions to astrochemistry. The spectroscopy described in this chapter, especially new laboratory techniques like the spectroscopy in traps, for which C_{60}^+ and the small molecular ions in the previous sections are just examples, lead to the detection of new molecules in space. Many of these species are not just another molecule; the presence of the deuterated versions of H_3^+ and the detection of C_{60}^+ are key molecules in the chemical networks that shape the universe in our local neighborhood, and perhaps also in other galaxies. Molecular spectroscopy is a prerequisite to obtain information on the molecular content in the universe, meaning this work is instrumental for the interpretation of the cosmic evolution.

6.9.4 Molecular Spectroscopy: a Molecular Physics Perspective

These interstellar molecules, and the many others discussed in this chapter, are also interesting from a molecular physics perspective, which guided us through this chapter. Molecular spectroscopy reveals many more details on the intermolecular forces and the various molecular structures that also make up our environment on Earth. Models for molecular spectra are of high predictive power, and therefore they can be used for the interpretation of ever richer spectra. However, current limits should also be overcome.

Molecular spectra, especially when observing low-lying vibrational motions (low barriers for internal motions), molecules with increasing levels of interactions, molecules at high levels of excitation (high temperatures), or just molecules that do not exhibit a clear molecular structure, are extremely complicated and many molecular parameters are needed to describe them. One direction of research is concerned with automated fitting procedures, which should help to reduce the cumbersome spectral analysis. Taking even a 1 THz spectrum of a molecule requires not more than 24 h. The analysis, however, often takes months, and thus astrophysics could make a quantum leap when automated fitting becomes as routine as finding data using an internet search engine, two processes that share some common algorithms.

Apart from these brute force approaches that obtain more and more molecular parameters to predict large fractions of spectra, there is also need for fundamental theories. Ab initio calculations do not reach the quality of experimental spectra, but they make substantial progress. For example, the inclusion of proper long-range forces[23] was a long-standing problem, but is now in much better shape. As a result,

[23] Or better stated as defining and using proper functionals for calculating molecular energies.

rotational constants for larger molecules where such forces, acting from one part of the molecule to another more distant part, are much improved. Therefore, constants from these calculations lead to faster convergence when trying to find the rotational constants from microwave spectra even for small molecules like propandiol. The derivation of new, more adequate Hamiltonians also seems necessary for molecules where unfavorable choices of coordinates—and thus molecular degrees of freedom —do not lead to a satisfactory model description of molecular spectra.

At last, developing new experimental tools with increased sensitivity, broader bandwidth, higher resolution, and better selectivity for the species studied, are leading to the sort of progress in spectroscopy that is needed to find formerly unexpected species like C_{60}^+. Synthesis of the molecule of interest and sample preparation are often the limiting step in recording their spectra. Here also, new routes are being taken. In one of the more counterintuitive—but presently more and more popular approaches—large fractions of spectra are taken from the many molecules that appear in plasmas (Martin-Drumel et al. 2016). Many molecules are present at once in those spectra. However, the large body of information on known spectra of known species allows those spectra to be reduced to lines of potential new species. The process called spectral taxonomy (Crabtree et al. 2016), and then further selection steps like changing the composition of the precursor molecules, thus adding or subtracting specific atoms, are undertaken to finally pin down the carrier of the recorded spectra. Molecular spectroscopy is an old branch of science with many new approaches. Due to the high demand for spectra of astrophysical relevant molecules, the field is always reinventing itself anew, which constantly leads to surprises like the detection of C_{60}^+ in space.

Acknowledgements

Examples for the outlines of molecular spectroscopy expanded in this chapter are based on work that has been conducted primarily by my research group in Cologne over the last ten-plus years. I am deeply grateful for my longtime collaboration with Oskar Asvany, who prepared the section on CH_3^+. This gratitude also holds for the hard and persistent work of the other senior members of my group, Sandra Brünken, Christian Endres, Pavol Jusko, Frank Lewen, and Holger Müller. Special thanks go to Sven Thorwirth for preparing the FTIR spectroscopy of N_2O and Nadine Wehres for work on the laboratory emission spectroscopy, as well as to Elena Zakharenko, who worked on the methacroline example in Lille during her thesis. Marius Herrmanns prepared work on NH_3 and built the chirped-pulse spectroscopy instruments in our group with initial help from Christian Endres. I am very grateful for the long-term support of and collaboration with our group through astrophysics colleagues at the Max-Planck-Institute for radio astronomy in Bonn, Karl Menten and Arnaud Belloche. Many of the works presented as examples of molecular spectroscopy were conducted in the PhD theses of a number of brilliant graduate students: Sabrina Gärtner, Monika Koerber, Lars Kluge, Matthias Ordu, and Hanno Schmiedt. Oliver Zingsheim was the leading scientist in the work on propanal. Jakob Maßen conducted the Lamb-dip measurements of OCS with help from Frank Lewen. Special thanks go to Sven Fanghänel, who prepared some of the graphs and tables in this work. I am very grateful to John Maier for supplying graphs for the C_{60}^+ work to this chapter, and to Kotaro Kohno for making the figure on the SED available for the introduction. I am also thankful to the workshops of our institute for all the in-house instrumental developments. Financial support came mainly from the German Science Foundation (DFG), in particular through Collaborative Research Center CRC 956 and the Cologne Center for THz spectroscopy, but also through individual research funds.

References

Amano, T. 2010, A&A, 516, L4

Amiot, C., & Guelachvili, G. 1976, JMoSp, 59, 171

Aoiz, F. J., González-Lezana, T., & Sáez Rábanos, V. 2007, JChPh, 127, 174109

Asvany, O., Brünken, S., Kluge, L., & Schlemmer, S. 2014, ApPhB, 114, 203

Asvany, O., Ricken, O., Müller, H. S. P., et al. 2008, PhRvL, 100, 233004

Asvany, O., Thorwirth, S., Redlich, B., & Schlemmer, S. 2018, JMoSp, 347, 1

Belloche, A., Garrod, R. T., Müller, H. S. P., & Menten, K. M. 2014, Sci, 345, 1584

Belloche, A., Menten, K. M., Comito, C., et al. 2008, A&A, 482, 179

Bergin, E. A., Phillips, T. G., Comito, C., et al. 2010, A&A, 521, L20

Bernath, P. F. 1995, Spectra of Atoms and Molecules (Oxford: Oxford Univ. Press)

Bishop, D. M. 1993, Group Theory and Chemistry (New York: Dover)

Bovino, S., Grassi, T., Schleicher, D. R. G., & Caselli, P. 2017, ApJL, 849, L25

Brünken, S., Sipilä, O., Chambers, E. T., et al. 2014, Natur, 516, 219

Bryant, M., Reeve, S. W., & Burns, W. A. 2008, JChEd, 85, 121

Bunker, P. R., & Jensen, P. 1998, Molecular Symmetry and Spectroscopy (2nd ed; Ottawa: NRC Research Press)

Bunker, P. R., & Jensen, P. 2005, Fundamentals of Molecular Symmetry (Bristol: Institute of Physics Publishing)

Campbell, E. K., Holz, M., Gerlich, D., & Maier, J. P. 2015, Natur, 523, 322

Chakrabarty, S., Holz, M., Campbell, E. K., et al. 2013, JPCL, 4, 4051

Crabtree, K. N., Martin-Drumel, M.-A., Brown, G. G., et al. 2016, JChPh, 144, 124201

Crofton, M. W., Jagod, M. F., Rehfuss, B. D., Kreiner, W. A., & Oka, T. 1988, JChPh, 88, 666

de Miranda, B. K. C., Alcaraz, C., Elhanine, M., et al. 2010, JPCA, 114, 4818

Demtröder, W. 1996, Laser Spectroscopy: Basic Concepts and Instrumentation (Berlin: Springer)

Endres, C. P., Schlemmer, S., Schilke, P., Stutzki, J., & Müller, H. S. P. 2016, JMoSp, 327, 95

Gärtner, S., Krieg, J., Klemann, A., et al. 2013, JPCA, 117, 9975

Geballe, T. R., & Oka, T. 1996, Natur, 384, 334

Gordy, W., & Cook, R. L. 1984, Microwave Molecular Spectra, Techniques of Chemistry (New York: Wiley)

Harju, J., Sipilä, O., Brünken, S., et al. 2017a, ApJ, 840, 63

Herzberg, G. 1989, Molecular Spectra and Molecular Structure: Spectra of Diatomic Molecules (Malabar, FL: Krieger)

Herzberg, G. 1991a, Molecular Spectra and Molecular Structure: Electronic Spectra and Electronic Structure of Polyatomic Molecules (Malabar, FL: Krieger)

Herzberg, G. 1991b, Molecular Spectra and Molecular Structure: Infrared and Raman Spectra of Polyatomic Molecules (Malabar, FL: Krieger)
Herzberg, G., & Spinks, J. W. T. 1945, Atomic Spectra and Atomic Structure (New York: Dover)
Hollas, J. M. 1996, Modern Spectroscopy (Chichester: Wiley)
Huang, J., & Oka, T. 2015, MolPh, 113, 2159
Huber, K. P., & Herzberg, G. in NIST Chemistry WebBook, NIST Standard Reference Database Number 69, ed. J. W. Gallagher, & R. D. Johnson, III (Gaithersburg, MD: National Institute of Standards and Technology)
Hugo, E., Asvany, O., & Schlemmer, S. 2009, JChPh, 130, 164302
Indriolo, N., & McCall, B. J. 2012, ApJ, 745, 91
Jagod, M.-F., Rösslein, M., Gabrys, C. M., & Oka, T. 1992, JMoSp, 153, 666
Jusko, P., Konietzko, C., Schlemmer, S., & Asvany, O. 2016, JMoSp, 319, 55
Jusko, P., Stoffels, A., Thorwirth, S., et al. 2017a, JMoSp, 332, 59
Jusko, P., Töpfer, M., Müller, H. S. P., et al. 2017b, JMoSp, 332, 33
Lisy, J. M. 2006, JChPh, 125, 132302
Martin-Drumel, M.-A., McCarthy, M. C., Patterson, D., McGuire, B. A., & Crabtree, K. N. 2016, JChPh, 144, 124202
Martin-Drumel, M. A., van Wijngaarden, J., Zingsheim, O., et al. 2015, JMoSp, 307, 33
McGuire, B. A., Carroll, P. B., Loomis, R. A., et al. 2016, Sci, 352, 1449
Müller, H. S. P., Schlöder, F., Stutzki, J., & Winnewisser, G. 2005, JMoSt, 742, 215
Müller, H. S. P., Thorwirth, S., Roth, D. A., & Winnewisser, G. 2001, A&A, 370, L49
Müller, H. S. P., Walters, A., Wehres, N., et al. 2016, A&A, 595, A87
Oepts, D., van der Meer, A. F. G., & van Amersfoort, P. W. 1995, InPhT, 36, 297
Roueff, E., Gerin, M., Lis, D. C., et al. 2013, JPCA, 117, 9959
Roueff, E., Loison, J. C., & Hickson, K. M. 2015, A&A, 576, A99
Schlemmer, S., Kuhn, T., Lescop, E., & Gerlich, D. 1999, IJMSp, 185, 589
Tielens, A. G. G. M. 2005, The Physics and Chemistry of the Interstellar Medium (Cambridge: Cambridge Univ. Press)
Töpfer, M., Jusko, P., Schlemmer, S., & Asvany, O. 2016, A&A, 593, L11
Toth, R. A. 1991, ApOpt, 30, 5289
Townes, C. H., & Schawlow, A. L. 1975, Microwave Spectroscopy (New York: Dover)
Vastel, C., Mookerjea, B., Pety, J., & Gerin, M. 2017, A&A, 597, A45
Wehres, N., Heyne, B., Lewen, F., & Hermanns, M. 2018a, in IAU Symp. S332, Astrochemistry VII: Through the Cosmos from Galaxies to Planets (Cambridge: Cambridge Univ. Press), 332
Wehres, N., Maßen, J., Borisov, K., et al. 2018b, PCCP, 20, 5530
Western, C. M. 2017, JQSRT, 186, 221
Wilson, E. B. Jr., Decius, J. C., & Cross, P. C. 1980, Molecular Vibrations: The Theory of Infrared and Raman Vibrational Spectra (New York: Dover)
Zakharenko, O., Motiyenko, R. A., Aviles Moreno, J.-R., et al. 2016, JChPh, 144, 024303
Zingsheim, O., Müller, H. S. P., Lewen, F., Jorgensen, J. K., & Schlemmer, S. 2017, JMoSp, 342, 125

Chapter 7

Excitation of Astrophysical Molecules

Paul J. Dagdigian

The column densities (integrals of densities over the line of sight) of molecules in the interstellar medium are determined from the intensities of spectroscopic transitions between individual molecular levels, either in absorption or emission. What is of most interest is the total molecular density, which requires knowledge of the size of the interstellar cloud, the radiation field, and very importantly, the fraction of molecules in the detected levels. If the cloud is in local thermodynamic equilibrium (LTE), then the fraction of molecules in the detected levels can be estimated, provided that the local temperature can be determined.

At equilibrium, the rate of production of a particular molecular level is equal to its rate of destruction. Population can be transferred between molecular levels by radiative and collisional transitions. Considering just a pair of levels and balancing the forward and backward transfer probabilities, we have in statistical equilibrium

$$n_l(R_{u\leftarrow l} + C_{u\leftarrow l}) = n_u(R_{u\rightarrow l} + C_{u\rightarrow l}) \tag{7.1}$$

where n_u and n_l are the densities of the lower and upper levels, respectively. The probabilities of radiative excitation and de-excitation are denoted in Equation (7.1) as $R_{u\leftarrow l}$ and $R_{u\rightarrow l}$, respectively. The rate of radiative excitation is given by

$$R_{u\leftarrow l} = B_{u\leftarrow l}\frac{cu_\nu}{4\pi} \tag{7.2}$$

where u_ν is the monochromatic energy density at frequency ν (units: erg cm^{-3} Hz^{-1}) of the radiation and $B_{u\leftarrow l}$ is the Einstein coefficient of absorption. The rate of radiative de-excitation is given by

$$R_{u\rightarrow l} = A_{u\rightarrow l} + B_{u\rightarrow l}\frac{cu_\nu}{4\pi} \tag{7.3}$$

where $A_{u\rightarrow l}$ and $B_{u\rightarrow l}$ are the Einstein coefficients for spontaneous and stimulated emission, respectively. The Einstein coefficients are defined in terms of molecular properties in Section 7.1.

doi:10.1088/2514-3433/aae1b5ch7

The quantities $C_{u\leftarrow l}$ and $C_{u\rightarrow l}$ in Equation (7.1) are the probabilities for collisional excitation and de-excitation, respectively, and are proportional to the density of the species responsible for the collisions. If there is one dominant collision partner (for example, H_2), then these collisional probabilities are given by the following—in this example, for the upward transition:

$$C_{u\leftarrow l} = nk_{u\leftarrow l} \tag{7.4}$$

where $k_{u\leftarrow l}$ is the bimolecular rate constant (often called the rate coefficient) for transfer from molecular level l to u, and n is the density of the collision partner. The collisional probability is a sum of terms given by the right-hand side of Equation (7.4) when there is more than one collision partner having a significant density.

When the molecular kinetic (translational) and radiative field temperatures both equal the same temperature T, the densities n_i of the molecular levels will, in this case, follow the Boltzmann distribution

$$n_u/n_l = (g_u/g_l)e^{-(\epsilon_u-\epsilon_l)/k_BT} \tag{7.5}$$

where ϵ_i and g_i are the energy and degeneracy of the ith level, and k_B is the Boltzmann constant. Here, the fraction of molecules in a given level is given by

$$n_i/N_{\rm mol} = g_i e^{-\epsilon_i/k_BT}/Q_{\rm mol} \tag{7.6}$$

where $N_{\rm mol}$ is the total molecular population and $Q_{\rm mol}$ is the molecular partition function.

In most instances, the clouds are not in local thermodynamic equilibrium. In this case, a collisional-radiative model must be applied. The ingredients for such a model are the excitation and de-excitation rates for radiative and collisional transitions between the molecular levels. If the particle density in the interstellar medium is sufficiently low, then collisional transitions can be ignored and transitions between molecular levels are governed entirely by radiative transitions.

This chapter is organized as follows. Section 7.1 presents a discussion of the rate of radiative transitions and introduces the concept of the line strength factor, which governs the intensity of radiative transitions. Section 7.2 considers two special cases governing the radiative intensity and molecular densities, namely the transfer equation and the LVG approximation, in which the coupling of radiative transfer and collisional excitation is considered. The former describes the situation where collisional transitions can be ignored in comparison with radiative transitions. Section 7.3 describes in detail the calculation of collisional excitation rate constants from first principles.

7.1 Radiative Transitions

Radiative transitions can be induced by absorption of microwave, infrared, or ultraviolet radiation, which induce pure rotational, vibrational, and electronic transitions, respectively. Molecules in excited levels can also be de-excited by stimulated emission, in a process that is the reverse of absorption, and by spontaneous radiative decay. Molecular clouds can also be exposed to ultraviolet radiation. For many molecules, the excited electronic states are dissociative, i.e., the excited state is not bound and the molecule falls apart upon absorption of a photon. In this section, we consider radiative transitions only between pairs of bound molecular levels, in particular rotational levels.

The Einstein coefficients for absorption and emission are given in terms of molecular properties as (Condon & Shortley 1951)

$$B_{u\leftarrow l} = \frac{32\pi^4}{3h^2c}\frac{S(j',j)}{g} \tag{7.7}$$

$$B_{u\rightarrow l} = \frac{32\pi^4}{3h^2c}\frac{S(j',j)}{g'} \tag{7.8}$$

$$A_{u\rightarrow l} = \frac{64\pi^4\nu^3}{3hc^3}\frac{S(j',j)}{g'} \tag{7.9}$$

where j' and j are the total angular momenta of the upper and lower levels, respectively, ν is the frequency of the transition, g is the degeneracy of the level, and $S(j',j)$ is the line strength factor:

$$\begin{aligned} S(j',j) &= \sum_{m'm} |\langle j'm'|\mu|jm\rangle|^2 \\ &= 3\sum_{m} |\langle j'm|\mu_Z|jm\rangle|^2. \end{aligned} \tag{7.10}$$

Here, μ is the electric dipole moment operator. The second line in Equation (7.10) follows for isotropic (de-)excitation. The primed and unprimed quantum numbers in Equations (7.7)–(7.10) refer to the upper and lower levels, respectively.

For rotational transitions, the line strength factor is proportional to the square of the molecular dipole moment $\mu_{\rm mol}$, and it is customary to factor out this quantity:

$$S(j',j) = \mu_{\rm mol}^2 S'(j',j) \tag{7.11}$$

where $S'(j',j)$ are the rotational line strength factors, which for symmetric tops equal

$$S'(j',k',j,k) = (2j'+1)(2j+1)\begin{pmatrix} j' & 1 & j \\ -k' & k'-k & k \end{pmatrix}^2. \tag{7.12}$$

Here, j and k are the angular momentum of the molecule and its projection on the molecular figure (symmetry) axis, and (:::) is a $3j$ symbol (Brink & Satchler 1993). The line strength factors in Equation (7.12) are the so-called Hönl–London factors. For closed-shell diatomic molecules, the allowed transitions involve $\Delta j = j' - j = \pm 1$, and Equation (7.12) reduces to

$$S'(j',j) = \max(j',j). \tag{7.13}$$

Many astrophysical diatomic molecules have partially filled orbitals and hence unpaired electrons. We consider here rotational line strength factors for linear molecules in $^2\Pi$ ground electronic states, such as the important OH and CH species. These molecules have electron occupation π and π^3, respectively, and have electron spin S equal to 1/2 and molecule-frame projection $\Lambda = 1$ of the electron orbital angular momentum L. A simplified Hamiltonian describing the rotational levels of such molecules would include the rotational energy and the spin–orbit interaction.

Figure 7.1 displays the lower rotational levels of the CH and OH molecules. For the CH molecule, the spin–orbit coupling is much less than the rotational spacings, and this molecule lies close to the Hund's case (b) limit (Herzberg 1950). The angular momentum n exclusive of spin has values 1, 2, etc. This angular momentum can couple with the electron spin to yield total molecular angular momentum j equal to $n + 1/2$ (F_1 fine-structure manifold) or $n - 1/2$ (F_2 fine-structure manifold). Because of the orbital degeneracy of a Π electronic state, there are two Λ-doublet levels, denoted e and f (Brown et al. 1975), for each rotational/fine-structure level n. The lower levels of $OH(X^2\Pi)$ lie close to the Hund's case (a) limit; in this case, the fine-structure manifolds correspond to definite values of the molecule-frame projection $\Omega = \Lambda + \Sigma$ of the angular momentum j, where Σ is the molecule-frame projection of the electron spin S. Because the spin–orbit constant of OH is negative (Huber & Herzberg 1979), the F_1 and F_2 manifolds are associated with $\Omega = 3/2$ and 1/2, respectively.

Because the rotational/fine-structure levels of a molecule in a $^2\Pi$ electronic state generally lie between the Hund's (a) and (b) limits, we must diagonalize the

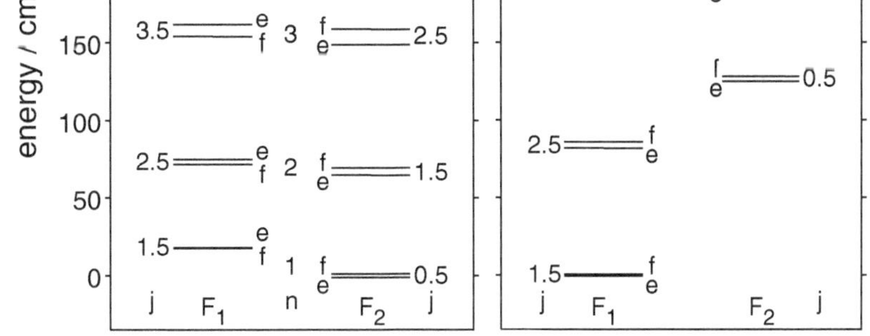

Figure 7.1. Energy of the lower rotational/fine-structure levels of (a) $CH(X^2\Pi)$ and (b) $OH(X^2\Pi)$. The levels are labeled by the total angular momentum j and the parity labels e and f (Brown et al. 1975). The two fine-structure manifolds are labeled F_1 and F_2 (Herzberg 1950). The Λ-doublet splittings have been exaggerated by a factor of 20, for clarity. The CH levels, which lie near the Hund's case (b) limit, are also labeled with n, the angular momentum exclusive of the electron spin.

rotational/fine-structure Hamiltonian. For the purposes of computing the line strengths, it is sufficient to consider the following Hamiltonian (see, for example, Zare 1988)

$$\hat{H} = B(\mathbf{j} - \mathbf{L} - \mathbf{S})^2 + A\mathbf{L} \cdot \mathbf{S}. \tag{7.14}$$

In Equation (7.14), B and A are the rotational constant and spin–orbit interaction constant, respectively, for the molecule. The nuclear rotational angular momentum is given by $\mathbf{j} - \mathbf{L} - \mathbf{S}$. Expressing the Hamiltonian in a case (a) basis $|j\Omega M\Lambda S\Sigma\rangle$, the Hamiltonian in Equation (7.14) is described by the 2×2 matrix

$$\begin{pmatrix} B[j(j+1) - \frac{7}{4} + \frac{1}{2}A & -B[j(j+1) - \frac{3}{4} \\ -B[j(j+1) - \frac{3}{4} & B[j(j+1) + \frac{1}{4} - \frac{1}{2}A \end{pmatrix}. \tag{7.15}$$

The eigenfunctions of Equation (7.15) are expressed as

$$|jF_i\rangle = \sum_{\Omega=\frac{1}{2}}^{\frac{3}{2}} c_{j,\Omega}^{F_i} |j\Lambda S\Sigma\rangle. \tag{7.16}$$

The rotational line strengths for a molecule in a $^2\Pi$ electronic state can be expressed as

$$\begin{aligned} S'(j', j) = (2j' + 1)(2j + 1)|\sum_{\Omega'}\sum_{\Omega} & c_{j',\Omega'}^{F_i'} c_{j,\Omega}^{F_i} \\ & \times (-1)^{j'-1+\Omega} \begin{pmatrix} j' & 1 & j \\ -\Omega' & \Omega' - \Omega & \Omega \end{pmatrix} |^2. \end{aligned} \tag{7.17}$$

The microwave and radio frequency spectra of many molecules show resolved hyperfine structure due to splittings caused by the nonzero spin of one or more nuclei. For a molecule with one nuclear spin I greater than zero, each rotational level of angular momentum j is split into several hyperfine levels, with angular momentum $\mathbf{F}$ equal to the vector sum of $\mathbf{j}$ and $\mathbf{I}$, i.e., $F = |j - I|, |j - I| + 1, \ldots, j + I - 1, j + I$. Figure 7.2 displays the hyperfine levels and transitions involving the $OH(X^2\Pi)$ $j = 3/2$ $F_1 e/f$ Λ-doublet levels. The allowed radiative transitions have $\Delta F = 0, \pm 1$, except $F' = 0 \rightarrow F = 0$ is forbidden. The line strength factors for hyperfine-resolved transitions are obtained by multiplying the standard line strength factors by

$$(2F' + 1)(2F + 1)\begin{Bmatrix} j & F & I \\ F' & j' & 1 \end{Bmatrix}^2. \tag{7.18}$$

There are two main publicly available databases that present transition frequencies, transition strengths or Einstein A factors, and statistical weight factors for astrophysical molecules, namely the Jet Propulsion Laboratory catalog (JPL; Pickett

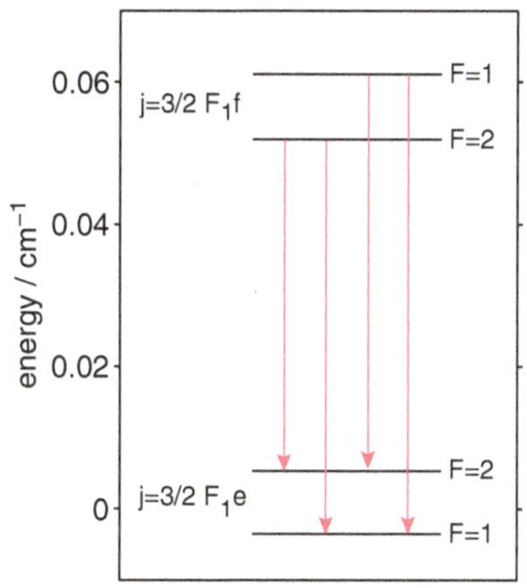

Figure 7.2. Hyperfine levels associated with the ground, $j = 3/2\ F_1e/f$ rotational/fine-structure levels of $OH(X^2\Pi)$. The red lines denote the allowed radiative transitions between the Λ-doublet levels. The hyperfine splittings have been exaggerated by a factor of 5 for clarity.

et al. 1998; spec.jpl.nasa.gov) and the Cologne Database for Molecular Spectroscopy (CDMS; Muller et al. 1998, 2005; http://www.astro.uni-koeln.de/cdms).

7.2 Non-LTE Situations

In low-density clouds, the collisional transition probabilities $C_{f\leftrightarrow i}$ can be neglected, compared to the radiative transition probabilities $R_{f\leftrightarrow i}$. Following Lequeux (2005), we outline the derivation of the transfer equation, which relates the radiation intensity incident on and transmitted through a molecular cloud at a given frequency ν.

We consider the intensity I_ν at a given frequency ν and solid angle incident on a cloud. Within a small slice ds along the direction of propagation of the light, this intensity is decreased by absorption and increased by emission, following Equations (7.2) and (7.3), respectively. Thus, we can express the evolution of the intensity at frequency ν as

$$\frac{dI_\nu}{ds} = \frac{h\nu}{4\pi}[n_u(\nu)A_{u\rightarrow l} - [n_l(\nu)B_{u\leftarrow l} - n_u(\nu)B_{u\rightarrow l}]I_\nu]. \tag{7.19}$$

Equation (7.23) applies to a single frequency within the line profile, and n_l and n_u are the densities of molecules in the lower and upper levels that interact with the light at that frequency.

It is often of interest to integrate the line profile. We thus define the absorption coefficient

$$\kappa_\nu = \frac{h\nu}{4\pi}[n_l(\nu)B_{u\leftarrow l} - n_u(\nu)B_{u\rightarrow l}] \tag{7.20}$$

and the optical depth (or optical thickness)

$$\tau_\nu = \int \kappa_\nu ds. \tag{7.21}$$

Defining the source function

$$S_\nu = \frac{n_u(\nu)A_{u\to l}}{n_l(\nu)B_{u\leftarrow l} - n_u(\nu)B_{u\to l}} \tag{7.22}$$

the transfer equation takes the simple form

$$\frac{dI_\nu}{d\tau_\nu} = S_\nu - I_\nu. \tag{7.23}$$

If the source function is uniform along the line of sight, the transfer equation can be integrated to yield

$$I_\nu(\tau_\nu) = I_\nu(0)e^{-\tau_\nu} + S_\nu(1 - e^{-\tau_\nu}). \tag{7.24}$$

We now assume that medium is in local thermodynamic equilibrium (LTE) at a temperature T. The ratio of the populations of the levels are thus given by the Boltzmann distribution:

$$n_u(\nu)/n_l(\nu) = (g_u/g_l)e^{-h\nu_0/k_BT} \tag{7.25}$$

where g_u and g_l are the statistical weights of the levels and ν_0 is the center frequency of the line. We are assuming for Equation (7.25) that the line has the same shape in emission and absorption.

If the optical depth is large, then the intensity I_ν will approach the blackbody intensity $B_\nu(T)$ at the temperature of the medium. The source function S_ν will then equal the Planck function B_ν, and the transfer equation (Equation (7.23)) can be written as

$$\frac{dI_\nu}{d\tau_\nu} = B_\nu(T) - I_\nu. \tag{7.26}$$

At line center, the Planck function equals

$$B_{\nu_0}(T) = \frac{2h\nu_0^3}{c^2}\frac{1}{e^{h\nu_0/k_BT} - 1}. \tag{7.27}$$

The absorption coefficient (Equation (7.20)) can be recast with the help of Equations (7.7) and (7.8), which define the Einstein coefficients for absorption and stimulated emission to yield

$$\kappa_\nu = \frac{c^2 n_l(\nu)g_u}{8\pi\nu_0^2 g_l}A_{u\to l}\left[1 - \frac{g_l n_u(\nu)}{g_u n_l(\nu)}\right]. \tag{7.28}$$

We see from Equation (7.28) that the absorption coefficient is proportional to the density. We can define the column density N of molecules along the line of sight, which equals $\int n\, ds$. The optical depth τ_ν (Equation (7.21)) can then be seen from Equation (7.28) to be proportional to N.

In the radio-frequency portion of the spectrum, the Planck function can be approximated by the Rayleigh–Jeans formula

$$B_\nu(T) = \frac{2k_B T\nu^3}{c^2}. \tag{7.29}$$

At LTE, the solution to the transfer equation (Equation (7.23)) becomes

$$I_\nu(\tau_\nu) = I_\nu(0) + \frac{2k_B T\nu^3}{c^2}(1 - e^{-\tau_\nu}). \tag{7.30}$$

With this form for the solution of the transfer equation, surface intensities are often described by a brightness temperature T_B, which is defined as

$$T_B = \frac{2k_B T\nu^3}{c^2} I_\nu. \tag{7.31}$$

With Equation (7.31), the solution of the transfer equation at LTE becomes

$$T_B(\tau_\nu) = T_B(0)e^{-\tau_\nu} + (1 - e^{-\tau_\nu})T. \tag{7.32}$$

We see from this equation that the line is seen in emission if the medium is at a higher temperature than that of the background, or is seen in absorption if the medium has a lower temperature. If the medium is optically thin (τ_νL1) and the background is weak, Equation (7.32) simplifies to

$$T_B(\tau_\nu) = \tau_\nu T. \tag{7.33}$$

For an optically thick medium, the brightness temperature T_B equals the kinetic temperature of the medium, i.e., the temperature characterizing the translational velocity distribution of the particles.

We now consider the situation where collisional transitions cannot be neglected. If the velocity distribution can be defined by a kinetic temperature T_K, then

$$k_{u\leftarrow l} = k_{u\rightarrow l}(g_u/g_l)e^{-h\nu/k_B T_K}. \tag{7.34}$$

In the general case, Equation (7.1) can be written as

$$\frac{n_u}{n_l} = \frac{g_u}{g_l}\frac{A_{u\rightarrow l}I_\nu c^2/2h\nu^3 + C_{u\rightarrow l}e^{-h\nu/k_B T_K}}{(1 + I_\nu c^2/2h\nu^3)A_{u\rightarrow l} + C_{u\rightarrow l}} \tag{7.35}$$

where we have related the Einstein $B_{u\rightarrow l}$ and $B_{u\leftarrow l}$ coefficients to the spontaneous radiative transition $A_{u\rightarrow l}$ through Equations (7.7), (7.8), and (7.9). There exists a critical density $n_{\rm crit}$ for which collisional and radiative transitions have equal importance, namely

$$n_{\rm crit} = \frac{1}{k_{u\leftarrow l}}\left(1 + \frac{I_\nu c^2}{2hc\nu^3}\right)A_{u\rightarrow l}. \tag{7.36}$$

We have seen that the balance of transitions between the upper and lower levels at a given location is determined by both the local molecular density and the radiation field at the transition frequency. By contrast, the radiation field is determined from the properties of the emitting molecular species in the cloud. In order to compute the populations of the levels, we thus must solve the equation of statistical equilibrium and the transfer equation at all points in the medium to obtain the steady-state values of the level populations. This calculation is thus an iterative process. An initial guess of the level populations following statistical equilibrium is made, and the optical depths of the lines then determined. The latter are then used to recalculate the molecular excitation. The calculation is complete when the level populations and radiation field are self-consistent.

These non-LTE radiative transfer calculations can be quite involved when the medium is not homogeneous. A conceptually simple method of calculation is to employ a Monte Carlo approach (Bernes 1979). Here, one follows "model" photons representing a larger number of real photons through the medium. The medium is represented by an ensemble of boxes containing the molecules, each with a different kinetic temperature and density. After all the photons have been followed and absorption and emission events tabulated, the level populations are adjusted, and the process is repeated until the level populations and radiation field are self-consistent. The Monte Carlo method is conceptually simple and easily implemented. However, it does suffer from numerical noise, and convergence can be slow when optically thick lines are modeled.

The treatment of a homogeneous medium is much simpler than for an inhomogeneous medium. In the former case, the radiative transfer problem can be solved on a personal computer. For example, van der Tak et al. (2007) have written a program (RADEX, which is available at http://home.strw.leidenuniv.nl/moldata) to carry out a non-LTE-analysis of interstellar line spectra in the case of an assumed homogeneous medium.

We consider next the coupling of excitation and transfer in the large velocity gradient (LVG) approximation (Goldreich & Kwan 1974; Surdej 1979). This situation applies to a locally narrow line of velocity width δv when the medium is undergoing rapid radial expansion or contraction. In this case, the emission at a given frequency arises only from a restricted portion of the line of sight whose velocity corresponds to this frequency. We assume here that the velocity gradient, as well as the density and the kinetic and excitation temperatures, are uniform in the medium. We follow the derivation given by Lequeux (2005).

The statistical equilibrium condition can be expressed, from Equations (7.1), (7.2), and (7.3), as

$$n_l\left(\frac{c}{4\pi}u(\mathbf{r})B_{u\leftarrow l} + C_{u\leftarrow l}\right) = n_u\left(A_{u\rightarrow l} + \frac{c}{4\pi}u(\mathbf{r})B_{u\rightarrow l} + C_{u\rightarrow l}\right). \tag{7.37}$$

Here, $u(\mathbf{r})$ is the local radiation density at the location $\mathbf{r}$. With a large velocity gradient, photons within the line profile can come only from a local region of size d, where

$$d \approx \delta v \ \mathrm{R}/\Delta V. \tag{7.38}$$

Here, R is the total length of the medium along the line of sight and ΔV is the total velocity width. If we assume that an absorbed photon is re-emitted without memory of the frequency, direction, or polarization of the incident photon, then we can relate $u(\mathbf{r})$ to the source function $S(\mathbf{r})$ (see Equation (7.22)) in the following way:

$$u(\mathbf{r}) = [1 - \beta(\mathbf{r})]S(\mathbf{r}) \tag{7.39}$$

where $\beta(\mathbf{r})$ is the escape probability of the photon at location $\mathbf{r}$.

The absorption probability $\alpha = 1 - \beta$ of an emitted photon of frequency ν in the direction $\mathbf{s}$ equals

$$\begin{aligned} \alpha(\nu, \mathbf{s}) = \int_0^{s(\nu)} & \exp\left[-\int_0^{s} \kappa(\nu)\phi\left(\nu - \nu_0 - \frac{\nu_0}{c}\frac{dv_s}{ds}s'\right)ds'\right]. \\ & \times \kappa(\nu)\phi\left(\nu - \nu_0 - \frac{\nu_0}{c}\frac{dv_s}{ds}s\right)ds \end{aligned} \tag{7.40}$$

The exponential factor in the integrand represents the probability that a photon of local frequency ν will reach the position s in the given direction without being absorbed. The function ϕ is the line profile, which is normalized such that the integral over the line is unity:

$$\int \phi(\nu)d\nu = 1. \tag{7.41}$$

The term $-(\nu_0/c)(dv_s/ds)s'$ in the argument of the line profile function expresses the Doppler shift due to the velocity between s' and $s' + ds'$. The factor $\kappa(\nu)\phi$ on the second line of Equation (7.40) represents the probability of absorption between s and $s + ds$. The upper limit of the integral is limited by the position $s(\nu)$ for which the photon is no longer within the line profile.

Equation (7.40) can be simplified with the following changes of variable:

$$x = \nu - \nu_0 - \frac{\nu_0}{c}\frac{dv_s}{ds}s' \tag{7.42}$$

$$y = \nu - \nu_0 - \frac{\nu_0}{c}\frac{dv_s}{ds}s. \tag{7.43}$$

With our assumptions about the medium, the quantities $\kappa(\nu)$ and dv_s/ds can be taken to be constant. The optical depth (Equation (7.21)) can be expressed here as

$$\tau_0(\mathbf{s}) = \frac{\kappa(\nu)}{\nu_0}\frac{c}{dv_s/ds}. \tag{7.44}$$

The absorption probability in Equation (7.40) can be simplified to

$$\alpha(\nu, \mathbf{s}) = \int_{y(s(\nu))}^{\nu_0-\nu} \frac{d}{dy}\left[\exp\left(\tau_0(s)\int_{\nu_0-\nu}^{y} \phi(x)dx\right)\right]$$
$$= 1 - \exp\left(\tau_0(s)\int_{\nu_0-\nu}^{y(s(\nu))} \phi(x)dx\right). \tag{7.45}$$

We now integrate over the line profile, which is assumed to cover the frequency range from $-\Delta\nu$ to $+\Delta\nu$ (or the local width of the line):

$$\alpha(\mathbf{s}) = \int_{\nu_0-\Delta\nu}^{\nu_0+\Delta\nu} \alpha(\nu, \mathbf{s})\phi(\nu - \nu_0)d\nu. \tag{7.46}$$

We change the frequency variable from ν to the displacement $f = \nu - \nu_0$ from line center and obtain

$$\alpha(\mathbf{s}) = \int_{-\Delta\nu}^{+\Delta\nu}\left[1 - \exp\left(\tau_0(\mathbf{s})\int_{f}^{y(\mathbf{s}(\nu))} \phi(x)dx\right)\right]df. \tag{7.47}$$

With the normalization of the line profile (Equation (7.41)), Equation (7.47) becomes

$$\alpha(\mathbf{s}) = 1 + \int_{-\Delta\nu}^{+\Delta\nu} \frac{d}{df}\left[\exp\left(\tau_0(\mathbf{s})\int_{f}^{y(\mathbf{s}(\nu))} \phi(x)dx\right)\right]df. \tag{7.48}$$

We have

$$y(s(\nu)) = \nu - \nu_0 \mp \frac{\nu_0}{c}\frac{dv_s}{ds}s(\nu) \tag{7.49}$$

depending upon the sign of the velocity gradient dv_s/ds. The Doppler shift at $s(\nu)$, after which the absorption is negligible, is related to the line width by

$$\frac{\nu_0}{c}\frac{dv_s}{ds} = \nu - (\nu_0 - \Delta\nu). \tag{7.50}$$

Hence, we have

$$y(s(\nu)) = \mp\Delta\nu. \tag{7.51}$$

With Equation (7.51), Equation (7.48) can be simplified to yield the final expression for the absorption probability:

$$\alpha(\mathbf{s}) = 1 - \frac{1 - \exp(-\tau_0(\mathbf{s}))}{\tau_0(\mathbf{s})}. \tag{7.52}$$

Hence, the escape probability equals

$$\beta(\mathbf{s}) = \frac{1 - \exp(-\tau_0(\mathbf{s}))}{\tau_0(\mathbf{s})}. \tag{7.53}$$

We now integrate over all directions to obtain the overall escape probability. We consider the plane parallel case. We define μ to be the cosine of the angle between the line of sight **s** and the normal **n** to the layers of the medium. The escape probability can then be expressed as

$$\beta = \int \frac{1 - \exp(-\tau_0(\mu))}{\tau_0(\mu)} \frac{d\Omega}{4\pi} = \frac{1}{2} \int \frac{1 - \exp(-\tau_0(\mu))}{\tau_0(\mu)} d\mu. \tag{7.54}$$

We integrate the optical depth (Equation (7.44)) over the line profile and define

$$\tau_0 = \frac{\kappa}{\nu_0} \frac{c}{dv/dz} \tag{7.55}$$

where z is normal to the layers of the medium. We can relate the velocity gradient dv_s/ds along the line of sight to dv/dz by

$$dv_s/ds = \mu^2(dv/dz). \tag{7.56}$$

Equation (7.54) can be integrated with the help of the Eddington approximation (see Huang 1968), with which we replace μ^2 with the average value of $\cos^2(\mathbf{s} \cdot \mathbf{n})$ integrated over all directions, or the value 1/3. We then obtain the final value for the escape probability:

$$\beta = \frac{1 - \exp(-3\tau_0)}{3\tau_0}. \tag{7.57}$$

We see from Equation (7.57) that β approaches unity when τ_0 goes to zero, and β approaches $1/3\tau_0$ when τ_0 goes to infinity. With the statistical equilibrium condition (Equation (7.37)), the brightness temperature (Equation (7.31)) in the low-frequency Rayleigh–Jeans approximation equals

$$T_B = \frac{T_K}{1 + (kT_K/h\nu_0)\ln[1 + (A_{u\to l}/3C_{u\to l}\tau_0)(1 - \exp(-3\tau_0))]}. \tag{7.58}$$

7.3 Collisional Transitions

Over the past few decades, advances in experimental methods, primarily involving molecular beams and laser-based detection methods, have facilitated the measurement of integral and differential cross sections for principally rotational transitions between specified levels of a molecule (Dagdigian 1995; Schiffman & Chandler 1995). For example, the development of a method to vary continuously the velocity of a state-selected molecular beam by Stark deceleration (Scharfenberg et al. 2009) has allowed measurement in a crossed-beam experiment (Schewe et al. 2015) of relative state-to-state cross sections as a function of collision energy for rotational transitions in $OH(X^2\Pi)$ in collisions with para-H_2 and normal-H_2. The measured cross sections were found to be in good agreement with cross sections using a new set of potentials describing the interaction of $OH(X^2\Pi)$ with H_2 (Ma et al. 2014).

Recent experiments have directly observed interference effects in inelastic molecular collisions at low energy. Long-lived complexes that correspond to quasi-bound states in the van Waals potential give rise to peaks in the energy-dependent state-to-state cross sections. Such shape and Feshbach resonances have very narrow energy widths and are difficult to observe. Resonances associated with the $j = 0 \rightarrow 1$ transition of CO in collisions with He and H_2 have been observed in crossed-beam experiments at collision energies of $<20\ \mathrm{cm}^{-1}$ with an acute beam intersection angle (Chefdeville et al. 2012; Bergeat et al. 2015). Another type of interference effect, namely diffraction oscillations in differential cross sections, have also recently been observed in collisions of $NO(X^2\Pi)$ with He and D_2 (de Jongh et al. 2017). In conjunction with these experiments, scattering calculations with ab initio potential energy surfaces show good agreement with the measured cross sections.

Along with advances in experimental techniques for the measurement of cross sections in molecular beam experiments, there have been advances in the direct measurement of rate constants for rotational energy transfer at low temperatures through the use of the CRESU (Cinétique de Réaction en Ecoulment Supersonique Uniforme) technique (Rowe et al. 1984; Rowe & Marquette 1987). This technique has been employed to determine rate constants for transitions between the CO ($X^2\Sigma^+$, $v = 2, j = 0$, 1, 4, and 6) rotational levels in collisions with He and temperatures between 294 and 15 K (Carty et al. 2004).

In concert with the advances in experimental techniques and the measurement of state-to-state cross sections and rate constants, there have been similar advances in the ability to compute reliable cross sections and rate constants for collisions of diatomic and larger molecules with rare gases and small molecules such as H_2 (Dagdigian 2013). This progress has been made possible in part by new theoretical methods and especially by the the availability of more powerful computational resources.

Comparison of the best experiments with scattering calculations employing state-of-the-art potential energy surfaces indicates that the rate constants important for astrophysical modeling can be computed reliably. It is important to note that these rate constants for astrophysically relevant conditions cannot generally be measured in the laboratory. Hence, it is reassuring that these can be accurately computed by theoretical methods.

The calculation of molecular collision properties begins with the Born–Oppenheimer approximation. Because the electrons are much lighter than the nuclei and have similar kinetic energies, the electrons move much faster than the nuclei. It is usually reasonable to express the total wave function as a product of an electronic wave function and a wave function describing nuclear motion. The motion of the nuclei, for example in a molecular collision, is then governed by the potential energy surface (PES) of interaction generated by the electrons in the ground electronic states of the collision partners. In some cases, one must consider more than one electronic state and the coupling between the states.

The first step in the determination of rate constants (sometimes called rate coefficients) for collision-induced transitions between specified molecular levels is the calculation of the potential energy surface of interaction between the molecular partners. This PES is then employed in a quantum scattering calculation to determine the cross sections, and after thermal averaging over the velocity distribution, the rate constants for collision-induced transitions. Section 7.3.1 describes the calculation of PESs and the fitting of the calculated interaction energies for use in the scattering calculation. Section 7.3.2 describes the use of the time-independent approach to quantum scattering calculations of cross sections, and after thermal averaging, rate constants for collision-induced molecular transitions. Section 7.3.3 discusses the calculation of cross sections for hyperfine-resolved transitions.

7.3.1 Calculation and Fitting of Potential Energy Surfaces

For nonreactive collisions, the potential energy surface of interaction between the collision partners is usually governed by the weak van der Waals interaction. Chałasiński & Szczęśniak (2000) have extensively reviewed the ab initio calculation of intermolecular interactions. Many references to earlier reviews are provided in this review, and the authors discuss the supermolecular approach to the calculation of interaction energies. The accuracy of the computed interaction energies depends upon both the level of theory employed and the basis set of orbitals used. Coupled-cluster theory can be employed for molecules whose electronic wave functions can be described reasonably well by a single determinant wave function. Crawford & Schaefer (2000) have written a tutorial introduction to coupled cluster theory. It should be noted that coupled cluster theory is applicable only for the lowest state of a given symmetry. When dealing with states for which a single determinant wave function is not an accurate description, multi-reference configuration interaction (MRCI) calculations must be carried out.

It is sufficient for most calculations to employ a coupled cluster theory that includes single, double, and (perturbatively) triple excitations (CCSD(T); Čížek 1969; Raghavachari et al. 1989; Hampel et al. 1992; Deegan & Knowles 1994). Coupled-cluster theory captures a significant fraction of the correlation energy, which is important in describing the nonbonding van der Waals dispersion interaction. Correlation-consistent basis sets at the quadrupole zeta (AVQZ) level (Kendall et al. 1992) are usually employed currently in calculations of intermolecular interactions.

To check for basis set convergence, one can carry out a series of calculations with triple zeta (AVTZ), AVQZ, and possibly larger basis sets, using an extrapolation algorithm (Helgaker et al. 1997; Helkier et al. 1999; Hill et al. 2009). To increase basis set convergence, it is often useful to add mid-bond functions (Tao & Pan 1992; Koch et al. 1998). Alternatively, coupled cluster theory with explicitly correlated basis sets (Hättig et al. 2012; Kong et al. 2012) can be employed in the calculation of intermolecular interactions, as illustrated by a recent calculation of the H_2O–H_2 potential energy surface (Valiron et al. 2008).

In the supermolecular approach, the interaction energy $V(\mathbf{R}, \mathbf{r})$, where $\mathbf{R}$ is the location of the center of mass of one collision partner with respect to the center of mass of the other and **r** describes the internal coordinates of the collision partners, is computed as follows:

$$V(\mathbf{R}, \mathbf{r}) = E_{\text{A-B}}(\mathbf{R}, \mathbf{r}) - E_{\text{A}}(\mathbf{r}) - E_{\text{B}}(\mathbf{r}) - \Delta E_{\text{CP}}(\mathbf{R}, \mathbf{r}). \quad (7.59)$$

Here, $E_{\text{A-B}}$ is the energy of the supermolecular system, and E_A and E_B are the energies of the separated collision partners. The counterpoise correction ΔE_{CP} (Boys & Benardi 2002) for basis set superposition error (BSSE), namely

$$\Delta E_{\text{CP}}(\mathbf{R}, \mathbf{r}) = E_{\text{A}}(\mathbf{R}, \mathbf{r}) + E_{\text{B}}(\mathbf{R}, \mathbf{r}) - E_{\text{A}}(R = \infty, \mathbf{r}) - E_{\text{B}}(R = \infty, \mathbf{r}) \quad (7.60)$$

adjusts for the lack of saturation of the orbital basis. The energies on the right-hand side of Equation (7.60) are computed with the basis of the supermolecule. The use of mid-bond functions has a tendency to increase the BSSE and to alter the electrostatic energy, and should thus be employed with caution.

Because energy transfer rate constants are of interest for astrophysical application mainly for collision-induced rotational transitions, it is usually sufficient to fix the bond lengths and angles in the collision partners. Jeziorska et al. (2000) discuss the optimum choice of bond length in calculations on the Ar–HF complex. Faure et al. (2005) advocate averaging over the ground-state vibrational probability distribution. Such averaging over the NH_3 inversion coordinate was found to be important for the NH_3–He PES (Gubbels et al. 2012b). Generally, the average nuclear geometry for the ground vibrational level is chosen.

The coordinates used to describe the geometry of the collision complex should be appropriate for the calculation of matrix elements of the rotation–translation scattering basis functions. Figure 7.3 displays the coordinates appropriate to describe complexes of a symmetric top with an atom and of the interaction of two diatomic molecules. The separation of the centers of mass of the collision partners is

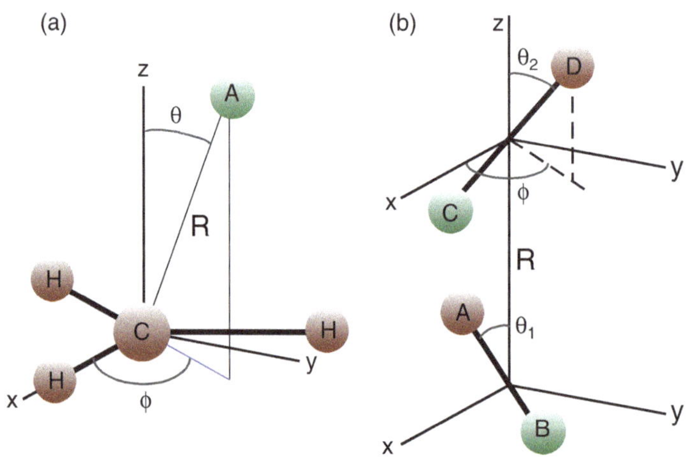

Figure 7.3. Body-frame coordinate systems defining the geometries of (a) the complex of a symmetric top (in this case, the methyl radical) with an atom and (b) of the interaction of two diatomic molecules, denoted AB and CD. The z axis in panel (a) is along the c inertial axis of CH_3; in panel (B), the z axis is along the vector **R** connecting the centers of mass of the molecules.

denoted by R. The location of the atom with respect to the molecule in the symmetric top–atom complex is described by the angles (θ, ϕ) and the separation R. For the diatom–diatom complex, the first molecule (AB) is defined to lie along the body-frame xz plane. The geometry of the complex is defined by the two polar angles θ_1 and θ_2, the dihedral angle ϕ, and the separation R.

For the symmetric top–closed-shell atom complex, the angular dependence of the potential can be expressed conveniently as

$$V(R, \theta, \phi) = \sum_{\lambda\mu} v_{\lambda\mu}(R)(1 + \delta_{\mu 0})^{-1}[C_{\lambda\mu}(\theta, \phi) + (-1)^{\mu} C_{\lambda,-\mu}(\theta, \phi)]] \tag{7.61}$$

because the xz plane is a plane of symmetry (see Figure 7.3(a)). In this case, $\lambda + \mu$ must be even. In Equation (7.61), the $C_{\lambda\mu}$ are Racah normalized spherical harmonics (Brink & Satchler 1993). For many calculations on systems involving a symmetric or asymmetric top, normalized spherical harmonics $Y_{\lambda\mu}$ (Brink & Satchler 1993) are often employed in the angular expansion; these are related to the $C_{\lambda\mu}$ functions by a normalization factor. For the methyl radical, μ must be a multiple of 3 because of the C_3 symmetry of the molecule. Similar symmetry restrictions will apply to other symmetrical molecules.

For an atom–diatom complex, Equation (7.61) simplifies to

$$V(R, \theta) = \sum_{\lambda} v_{\lambda}(R) P_{\lambda}(\cos\theta). \tag{7.62}$$

The ϕ coordinate is not needed to describe the geometry of this complex. The two halves of a homonuclear diatomic molecule (or any symmetrical polyatomic molecule such as CO_2) are the same. In this case, only even values of λ are present in the sum in Equation (7.62).

For the diatom–diatom complex where both molecules are closed-shell species, the angular dependence of the potential can be expanded as

$$V(R, \theta_1, \theta_2, \phi) = \sum_{\lambda_1\lambda_2\lambda} \sum_{mu>0} (\lambda_1\mu\lambda_2, -\mu|\lambda 0)(2 - \delta_{\mu 0}) d^{\lambda_1}_{\mu 0}(\theta_1) d^{\lambda_2}_{-\mu 0}(\theta_2)\cos(\mu\phi) \tag{7.63}$$

where (....|..) is a Clebsch–Gordan coefficient and $d^{\lambda}_{\mu,\nu}$ is a reduced rotation matrix element (Brink & Satchler 1993).

Because some important astrophysical molecules, such as OH and CH, have $^2\Pi$ ground electronic states, the angular expansion of the interaction of a $^2\Pi$ diatomic molecule with a structureless target is described here. The $^2\Pi$ state is orbitally degenerate and has Cartesian components $|\Pi_x\rangle$ and $|\Pi_y\rangle$, where the z axis lies along the internuclear axis. With the approach of the structureless collision partner in, say, the xz plane, the wave function of the supermolecular system has two states of A'' and A' reflection symmetry, respectively. Alexander (1985) has shown that it is convenient to take the average and half-difference of the interaction energies of these two states, namely

$$V_{\text{sum}} = \frac{1}{2}[V(A'') + V(A')] \tag{7.64}$$

$$V_{\text{dif}} = \frac{1}{2}[V(A'') - V(A')]. \tag{7.65}$$

This transformation is carried out so that the electronic basis is appropriate to describe the isolated $^2\Pi$ molecule, namely in terms of the two functions with projections $\Lambda = \pm 1$ along the internuclear axis.

The angular expansions of these functions are given by

$$V_{\text{sum}}(R, \theta) = \sum_{\lambda} v_{0,\lambda}(R) d_{00}^{\lambda}(\theta) \tag{7.66}$$

$$V_{\text{dif}}(R, \theta) = \sum_{\lambda} v_{2,\lambda}(R) d_{02}^{\lambda}(\theta). \tag{7.67}$$

The reader is referred to the original literature for the form of the angular expansion describing the interaction of a diatomic molecule in a $^2\Pi$ state with a closed-shell diatomic molecule (Wormer et al. 2005; Ma et al. 2014; Dagdigian 2016).

The set of nuclear geometries for which interaction energies (Equation (7.59)) are computed is usually governed by the strategy by which the angular expansion coefficients ($v_{\lambda\mu}$ or v_λ coefficients in Equations (7.61), (7.62), (7.63)) will be determined. A common approach is to compute the interaction energy on the same angular grid for each value of R in the radial grid. The expansion coefficients for a given R are then obtained by a least-squares fit to the computed interaction energies. The number of expansion coefficients employed is dictated by the desire to fit the energies with small residual errors, but not with so many expansion coefficients so that there are nonphysical oscillations in the fitted interaction energy as a function of the angles.

For highly anisotropic PESs, a large number of coefficients will be required. Care must be taken in the fitting Equation (7.62) to calculated points in the PES of a strongly anisotropic system; for example, one involving a long linear polyatomic molecule such as HC_3N. Wernli et al. (2007) describe a modification of the procedure described here for determining the expansion coefficients. The quality of the fits will degrade for small R. For computational simplicity in the scattering calculations, the same angular expansion should be employed for all values of R.

In choosing the angular grid, Rist & Faure (2012) advocate uniform sampling of the differential solid angle. For the interaction of two diatomic molecules, the angular grid should be chosen with equal spacing between -1 and 1 for $\cos\theta_1$, and $\cos\theta_2$ and equal spacing of ϕ between 0° and 180°. For one or more collision partners with symmetry, the full angular grid does not have to be sampled.

In their fit to computed energies for the H_2O–H_2 system, Valiron et al. (2008) employed a slightly different procedure for determining the angular expansion coefficients. In this case, four angles are required to determine the geometry of the complex. They computed the interaction energy for each value of R at a large, random set of angles and carried out a linear least-squares fit to determine the expansion coefficients. To judge the quality of the fit, Monte Carlo sampling of the angles was used to estimate the error in the expansion coefficients (Rist & Faure 2012).

The inner range of the radial grid should be small enough that the potential is more repulsive than the expected highest collision energy for which scattering calculations will be carried out. The outer range of R is dictated by the reliability of the calculations of the interaction energies. This will usually be smaller than $20a_0$. However, the scattering calculations must be extended to larger R for the purpose of matching the scattering wave function to the asymptotic boundary conditions. The expansion coefficients for the lowest-order terms (particularly V_{00}) should be extrapolated to large R with a R^{-n} dependence. A switching function is typically used to connect the fitted coefficients at smaller R with the extrapolated radial dependence. The higher-order terms (van der Avoird et al. 1980) are damped to zero with a switching function. If cross sections at very low energies are desired, then more elaborate procedures for fixing the long-range behavior of the PES should be carried out (Janssen et al. 2011; Gubbels et al. 2012a).

7.3.2 Time-independent Quantum Scattering Calculations

For calculation of cross sections at the low collision energies appropriate for the low temperatures in most molecular clouds, time-independent quantum scattering calculation of the cross sections is the method of choice. This theory for the scattering of a rigid diatomic molecule was first given by Arthurs & Dalgarno (1960).

In the close-coupled treatment of the collision dynamics, the scattering wave function is expanded in (ideally) a complete set of the internal states of the system. These are constructed as direct products of the internal states of one (or both) of the collision partners and the angular functions describing the rotation of the collision partners about each other. These functions are then coupled so that the scattering basis functions are eigenfunctions of the total angular momentum J. The set of these states (usually called channels) is designated as the column vector $\Phi(\hat{\mathbf{R}}, \mathbf{r})$, where $\mathbf{r}$ designates the internal coordinates and $\hat{\mathbf{R}} = (\alpha, \beta)$ denotes the orientation of $\mathbf{R}$. The internal coordinates of the complex have usually been expressed in the space frame (SF), as in the original treatment (Arthurs & Dalgarno 1960). Alternatively, the coordinates can alternatively be expressed in the body frame (BF). The advantage of the latter is that the potential is more naturally expressed in BF coordinates.

The SF scattering basis functions are defined as follows. We denote the SF rotational wave function of a molecular collision partner as $\psi_{jm_j}(\hat{\mathbf{r}}_{\text{SF}})$, where $\hat{\mathbf{r}}_{\text{SF}}$ denotes the orientation of the molecule with respect to the SF axes, and j and m_j are the rotational angular momentum and its projection along the SF Z axis. The molecule is usually assumed to be rigid. The rotational wave functions of a diatomic molecule and symmetric top, as well as the rotational electronic wave functions of a diatomic molecule in a $^2\Pi$ state, are given below.

For a diatomic molecule, we have

$$\psi_{jm_j}(\hat{\mathbf{r}}_{\mathrm{SF}}) = Y_{jm_j}(\theta_{\mathrm{SF}}, \phi_{\mathrm{SF}}) \tag{7.68}$$

where θ_{SF} and ϕ_{SF} denote the SF orientation of the diatomic molecule. The parity of the functions in Equation (7.68) is $p = (-1)^j$.

For a symmetric top, the SF rotational wave function can be written as

$$\psi_{jm_jk\epsilon}(\hat{\mathbf{r}}_{\mathrm{SF}}) = \left(\frac{2j+1}{8\pi^2}\right)^{1/2} [2(1+\delta_{k0})]^{-1/2}\left[D^{j*}_{jm_jk}(\hat{\mathbf{r}}_{\mathrm{SF}}) + \epsilon D^{j*}_{jm_j,-k}(\hat{\mathbf{r}}_{\mathrm{SF}})\right] \tag{7.69}$$

where $\hat{\mathbf{r}}_{\mathrm{SF}} = (\phi_{\mathrm{SF}}, \theta_{\mathrm{SF}}, \chi_{\mathrm{SF}})$ denotes the orientation of the symmetric top with respect to the SF axes. In Equation (7.69), j, k ($\geqslant 0$), and m_j are the rotational angular momentum, its projection along the BF z axis (figure axis), and its projection along the SF Z axis, respectively. The symmetry index $\epsilon = \pm 1$ except for $k = 0$, for which only $\epsilon = +1$ is allowed. The parity of the functions in Equation (7.69) is $p = \epsilon(-1)^{j+k}$.

Finally, we consider a diatomic molecule in a $^2\Pi$ electronic state. In this case, the SF rotational wave function in the Hund's case (a) limit (Zare 1988) can be written as

$$\psi_{jm_j\Omega\epsilon}(\hat{\mathbf{r}}_{\mathrm{SF}}) = \left(\frac{2j+1}{4\pi}\right)^{1/2}\left[D^{j*}_{m_j\Omega}(\hat{\mathbf{r}}_{\mathrm{SF}}) + \epsilon D^{j*}_{m_j,-\Omega}(\hat{\mathbf{r}}_{\mathrm{SF}})\right] \tag{7.70}$$

where $\hat{\mathbf{r}}_{\mathrm{SF}} = (\phi_{\mathrm{SF}}, \theta_{\mathrm{SF}}, 0)$ denotes the orientation of the diatomic molecule with respect to the SF axes. The projection of the angular momentum along the molecular axis is denoted Ω and can have the values 1/2 and 3/2, and the symmetry index $\epsilon = \pm 1$ in Equation (7.70). The BF projection quantum numbers Λ, Σ, and $\Omega = \Lambda + \Sigma$ have been defined earlier. The parity of the functions in Equation (7.70) is $p = \epsilon(-1)^{j-s}$, where s is the electronic spin angular momentum and equals 1/2. In this case, J takes on half-integral values.

The Hamiltonian for a $^2\Pi$ diatomic molecule includes the rotational energy, the spin–orbit interaction, and the Λ-doublet splitting:

$$H_{\mathrm{mol}} = H_{\mathrm{rot}} + H_{\mathrm{so}} + H_{\Lambda} \tag{7.71}$$

The matrix elements of H_{mol} can be found in the literature (Lefebvrre-Brion & Field 1986; Zare 1988). The true wave functions for a diatomic molecule in a $^2\Pi$ electronic state are a linear combination of the $\Omega = 1/2$ and 3/2 states in Equation (7.70).

The total SF wave function can be constructed from the molecular rotational wave function and the wave function for end-over-end rotation of the complex. We take linear combinations of the products of these two types of functions so that the total wave function is an eigenfunction of the total angular momentum J to describe a complex consisting of a molecule and a structureless atom. In the case of a collision of a closed-shell $^1\Sigma^+$ linear molecule with a structureless atom, we have

$$\Phi_{jlJM}(\hat{\mathbf{R}}, \hat{\mathbf{r}}_{\mathrm{SF}}) = \sum_{m_jm_l}(jm_j\, lm_l|JM)\psi_{jm_j}(\hat{\mathbf{r}}_{\mathrm{SF}})Y_{lm_l}(\hat{\mathbf{R}}) \tag{7.72}$$

where $(jm_j lm_l|JM)$ is a Clebsch–Gordan coefficient (Brink & Satchler 1993), $\hat{\mathbf{R}} = (\alpha, \beta)$ is the orientation of the vector R connecting the centers of mass of the collision partners, and M is the projection of J on the SF Z axis. The double sum in Equation (7.72) can be reduced to a single sum because $m_j + m_l = M$. The quantum numbers specifying the internal state of the molecule beyond j and m_j have been dropped in Equation (7.72) for notational simplicity.

If both collision partners have internal structure, then their rotational wave functions must first be coupled:

$$\psi_{j_1j_2j_{12}m_{12}}(\hat{\mathbf{r}}_{SF}) = \sum_{m_1m_2} (j_1m_1j_2m_2|j_{12}m_{12})\psi_{j_1m_1}(\hat{\mathbf{r}}_{SF1})\psi_{j_2m_2}(\hat{\mathbf{r}}_{SF2}). \tag{7.73}$$

Here, $\hat{\mathbf{r}}_{SF1}$ and $\hat{\mathbf{r}}_{SF2}$ are the SF orientations of the two collision partners. The total SF wave function in this case then equals

$$\Phi_{j_1j_2j_{12}lJM}(\hat{\mathbf{R}}, \hat{\mathbf{r}}_{SF}) = \sum_{m_{12}m_l} (j_{12}m_{12}lm_l|JM)\psi_{j_{12}m_{12}}(\hat{\mathbf{r}}_{SF}) Y_{lm_l}(\hat{R}). \tag{7.74}$$

The total BF wave function is given by the product of the wave function describing the overall rotational motion of the complex and the rotation of the molecule within the complex. The total BF wave function for an atom–$^1\Sigma^+$ molecule collision pair is given by

$$\Phi_{JMKj}(\hat{\mathbf{R}}, \hat{\mathbf{r}}_{BF}) = \left(\frac{2J+1}{4\pi}\right)^{1/2} D^{J*}_{MK}(\alpha, \beta, 0) Y_{jK}(\theta_{BF}, \phi_{BF}) \tag{7.75}$$

where θ_{BF} and ϕ_{BF} denote the BF orientation of the diatomic molecule, and K is the projection of J and j along the Jacobi vector $\mathbf{R}$. BF wave functions of definite parity can be obtained by taking linear combinations of the wave functions in Equation (7.75):

$$\Phi_{JMKj\rho}(\hat{\mathbf{R}}, \hat{\mathbf{r}}_{BF}) = 2^{-1/2}[\Phi_{JM,+K,j\rho}(\hat{\mathbf{R}}, \hat{\mathbf{r}}_{BF}) + \rho\Phi_{JM,-K,j\rho}(\hat{\mathbf{R}}, \hat{\mathbf{r}}_{BF})] \tag{7.76}$$

where $\rho = \pm 1$ is the parity index. For $K = 0$, only the $\rho = +1$ function exists:

$$\Phi_{JM,K=0,j,\rho=+1}(\hat{\mathbf{R}}, \hat{\mathbf{r}}_{BF}) = \Phi_{JM,K=0,j}(\hat{\mathbf{R}}, \hat{\mathbf{r}}_{BF}). \tag{7.77}$$

The SF wave functions can be expressed in terms of the BF wave functions as

$$\Phi_{jlJM}(\hat{\mathbf{R}}, \hat{\mathbf{r}}_{SF}) = \sum_K \left(\frac{2l+1}{2J+1}\right)^{1/2} (jKl0|JK)\Phi_{JMKj}(\hat{\mathbf{R}}, \hat{\mathbf{r}}_{BF}). \tag{7.78}$$

Equation (7.78) can be employed for the BF–SF transformation for collision complexes involving any type of molecule.

The full scattering wave function can be written as

$$\Psi(\mathbf{R}, \mathbf{r}) = R^{-1}\mathbf{c}(R)\Phi(\hat{\mathbf{R}}, \mathbf{r}) \tag{7.79}$$

where $\mathbf{c}(R)$ is the row vector of radial expansion coefficients. Substitution of Equation (7.79) into the Schrödinger equation, premultiplication by one of the internal states, and integration over $\mathbf{r}$ yields a set of coupled ordinary differential equations for the radial expansion coefficients. In practice, a finite set of basis functions, chosen large enough so as not to affect the convergence of the calculation, is employed.

These coupled second-order differential equations can be written in matrix notation as

$$\left[\mathbf{I}\frac{d^2}{dR^2} + \mathbf{W}(R)\right]\mathbf{C}(R) = 0 \tag{7.80}$$

where $\mathbf{C}$ represents the matrix of column vectors of the expansion coefficients for each solution. In Equation (7.80), $\mathbf{I}$ denotes the identity matrix and the matrix $\mathbf{W}(R)$ is given by

$$\mathbf{W}(R) = \mathbf{k}^2 - \frac{\mathbf{l}^2}{R^2} + \frac{2\mu}{\hbar^2}\mathbf{V}(R). \tag{7.81}$$

Here, $\mu = m_A m_B/(m_A + m_B)$ is the reduced mass of the collision system, and m_A and m_B are the masses of the collision partners. In Equation (7.81), $\mathbf{V}(R)$ is the (symmetric) matrix of the potential in the scattering basis, and $\mathbf{k}^2$ represents the (diagonal) matrix of the wavevector, with matrix elements

$$k_{ii}^2 = \frac{2\mu}{\hbar^2}(E - \epsilon_i) \tag{7.82}$$

where E is the total energy and ϵ_i is the internal energy of the ith channel.

The internal energies ϵ_i in Equation (7.82) can usually be taken as the rigid rotor energies

$$\epsilon_i = Bj(j + 1) \tag{7.83}$$

for a closed-shell diatomic molecule, or for an oblate (like methyl) or prolate (like methyl chloride) symmetric top, respectively,

$$\epsilon_i = Bj(j + 1) + (C - B)k^2 \tag{7.84}$$

$$\epsilon_i = Bj(j + 1) + (A - B)k^2. \tag{7.85}$$

The rotational/fine-structure energies of a $^2\Pi$ diatomic molecule are given by the eigenvalues of H_{mol} (Equation (7.71)) as a function of the total diatomic angular momentum j.

In the space frame, the matrix of $\mathbf{l}^2$ is diagonal, with matrix elements

$$l_{ii}^2(R) = l_i(l_i + 1) \tag{7.86}$$

where l_i is the orbital angular momentum quantum number in the ith channel. The $\mathbf{l}^2$ operator is not diagonal in the body frame. The BF matrix elements of $\mathbf{l}^2$ are diagonal in J, M, and the quantum numbers describing the internal motion in the complex. The nonzero matrix elements between the signed-K BF functions defined in Equation (7.75) are

$$\langle JMKj|\mathbf{l}^2|JMKj\rangle = J(J+1) + j(j+1) - 2K^2 \tag{7.87}$$

$$\langle JM, K\pm 1, j|\mathbf{l}^2|JMKj\rangle = \{[J(J+1) - K(K\pm 1)][j(j+1) - K(K\pm 1)]\}^{1/2}. \tag{7.88}$$

We require matrix elements of the $\mathbf{V}(R)$ for the solution of Equation (7.80); these matrix elements are diagonal in J and M. For a symmetric top, the SF matrix element of one term in the angular expansion in Equation (7.69) involving the SF signed-k wave functions in Equation (7.69) is

$$\begin{aligned}\langle JMj'k'l' \; |C_{\lambda\mu}(\theta, \phi)|JMjkl\rangle &= (-1)^{j+j'+k+J} \\ &\times [(2j+1)(2j'+1)(2l+1)(2l'+1)]^{1/2} \\ &\times \begin{pmatrix} l & \lambda & l' \\ 0 & 0 & 0 \end{pmatrix} \begin{pmatrix} j & \lambda & j \\ k & \mu & -k \end{pmatrix} \begin{Bmatrix} l & \lambda & l' \\ j' & J & j \end{Bmatrix}\end{aligned} \tag{7.89}$$

where (:::) is a $3j$ symbol and {:::} is a $6j$ symbol (Brink & Satchler 1993).

For the atom–diatomic molecule system, this expression simplifies to

$$\begin{aligned}\langle JMj'l' \; |P_\lambda(\cos\theta)|JMjl\rangle &= (-1)^{j+j'+J} \\ &\times [(2j+1)(2j'+1)(2l+1)(2l'+1)]^{1/2} \\ &\times \begin{pmatrix} l & \lambda & l' \\ 0 & 0 & 0 \end{pmatrix} \begin{pmatrix} j & \lambda & j \\ 0 & 0 & 0 \end{pmatrix} \begin{Bmatrix} l & \lambda & l' \\ j' & J & j \end{Bmatrix}.\end{aligned} \tag{7.90}$$

The BF matrix elements are much simpler; these are diagonal in K. For a symmetric top, the BF matrix element corresponding to Equation (7.89) equals

$$\begin{aligned}\langle JMKj'k' \; |C_{\lambda\mu}(\theta, \phi)|JMKjk\rangle &= (-1)^{K-k}[(2j+1)(2j'+1)]^{1/2} \\ &\times \begin{pmatrix} j & \lambda & j' \\ -k & \mu & k \end{pmatrix} \begin{pmatrix} j & \lambda & j' \\ -K & 0 & K \end{pmatrix}.\end{aligned} \tag{7.91}$$

Note that k is the projection of j on the figure axis and K is the projection of J and j along $\mathbf{R}$. In contrast to the SF matrix elements, the BF matrix elements do not depend upon the total angular momentum J.

The SF and BF matrix elements of the $\mathbf{V}(R)$ for the interaction of a diatomic molecule in a $^1\Sigma^+$ electronic state can be obtained from Equations (7.89) and (7.91), respectively, because $d^{\lambda}_{0\mu}(\theta) = C_{\lambda\mu}(\theta, 0)$.

It is advantageous to employ scattering basis functions that have a definite parity because the matrix $\mathbf{V}(R)$ in Equation (7.81) is block diagonal in the parity and also the total angular momentum J and space-frame projection M. Because the computational effort in solving the Schrödinger equation scales roughly as the number of basis functions cubed, the solution of a number of smaller blocks (defined by the total angular momentum J and the parity) of the Hamiltonian requires less effort than the solution of the full matrix.

The matrix of solutions $\mathbf{C}(R)$ is obtained numerically by outward propagation of Equation (7.80) from the classically forbidden region at small R to a large value of R, denoted here as R_{end}. Equation (7.80) must be solved for each parity and a sufficient number of partial waves (total angular momentum J) to converge the cross sections. It takes fewer partial waves to converge the inelastic cross sections than for the elastic cross section.

There are two widely used, freely available packages for the integration of these equations and the determination of cross sections: HIBRIDON[1] and MOLSCAT[2]. In the former, the scattering equations are integrated with a hybrid integrator (Alexander 1984), in which Equation (7.80) is integrated with the log-derivative method (Johnson 1973; Mrugala & Secresty 1983; Manolopoulos 1986) for small R where the potential is varying rapidly, and at large R, where the potential is slowly varying, with an Airy function algorithm based in the original algorithm of Gordon (1969). The preferred method of solution in MOLSCAT is the hybrid integrator. In both of these packages, the scattering wave function is expressed in the space frame.

If the integration is carried out with the wave function defined in the body frame, then the wave function at R_{end} must be transformed to the space frame, employing Equation (7.78). This procedure is described by Skouteris et al. (2000).

All the information observable about the collision can be derived from the S matrix, which can be obtained by matching the expansion coefficients $\mathbf{C}(R)$ to the asymptotic form

$$\lim_{R\to\infty} \mathbf{C}(R) = \hat{\mathbf{h}}^{(2)}(R) - \mathbf{S}\hat{\mathbf{h}}^{(1)}(R) \tag{7.92}$$

where $\hat{\mathbf{h}}^{(1)}$ and $\hat{\mathbf{h}}^{(2)}$ are diagonal matrices with elements

$$\hat{h}_{ii}^{(1,2)}(R) = k_i^{-1/2} h_{l_i}^{(1,2)}(k_i R) \tag{7.93}$$

and $h_l^{(1,2)}$ are complex spherical Hankel functions (Abramowitz & Stegun 1965), namely

$$h_l^{(1,2)}(x) = j_l(x) \pm i y_l(x). \tag{7.94}$$

[1] HIBRIDON is a package of programs for the time-independent quantum treatment of inelastic collisions and photodissociation written by M. H. Alexander, D. E. Manolopoulos, H.-J. Werner, B. Follmeg, P. J. Dagdigian, and others. More information and/or a copy of the code can be obtained from the website https://www2.chem.umd.edu/groups/alexander/hibridon.

[2] MOLSCAT is a general-purpose package for performing quantum nonreactive molecular scattering calculations. It was written by Jeremy Hutson and Sheldon Green in near-standard FORTRAN 77 but is no longer being maintained. More information and/or a copy of the code can be obtained from the website https://www.giss.nasa.gov/tools/molscat/.

In Equation (7.94), the first (second) Hankel function corresponds to the plus (minus) sign. The functions $j_l^{(1,2)}$ and $y_l^{(1,2)}$ are spherical Bessel functions and are defined in terms of Bessel functions of integral order as

$$j_n(x) = (\pi/2x)^{1/2} J_{n+\frac{1}{2}}(x) \tag{7.95}$$

$$y_n(x) = (\pi/2x)^{1/2} Y_{n+\frac{1}{2}}(x). \tag{7.96}$$

Within the close-coupled formulation, the integral cross sections for a transition from an initial level i to a final level f are given by

$$\sigma_{f\leftarrow i} = \frac{\pi}{(2j+1)k_i^2} \sum_{Jll'} (2J+1) |T^J_{j'\alpha' l', j\alpha l}|^2. \tag{7.97}$$

In Equation (7.97), J is the total angular momentum, j is the rotational angular momentum, l is the orbital angular momentum, and α is a collective index defining the initial or final level. The unprimed and primed quantum numbers refer to the initial and final levels, respectively. The T in Equation (7.97) are elements of the transition matrix $\mathbf{T}$, which is defined as

$$\mathbf{T} = \mathbf{I} - \mathbf{S}. \tag{7.98}$$

With modern computational resources, it is generally possible to solve the close-coupled equations without approximations. In special cases in which many scattering basis functions need to be included, the coupled-states approximation can be employed. Kouri has written a review of the approximation methods that have been used to reduce the computational effort in solving the scattering equations (Kouri 1979).

State-to-state rate constants can be computed by taking the thermal average of the corresponding cross sections computed over a range of energies (Greene & Kuppermann 1968):

$$k_{f\leftarrow i}(T) = \left[\frac{8}{\pi\mu(k_B T)^3}\right]^{1/2} \int_0^\infty E_c \sigma_{f\leftarrow i}(E_c) e^{-E_c/k_B T_K} dE_c \tag{7.99}$$

where T_K is the translational (kinetic) temperature and E_c is the collision energy. It should be noted that these rate constants satisfy detailed balance (Equation (7.34)).

There are publicly available databases that present rate constants for selected astrophysically relevant collision pairs. These include the Leiden Atomic and Molecular Database (LAMDA; http://home.strw.leidenuniv.nl/moldata) and BASECOL, which is now part of the Virtual Atomic and Molecular Data Centre (VAMDC; http://portal.vamdc.org/vamdc_portal/home.seam).

In this section, the theoretical treatment of collisional rotational excitation has been considered. While of lesser importance in astrophysical applications, collisions can also induce vibrational excitation in a molecule. In this case, the potential energy surface would need to be computed also as a function of the vibrational coordinate. Transitions between vibrational levels would be governed by the matrix element of the potential between the vibrational levels. Collision pairs for which recent calculations of rovibrational excitation have been carried out include CO–H_2 (Castro et al. 2017), CN–H_2 (Yang et al. 2016), HCN–He (Denis-Alpizar et al. 2013b), C_3–He (Stoecklin et al. 2015), and CH_3–He (Ma et al. 2013).

7.3.3 Hyperfine Transitions

For many astrophysical molecules, the hyperfine structure of a rotational transition is often resolved by the high spectral resolution of observational spectra, especially in the radio-frequency portion of the spectrum. For example, the abundant nuclei ^{1}H and ^{14}N have nonzero nuclear spin, and molecules containing these nuclei, especially ^{14}N, i.e., CN, HCN, NH_3, and N_2H^+, show hyperfine-resolved rotational spectra.

Resolution of the hyperfine structure of a rotational transition can be very helpful because the optical depth of the transition can be determined directly by comparing the relative intensities of the hyperfine lines. For optically thin lines, the intensity ratios follow the ratios of the line strength factors, whereas these intensity ratios approach unity for optically thick lines.

In cold astronomical media, the main collision partners are He and para-$H_2(j = 0)$. Rate constants for hyperfine-resolved rotational transitions have been computed for a number of molecules in collisions with He or H_2, including CN (Lique 2011; Kalugina & Lique 2015), HCN (Lique 2012), HCl (Neufeld & Green 1994), NH_3 (Chen et al. 1998), N_2H^+ (Daniel et al. 2005), C_2H (Spiefiedel et al. 2012), OH (Marinakis et al. 2016), OH^+ (Lique et al. 2016), and DCO^+ (Buffa 2012).

It is beyond the scope of this chapter to describe in detail the rigorous and various approximate methods for the calculation of rate constants for hyperfine-resolved collision-induced rotational transitions. Faure & Lique (2012) have written a brief review on different methods of computing hyperfine-resolved rate constants for molecular collisions. Unless one is interested in rate constants at extremely low temperatures, it is sufficient to carry out nuclear-spin-free scattering calculations and then to determine the hyperfine-resolved cross sections and rate constants by means of a recoupling method (Alexander & Dagdigian 1985; Corey & Alexander 1988; Offer et al. 1994). This approach greatly simplifies the scattering calculation. Several approximate methods have also been employed, and their accuracy has been tested for some collision pairs (Faure & Lique 2012).

7.4 Excitation of Interstellar Molecules

We conclude this chapter with several examples of the importance of employing accurate collisional rates in solving a non-LTE radiative transfer problem to deduce molecular densities.

Hydrogen cyanide (HCN) and hydrogen isocyanide (HNC) are two important and abundant molecules in the interstellar medium and are useful as star formation indicators. Despite the fact that HCN is by far the more thermodynamically stable isomer, both molecules have been detected, and the ratio of their concentrations depends upon the kinetic temperature. Until recently, collisional data for the two isomers were not available, and the abundance ratio was estimated from line intensity ratios. This is not expected to be a reliable method for estimating the abundance ratio, given that level populations of interstellar molecules are usually out of thermodynamic equilibrium.

Reliable determination of abundance ratios requires radiative transfer calculations—and these, in turn, require radiative and collisional rates. Recently, Hernández-Vera et al. (2017) have calculated excitation rate constants for collisions of HCN and HNC with para-$H_2(j = 1)$ and ortho-$H_2(j = 1)$. They employed potential energy surfaces computed by Denis-Alpizar et al. (2013a) for HCN–H_2 and by Dumouchel et al. (2011) for HNC–H_2. The rate constants for the two systems differ significantly, especially for collisions with para-H_2. This indicates that radiative transfer calculations must be carried out to determine the HNC/HCN abundance ratio.

Isotope ratios are very useful tools for studying the origin of the solar system and the possible link with interstellar chemistry. Aside from hydrogen, the second largest variation of isotope ratios is found for nitrogen. The N_2H^+ ion is a useful species and shows the largest variation in the $^{14}N/^{15}N$ ratio of nitrogen-containing species studied. Daniel et al. (2016) detected the $j = 1 - 0$ line of $N^{15}NH^+$. Also available were data on the N_2H^+ $j = 1 - 0$ and $j = 3 - 2$ lines. To derive the isotope ratio from these data, the molecular excitation was solved with a nonlocal radiative transfer code. The required hyperfine-resolved collisional rates for N_2H^+ were calculated using a highly correlated potential energy surface (Spiefiedel et al. 2015). As a first approximation, rate constants for N_2H^+ could be taken as equal to those for N_2H^+ (when hyperfine splittings are ignored). However, even the small shift of the center of mass for $^{14}N \rightarrow {}^{15}N$ substitution is problematic (Flower & Lique 2015). The adiabatic-hindered-rotor treatment with the N_2H^+–H_2 potential energy surface was used to compute rate constants for collisions of $N^{15}NH^+$ with spherical $H_2(j = 0)$. With these rate constants employed in the radiative transfer calculation, a $N_2H^+/N^{15}NH^+$ ratio of ~330 was found, which agrees with typical values of the elemental isotope ratio in the local interstellar medium.

The ortho–para ratio (OPR) of molecular hydrogen is a fundamental parameter for the physics and chemistry in interstellar molecular clouds. In protostellar environments, H_2 forms on grain surfaces, and it is usually assumed that the initial OPR equals the statistical value of 3. It is generally accepted that ortho–para conversion occurs subsequently through proton exchange reactions with H^+ and H_3^+, and the OPR is expected to decrease toward the very small equilibrium value (which equals 3.6×10^{-7} at 10 K). However, the equilibrium value is not expected to be reached, because of recycling of hydrogen through gas-phase reactions and catalysis on grains. Direct observation of H_2 rovibrational lines in the warm gas associated with molecular shocks have found the OPR to be in the range 0.5–2 (see Maret et al. 2009, and references therein). However, H_2 is not directly observable in dark and cold regions.

The OPR of H_2 is also important for collision excitation at low temperature because the interaction potential is different for para-$H_2(j = 0)$, which is spherical, versus ortho-$H_2(j = 1)$, and the permanent quadrupole moment vanishes for $j = 0$.

Consequently, rate constants for collisions of $H_2(j = 1)$ are usually larger than for $H_2(j = 1)$, and the collisional propensity rules can be different for these two colliders. Through direct observation of the 6 cm absorption line ($1_{10} \leftarrow 1_{11}$) of H_2CO and radiative transfer calculations, Troscompt et al. (2009) have been able to constrain the OPR of H_2 to be close to zero in the observed dark molecular cloud. In their radiative transfer model, they employed H_2CO–H_2 rate constants computed with a recently determined potential energy surface for this system (Troscompt et al. 2009). They were also able to explain why the brightness temperature of the 6 cm line is below the cosmic background temperature (2.7 K), and they confirmed the propensity rule proposed by Townes & Cheung (1969) for collisional transfer in H_2CO.

References

Abramowitz, M., & Stegun, I. A. 1965, Handbook of Mathematical Functions, Vol. 55 (Washington, DC: US Govt Printing Office)

Alexander, M. H. 1984, JChPh, 81, 4510

Alexander, M. H. 1985, CP, 92, 337

Alexander, M. H., & Dagdigian, P. J. 1985, JChPh, 83, 2191

Arthurs, A. M., & Dalgarno, A. 1960, RSPSA, 256, 540

Bergeat, A., Onvlee, J., Naulin, C., van der Avoird, A., & Costes, M. 2015, NatCh, 7, 349

Bernes, C. 1979, A&A, 73, 67

Boys, S. F., & Benardi, F. 2002, MolPh, 100, 65

Brink, D. M., & Satchler, G. R. 1993, Angular Momentum (3rd ed; Oxford: Clarendon Press)

Brown, J. M., Hougen, J. T., Huber, K.-P., et al. 1975, JMoSp, 55, 500

Buffa, G. 2012, MNRAS, 421, 719

Carty, D., Goddard, A., Sims, I. R., & Smith, I. W. M. 2004, JChPh, 121, 4671

Castro, A., Doan, K., Klemka, M., et al. 2017, MolAs, 6, 47

Chałasiński, G., & Szczęśniak, M. M. 2000, ChRv, 100, 4227

Chefdeville, S., Stoecklin, T., Bergeat, A., et al. 2012, PhRvL, 109, 023201

Chen, J., Zhang, Y.-H., & Zhou, T.-C. 1998, ChA&A, 22, 113

Condon, E. U., & Shortley, G. H. 1951, The Theory of Atomic Spectra (London: Cambridge Univ. Press)

Corey, G. C., & Alexander, M. H. 1988, JChPh, 88, 6931

Crawford, T. D., & Schaefer, H. F. 2000, RvCoCh, 14, 33

Čížek, J. 1969, AdChP, 14, 35

Dagdigian, P. J. 1995, The Chemical Dynamics and Kinetics of Small Radicals, Part I, ed. K. Liu, & A. F. Wagner (Singapore: World Scientific), 315

Dagdigian, P. J. 2013, IRPC, 32, 229

Dagdigian, P. J. 2016, JChPh, 145, 114301

Daniel, F., Dubernet, M.-L., Meuwly, M., Cernicharo, J., & Pagani, L. 2005, MNRAS, 363, 1083

Daniel, F., Faure, A., Pagani, L., et al. 2016, A&A, 592, A45

de Jongh, T., Karman, T., Vogels, S., et al. 2017, JChPh, 147, 013918

Deegan, M. J. O., & Knowles, P. J. 1994, CPL, 227, 321

Denis-Alpizar, O., Kalugina, Y., Stoecklin, T., Vera, M. H., & Lique, F. 2013a, JChPh, 139, 224301

Denis-Alpizar, O., Stoecklin, T., Halvick, P., & Dubernet, M.-L. 2013b, JChPh, 139, 034304

Dumouchel, F., Kłos, J., & Lique, F. 2011, PCCP, 13, 8204
Faure, A., & Lique, F. 2012, MNRAS, 425, 740
Faure, A., Valiron, P., Wernli, M., et al. 2005, JChPh, 122, 221102
Flower, D. R., & Lique, F. 2015, MNRAS, 446, 1750
Goldreich, P., & Kwan, J. 1974, ApJ, 189, 441
Gordon, R. G. 1969, JChPh, 51, 14
Greene, E. F., & Kuppermann, A. 1968, JChEd, 45, 361
Gubbels, K. B., Ma, Q., Alexander, M. H., et al. 2012a, JChPh, 136, 144308
Gubbels, K. B., van de Meerakker, S. Y. T., Groenenboom, G. C., Meijer, G., & van der Avoird, A. 2012b, JChPh, 136, 074301
Hampel, C., Peterson, K. A., & Werner, H.-J. 1992, CPL, 190, 1
Hättig, C., Klopper, W., Köhn, A., & Tew, D. P. 2012, ChRv, 112, 4
Helgaker, T., Klopper, W., Koch, H., & Noga, J. 1997, JChPh, 106, 9639
Hernández-Vera, M., Lique, F., Dumouchel, F., Hily-Blant, P., & Faure, A. 2017, MNRAS, 468, 1084
Helkier, A., Helgaker, T., Jorgensen, P., Klopper, W., & Olsen, J. 1999, CPL, 302, 437
Herzberg, G. 1950, Molecular Spectra and Molecular Structure I, Spectra of Diatomic Molecules (Princeton, NJ: Van Nostrand-Reinhold)
Hill, J. G., Peterson, K. A., Knizia, G., & Werner, H.-J. 2009, JChPh, 131, 194105
Huang, S.-S. 1968, ApJ, 152, 841
Huber, K. P., & Herzberg, G. 1979, Molecular Spectra and Molecular Structure IV, Constants of Diatomic Molecules (Princeton, NJ: Van Nostrand-Reinhold)
Janssen, L. M. C., Zuchowski, P. S., van der Avoird, A., et al. 2011, JChPh, 134, 124309
Jeziorska, M., Jankowski, P., Szalewicz, K., & Jeziorski, B. 2000, JChPh, 113, 2957
Johnson, B. R. 1973, JCoPh, 13, 445
Kalugina, Y., & Lique, F. 2015, MNRAS, 446, L21
Kendall, R. A., Dunning, T. H., & Harrison, R. J. 1992, JChPh, 96, 6796
Koch, H., Fernandez, B., & Christiansen, O. 1998, JChPh, 108, 2784
Kong, L., Bischoff, F. A., & Valeev, E. F. 2012, ChRv, 112, 75
Kouri, D. J. 1979, Atom-Molecule Collision Theory, ed. R. B. Bernstein, (New York: Plenum), 301
Lefebvrre-Brion, H., & Field, R. W. 1986, Perturbations in the Spectra of Diatomic Molecules (Orlando, FL: Academic)
Lequeux, J. 2005, The Interstellar Medium (Berlin: Springer)
Lique, F. 2011, MNRAS, 413, L20
Lique, F. 2012, MNRAS, 419, 2441
Lique, F., Bulut, N., & Roncero, O. 2016, MNRAS, 461, 4477
Ma, Q., Dagdigian, P. J., & Alexander, M. H. 2013, JChPh, 138, 104317
Ma, Q., Kłos, J., Alexander, M. H., van der Avoird, A., & Dagdigian, P. J. 2014, JChPh, 141, 174309
Manolopoulos, D. E. 1986, JChPh, 85, 6425
Maret, S., Bergin, E. A., Neufeld, D. A., et al. 2009, ApJ, 698, 1244
Marinakis, S., Kalugina, Y., & Lique, F. 2016, EPJD, 70, 97
Mrugala, F., & Secresty, D. 1983, JChPh, 78, 5954
Muller, H. S. P., Schloder, F., Stutzki, J., & Winnewisser, G. 2005, JMoSt, 742, 215
Muller, H. S. P., Thorwith, S., Roth, D. A., & Winnewisser, G. 1998, A&A, 370, L49

Neufeld, D., & Green, S. 1994, ApJ, 432, 158
Offer, A. R., van Hemert, M. C., & van Dishoeck, E. F. 1994, JChPh, 100, 362
Pickett, H. M., Poynter, R. L., Cohen, A. A., et al. 1998, JQSRT, 60, 883
Raghavachari, K., Trucks, G. W., Pople, J. A., & Head-Gordon, M. 1989, CPL, 157, 479
Rist, C., & Faure, A. 2012, JMaCh, 50, 588
Rowe, B. R., Dupyrat, G., Marquette, J. B., & Gaucherel, P. 1984, JChPh, 80, 4915
Rowe, B. R., & Marquette, J. B. 1987, IJMSI, 80, 239
Scharfenberg, L., Haak, H., Meijer, G., & van de Meerakker, S. Y. T. 2009, PhRvA, 79, 023410
Schewe, H. C., Ma, Q., Vanhaecke, N., et al. 2015, JChPh, 142, 204310
Schiffman, A., & Chandler, D. W. 1995, IRPC, 14, 371
Skouteris, D., Castillo, J. F., & Manolopoulos, D. E. 2000, CoPhC, 133, 128
Spiefiedel, A., Senent, M. L., Kalugina, Y., et al. 2015, JChPh, 143, 024301
Spiefiedel, N., Feautrier, A., Najar, D., et al. 2012, MNRAS, 421, 1891
Stoecklin, T., Denis-Alpizar, O., & Halvick, P. 2015, MNRAS, 449, 3420
Surdej, J. 1979, A&A, 60, 303
Tao, F. M., & Pan, Y. K. 1992, JChPh, 97, 4989
Townes, C. H., & Cheung, A. C. 1969, ApJ, 157, L103
Troscompt, N., Faure, A., Maret, S., et al. 2009, A&A, 506, 1243
Valiron, P., Wernli, M., Faure, A., et al. 2008, JChPh, 129, 134306
van der Avoird, A., Wormer, P. E. S., Mulder, F., & Berns, R. M. 1980, TCCh, 93, 1
van der Tak, F. F. S., Black, J. H., Schöier, D. J., Jansen, D., & van Dishoeck, E. F. 2007, A&A, 468, 627
Wernli, M., Wiesenfeld, L., Faure, A., & Lique, F. 2007, A&A, 464, 1147
Wormer, P. E. S., Kłos, J., Groenenboom, G. C., & van der Avoird, A. 2005, JChPh, 122, 244325
Yang, B., Wang, X. H., Stancil, P. C., et al. 2016, JChPh, 145, 224307
Zare, R. N. 1988, Angular Momentum (New York: Wiley)

Chapter 8

Applications: the Molecular Viewpoint of Interstellar Observations

Evelyne Roueff

8.1 Introduction

Observing molecular spectra is a relatively young research field for astrophysicists. Whereas visible emission observed in the solar atmosphere or in planetary environments can be attributed to a small number of well-known stable molecules, the opening of the millimeter window in the early 1980s has provided the opportunity to discover a completely new and unexpected chemistry involving radicals, molecular ions, and isomers, some of which were not even known in terrestrial laboratories, toward interstellar environments, presumably empty regions of the sky. Still, unknown narrow visible absorption features detected in front of the bright star ξ Per, were reported in early 1937 (Dunham 1937) at 3957.69 Å, 4232.54 Å, and 4300.27 Å as being of possible molecular origin, rapidly identified as CH^+ for the first two transitions and CH for the latter (Douglas & Herzberg 1941; Adams 1941). The first astrochemical analysis was performed a few years later by Bates & Spitzer (1951), who conceived the different chemical processes hypothetically occurring in the gas phase, as well the possible influence of grains, that could explain such observations. Visible and ultraviolet molecular absorption studies have subsequently been conducted, leading to the detection of various diatomics and bringing important constraints to our understanding of the so-called diffuse and translucent clouds (for a review, see van Dishoeck & Black 1986 and Snow & McCall 2006). These observations are performed within a very narrow direction defined by the fortuitous location of the background star, providing the visible and ultraviolet continuum radiation in front of which absorption due to intersecting interstellar clouds can take place.[1] Remarkably, such absorption studies can also be performed

[1] In addition, diffuse interstellar bands (DIBs) observed in stellar spectra designate a very well-documented list of visible and infrared spectral absorption features that have been known for several decades but are still not identified.

doi:10.1088/2514-3433/aae1b5ch8

in the radio and submillimeter domain, thanks to continuum sources provided by quasars. Pioneering radio-millimeter studies with the Plateau de Bures interferometer by Lucas & Liszt (1993) have been pursued in the submillimeter region thanks to Herschel,[2] and have provided many successful new detections as reviewed by Gerin et al. (2016). By contrast, radio, millimeter, and submillimeter emission can be observed in any direction within a certain beam size (constrained by the transition frequency and the diameter of the antenna) by changing the pointing of telescope, or even mapped thanks to interferometers. Analyzing the spatial distribution of the emission allows the density and temperature distribution of the interstellar matter to be constrained.

The combination of spectral and spatial resolution is thus decisive in probing these interstellar environments where pre-stellar conditions are fulfilled and allowing the star formation history to be constrained. Complementary continuum mapping performed with bolometers at infrared, submillimeter, and millimeter wavelengths traces the thermal emission of tiny dust particles, which are mixed with the gas, providing additional constraints on the protostellar environments.

The present amount of identified molecules in the gas phase is of the order of 200 in the close galactic environment, whereas only less than a dozen dust mantle molecules have been definitively found. This fact is principally due to the difficulties in unequivocally assigning solid mantle species that display broad and faint spectral features. This pitfall will probably be considerably reduced with the advent of the *JWST*[3] instrument.

We report the inventory of detected interstellar and circumstellar gas-phase molecules in three different tables.

Table 8.1 displays the list of neutral, stable molecules, which have a singlet symmetry electronic ground state. Table 8.2 reports the open-shell neutral species, which may be referred as radicals, whereas Table 8.3 exhibits the molecular ions, where isotopic substituted components are referred as well, which include both positive and negative species. More than sixty molecules are even found in extragalactic sources, where the redshifted molecular transitions are detected at frequencies $\nu = \nu_0 \times \frac{1}{1+z}$, where z is the redshift of the intervening extragalactic cloud. The list of these extragalactic detected molecules is displayed in the Cologne Spectroscopic Molecular database (see Molecules in Space at http://cdms.ph1.uni-koeln.de/cdms/portal/; Endres et al. 2016). The display of this molecular population is already instructive, showing the very large variety of interstellar compounds, the presence of a significant number of radicals, reactive molecules, different isomers, and the occurrence of positive and negative molecular ions. Such a molecular diversity was completely unpredicted only fifty years ago, with the first astrochemical models devoted to diatomic molecules settled in the 1970s by Solomon and Klemperer (Solomon & Klemperer 1972) and cautiously extended to account for the observed stable HCN, H_2O, NH_3, and H_2CO molecules and their protonated ions by Herbst and Klemperer (Herbst & Klemperer 1973). The molecular complexity of

[2] Herschel is a space observatory launched by the European Space Agency in 2009, which covered the spectral range from the far-infrared to submillimeter region

[3] *JWST* stands for the *James Webb Space Telescope*.

Table 8.1. Neutral Saturated Molecules in a Singlet Electronic State Found in the ISM and CSEs

2 atoms	3 atoms	4 atoms	5 atoms	6 atoms	7 atoms	8 atoms	9 atoms
H_2	C_3	NH_3	CH_4	C_2H_4	CH_3CHO	$HCOOCH_3$	CH_3OCH_3
C_2	H_2O	H_2CO	c-C_3H_2	CH_3OH	CH_2CHCN	CH_3COOH	CH_3CH_2OH
CO	HCN	HNCO	l-C_3H_2	CH_3CN	CH_3NH_2	CH_2OHCHO	CH_3CH_2SH
HCl	HNC	HOCN	HCOOH	CH_3NC	CH_3C_2H	C_6H_2	CH_3CH_2CN
KCl	H_2S	HCNO	H_2CNH	CH_3SH	CH_2CHCN	CH_3C_3N	CH_3C_4H
NaCl	HNO	H_2CS	HC_3N	c-C_3H_2O	c-C_2H_4O	CH_2CCHCN	C_3H_6
SiO	N_2O	PH_3	H_2CCO	$HC\equiv CCH{=}O$	CH_3NCO	NH_2CH_2CN	HC_7N
SiS	CO_2	C_3O	NH_2CN	NH_2CHO	CH_2CHOH	CH_3CHNH	CH_3CONH_2
CS	SO_2	C_3S	CNCHO	HNCHCN	HC_5N		
HF	OCS	C_2H_2	HNCNH	SiH_3CN			
N_2	HCP	HNCS	SiH_4				
PN	AlNC	H_2O_2					
AlCl	AlOH	HMgNC					
AlF	KCN						
TiO	c-SiC_2						
O_2[1]							

10 atoms	11 atoms	12 atoms	>12 atoms
CH_3COCH_3	$CH_3CH_2CH_3$	c-C_6H_6	c-C_6H_5CN
$(CH_2OH)_2$	CH_3C_6H	$C_2H_5OCH_3$	C_{60}
CH_3CHCH_2O	C_2H_5OCHO	n-C_3H_7CN	C_{70}
CH_3CH_2CHO	CH_3OCOCH_3	i-C_3H_7CN	
CH_3C_5N			

Note. [1]O_2 has a $^3\Sigma^-$ ground state but is reported here as it is a very stable paramagnetic molecule.

the galactic and extragalactic interstellar clouds is now well-recognized and represents an important issue within the field of astrochemistry where elementary processes involved with molecular formation, destruction, and excitation are studied. The link between the interstellar environment and the cometary atmospheres is another valuable question, as important molecules—such as water H_2O, hydrogen cyanide HCN, formaldehyde H_2CO, and methanol CH_3OH—are found in cometary environments. The *Rosetta* mission has yielded a wealth of unique in situ measurements on comet 67P/Churyumov–Gerasimenko (67P/C-G), making it the best probe to date. The presence of molecular oxygen (Keeney et al. 2017) and multiply sulfuretted molecules (S_2, CS_2 S_3, S_4; Calmonte et al. 2016) in the coma of comet 67P/C-G measured by the *Rosetta* Orbiter Spectrometer represents a great opportunity to probe the connection between protostars and comets (Dulieu et al. 2017; Drozdovskaya et al. 2018). The comparison of isotopic ratios such as D/H in H_2O, HCN, H_2CO, and $^{14}N/^{15}N$ in HCN in cometary and interstellar environments may serve as an Ariadne's thread to trace the evolution from the precollapse phase to meteorites and comets today (Ceccarelli et al. 2014). Very recently, Bertaux &

Table 8.2. Molecular Radicals and Carbon Chains Found in the ISM and CSEs

2 atoms	3 atoms	4 atoms	5 atoms	6 atoms	7 atoms	8 atoms	9 atoms
CH	CH_2	c-C_3H	C_5	C_5H	C_6H	C_7H	C_8H
OH	C_2H	l-C_3H	H_2C_3	C_5N	HC_5O		
NH	NH_2	HC_2N	CH_2CN	HC_4N			
CN	HCO	H_2CN	C_4H				
NO	HNO	CH_3	CH_3O				
SH	C_2O	C_3N					
SO	HO_2	HCCO					
NS	NCO						
SiC	HCS						
SiN	HSC						
FeO	C_2N						
AlO	C_2S						
CP	S_2H[1]						
	SiCN						
	SiNC						
	MgCN						
	MgNC						
	FeCN						
	CCP						

Note. [1]S_2H is the first doubly sulfuretted molecule recently detected in the ISM (Fuente et al. 2017).

Table 8.3. Molecular Ions Found in the ISM and CSEs

2 atoms	3 atoms	4 atoms	5 atoms	6 atoms	7 atoms	9 atoms and more
CH^+	H_3^+	H_3O^+	H_2COH^+	HC_3NH^+	C_6H^-	C_{60}^+
$^{13}CH^+$	H_2D^+	$HOCO^+$	NH_3D^+	C_5H^-		C_8H^-
OH^+	D_2H^+	$HCNH^+$	H_2NCO^+	C_5N^-		
SH^+	H_2O^+	CH_2D^+	$C_2N_2H^+$			
CO^+	HCO^+	C_3H^+	C_4H^-			
SO^+	$H^{13}CO^+$	C_3N^-				
CF^+	DCO^+					
HCl^+	$HC^{18}O^+$					
NO^+	$HC^{17}O^+$					
NS^+	HOC^+					
ArH^+	HCS^+					
CN^-	N_2H^+					
	N_2D^+					
	$^{14}N^{15}NH^+$					
	$^{15}N^{14}NH^+$					
	$H_2{}^{35}Cl^+$					
	$H_2{}^{37}Cl^+$					

Note: Negative molecular ions are labeled in blue whereas recent detections in the preceding five years are labeled in red.

Lallement (2017) suggested that the large organic molecules found by *Rosetta* in the dust of comet 67P/CG originate from the interstellar medium (ISM) and could be the source of the DIBs. We shall not further address the cometary environments; instead, we focus now on the microscopic mechanisms occurring in cold and low-density media that are representative of interstellar galactic and extragalactic clouds.

The early identification of these exotic interstellar molecules may be compared to a Hercule Poirot investigation where each tiny detail matters. Translated into a more scientific vocabulary, the open shell structure of these radicals, combined with the presence of the nuclear spin of H, an almost ubiquitous component, allows an impressive number of constraints to be derived. Nevertheless, the decisive confirmation of a molecular candidate is only achieved when the experimental spectrum provides a perfect match to the observational spectrum. The step of numbering and classifying the molecular content is, however, only a prelude and must be completed by different tasks. Spectra are often observed in emission. It is then mandatory to understand the excitation processes, which in turn allow us to derive diagnostics of density and temperature, major physical characteristics of the medium which can be compared to the diagnostics derived from dust continuum emission, when available. Additionally, the presence of the different molecular species should be understood in terms of the possible chemical reactions taking place in the environment. The field of astrochemistry, including the consideration of the different mechanisms at work, either in the gas phase or via gas–solid interactions, has developed significantly over the last four decades, as testified by the increased attendance of the IAU symposia dedicated to astrochemistry. Numerical chemical models have been built in parallel in order to compare against observations and check the physical conditions. Both time-dependent and steady-state models are available, allowing the inclusion of several hundreds of species linked by several thousands of molecular reactions. Finally, I want to stress the importance of the ionic component that contains both positive and even negative molecular ions, as the corresponding fluid is coupled to the ambient magnetic field. Magnetic fields indeed play a critical role on all scales, from the cosmological debate of structure formation to the shapes of dusty filaments in the interstellar medium (Boulanger et al. 2018).

This chapter is intended to describe some applications of fundamental molecular studies reported in the other chapters of this book, so the possibilities are very extended. The issues related to the finally accepted identification of the C_3H^+ molecular ion will be the focus of Section 8.2, which allows the problems linked to the accuracy requirements of molecular data to be outlined. Success and limitations of gas-phase chemical models will be described in Section 8.3.1, and the need for invoking alternative gas–grain chemical processes is the subject of Section 8.4.

8.2 Importance of Accurate Molecular Data

Spectroscopic identification obviously requires not only the frequencies to be known with great precision, but also the associated transition intensities, in order to check

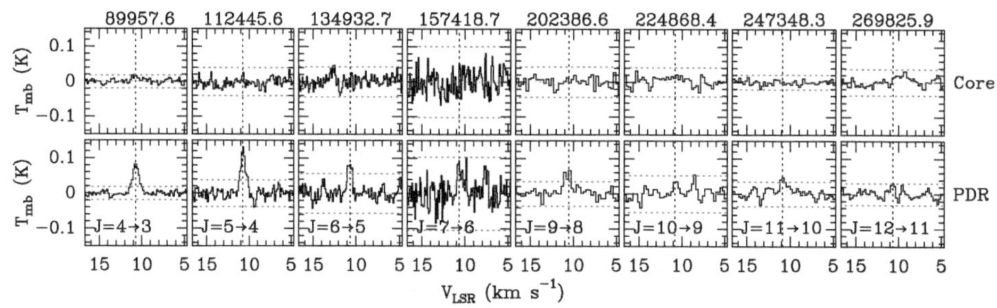

Figure 8.1. First spectrum of the C_3H^+ candidate obtained at two positions in the Horsehead Nebula. (Credit: Pety et al. 2012. Reproduced with permission. Copyright ESO.)

the observability conditions. The recent debate about the C_3H^+ identification illustrates very well the diverse channels through which one has to "slalom" to finally obtain a correct answer.

8.2.1 Spectroscopy

A sequence of as-yet unknown millimeter transitions ranging from 89 GHz up to 270 GHz in the so-called Horsehead PDR (photon-dominated region), thanks to the wideband availability of the heterodyne receivers offered at the IRAM 30 m telescope, were reported in Pety et al. (2012). These detections came from a complete unbiased survey of the 3 mm, 2 mm, and 1 mm bands at two positions of the Horsehead environment (WHISPER project; standing for "Wideband High-resolution Iram-30 m Surveys at two Positions with Emir Receivers"), as shown in Figure 8.1 and labeled respectively the "Core" and the "PDR" regions. Three transitions at 89.957, 112.445, and 134.932 GHz were apparent in the first analysis of the spectra at a level of about 100 mK, as shown in the bottom left part of Figure 8.1, which did not belong to any spectroscopic catalog currently used by radio-astronomers (Pickett et al. 2010),[4] (Endres et al. 2016).[5] Remarkably, these transitions were only present in the single PDR region of the Horsehead Nebula, whereas the two positions had been studied on equal footing. As a first comment, the presence of the carrier is clearly favored by the presence of photons, which are more prominent in the PDR. In addition, by carefully analyzing the frequencies of these three transitions, it was possible to derive that they were approximately multiple of a common value 22.489 GHz (respectively 4, 5, and 6). Such an occurrence reflects the properties of a linear rotor spectrum where rotation energy levels are given, at a first approximation, by $E(J) = BJ(J + 1)$, so that the frequencies of the electric dipole spectrum governed by a ($\Delta J = 1$) selection rule are given by an harmonic sequence $\nu(J \rightarrow J - 1) = 2BJ$. We thus explored the broadband available spectra for the other possible transitions lying at higher frequencies that escaped our first attention. As shown in Figure 8.1, some spectral hints were indeed present at the predicted

---

[4] Available on https://spec.jpl.nasa.gov/.

[5] Available at http://www.astro.uni-koeln.de/f/.

frequencies, albeit at a much lower signal-to-noise detection ratio. These findings allow us to affirm the presence of a closed shell linear molecule with a rotational constant B = 11.245 GHz, and even derive a rotational distortion constant D of 7.652 kHz when introducing the corresponding perturbation term $-DJ^2(J + 1)^2$ to the rigid rotor formula of the energy levels. How can one go further? A relative straightforward remark concerns the relatively small value of the derived rotational constant, indicating the presence of a heavy molecule. A further clue is provided by the comparison with known molecules, and in this specific case, the value of the rotation constant of the C_3H radical of 11.189 GHz is noticeably close to the value of 11.245 GHz of the potential candidate. Analyzing the electronic structure, it is easily found that the ionic counterpart, C_3H^+, is closed-shell and it was possible to recover in previous quantum ab initio studies (Radom et al. 1976; Wilson and Green 1980, 1982) that the ground electronic state is $^1\Sigma^+$ and possesses a permanent dipole moment of 2.6. D. Pety et al. (2012) noticed the unusually large value of the distortion constant, which traces the floppiness of the molecule. However, no experimental spectra was available at that time and the identification of C_3H^+, a new molecular ion in the Horsehead Nebula, was reported based on the compelling availability of eight harmonically related different rotational transitions. The derived column density was given within a factor of about 2, given the uncertainties in the beam filling factor, so the derived C_3H^+/C_3H ratio is about 0.05–0.1, when the column density of C_3H includes the cyclic *and* linear forms, both detected in the Horsehead Nebula. The proposed identification was shortly followed by an observational report at lower frequencies where data taken by McGuire & Carroll (2013) from a previous PRIMOS (PRebiotic Interstellar MOlecular Survey) project were analyzed at the expected frequencies of the $J = 1 - 0$ and $J = 2 - 1$ transitions, which were not available with the IRAM 30 m telescope. These authors were able to detect these transition in the Sgr B2(N) cloud, whereas only an upper limit was derived for SgrB2(OH). Examination of Turner's (Pulliam et al. 2012) for SgrB2 (OH) and Kaifu's (Kaifu et al. 2004) for TMC1 archives allowed McGuire and coworkers (McGuire & Carroll 2013) to derive upper limit on the abundances of this same ion.

However, the proposed identification was rapidly questioned by quantum chemists (Huang et al. 2013). Indeed, Huang et al. (2013) and Fortenberry & Lee (2013) revisited the C_3H^+ molecular ion structure, undertook new ab initio quantum calculations with the state-of-the-art technique developed in their group, and came to the conclusion that the astrophysical identification was not correct, as their highly accurate techniques predicted a much smaller distortion constant, which disagreed with the astrophysical value by more than 50%. These authors suggested alternatively (Huang et al. 2013) that the unknown candidate was instead the C_3H^- negative molecular ion, and even the first electronically metastable excited state which was predicted to be slightly nonlinear, such that its rotational electronic spectrum of the μ_a component of the dipole moment could mimic that of a linear species with a large distortion constant. Such a case, which was analytically developed by Polo (1957) for the rotational eigenvalues and corresponding rotational frequencies, can be written as

$$\nu_{J+1\to J} = \left\{(B + C) \pm \frac{1}{2}(B - C)\delta_{K,1} - 2D_{JK}K^2\right\}(J + 1) - \left\{4D_J + \frac{(B - C)^2}{c[A - (B + C)/2]}\right\}(J + 1)^3. \quad (8.1)$$

That expression represents the leading terms of an asymptotic series expansion of a near-prolate top where c is equal to 32 for $K = \pm 1$ and 8 for $K = 0$. The transitions detected by Pety et al. (2012) would then correspond to so-called a-type transitions with $K = 0$ and that frequency expression is then equivalent to that of a linear rotor with a corresponding rotational constant value of $(B + C)/2$ and a distortion constant of $D_J + \frac{(B - C)^2}{32[A - (B + C)/2]}$, determined by the observations of Pety et al. (2012). The role of quantum chemistry in the understanding of the structure of negative molecular ions was further emphasized by Fortenberry (2015).

In answer to that suggestion, the astrophysical team searched, however unsuccessfully, for the presence of other carbon-chain negative ions in the same source, which were indeed previously detected in dark clouds such as TMC1 and the IRC +10216 circumstellar envelope. An additional attempt (McGuire et al. 2014a) to satisfy the suggestion of Huang et al. (2013) was to search for b-type transitions corresponding to $K = 1$, but instead involving the rotational A, B, C constants, predicted only through quantal calculations with a significant amount of uncertainty. No hint of such transitions was spotted. In parallel, McGuire et al. (2014b) undertook a search of the available frequencies in other environments, including complex molecular sources and PDRs, in order to further constrain the understanding of this elusive species. The spectral signature was found in a single source, the Orion Bar, a well-known PDR, among the 39 investigated.

The unknown molecule was then referred as B11244, from the value of the rotational B constant.

8.2.2 Chemical Reactivity

Whereas the spectroscopic identification of the negative molecular ion may appear reasonable, the important question of the likelihood of the presence of such a molecular compound should be posed. Given the value of the transition intensities, Huang et al. (2013) further derived the corresponding column density and found that the negative C_3H^- over neutral C_3H ratio was about 57%, a surprisingly high value when compared to those of other negative molecular ions, as shown in Table 8.4.

On the other hand, the proposed reaction network to explain the presence of the C_3H^+candidate was set up in Guzmán et al. (2015), slightly revised in Figure 8.2, and introduced in a chemical model in order to compare its predictions to the observations.

8.2.3 Quantum Chemistry

The debate was subsequently reinvigorated by quantum chemists themselves when Botschwina et al. (2014) addressed the accuracy of the potential energy calculations

Table 8.4. Fractional Abundances of Interstellar Neutral Carbon Chains and Their Corresponding Anions

	Horsehead Nebula	TMC1	Orion Bar	SgB2
C_2H	$1.1 \pm 0.4 \times 10^{-8}$	7.2×10^{-9}	$(0.7 - 2.7) \times 10^{-8}$	
c-C_3H	$2.7 \pm 1 \times 10^{-10}$	3×10^{-9}	$(0.3 - 1.6) \times 10^{-10}$	3×10^{-9}
l -C_3H	$1.4 \pm 0.7 \times 10^{-10}$	5.6×10^{-10}	$(1.1 - 5.4) \times 10^{-11}$	
c-C_3H_2	$6.4 \pm 2.1 \times 10^{-10}$	1.9×10^{-9}	$(2.1 - 9.4) \times 10^{-10}$	
l-C_3H_2	$1.9 \pm 1.1 \times 10^{-10}$	5.9×10^{-11}	$(0.6 - 3.7) \times 10^{-11}$	
C_4H	—	7.1×10^{-8}	$(0.6 - 3.2) \times 10^{-9}$	—
C_3H^+ (B11244)	$(1.9 - 4.2) \times 10^{-11}$	6×10^{-11}	$(0.8 - 4.0) \times 10^{-11}$	2.4×10^{-10}
C_4H^-	—	$\leqslant 3.7 \times 10^{-12}$		
C_6H^-	—	1.0×10^{-11}		
C_8H^-	—	2.1×10^{-12}		

Note. The values reported for the Horsehead, TMC1, Orion Bar, and SgB2 are respectively from Pety et al. (2012), Agúndez & Wakelam (2013), Cuadrado et al. (2015), and McGuire et al. (2014a).

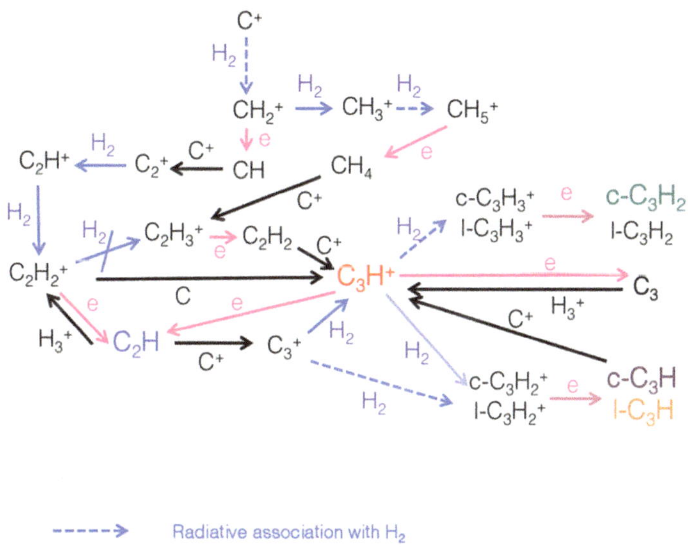

Figure 8.2. Reaction network for C_3H^+.

and found a remarkably close to observational rotational constant for the C_3H^+ molecular ion. Such an achievement was made possible through careful consideration of inner-shell electron correlations, which may be considered as the difference in energies of calculations with either *all* electrons or only the valence electrons being correlated. This contribution, called CV by Botschwina et al. (2014), was apparently underestimated in the computations of Huang et al. (2013). Additionally, the difficulty of theoretically recovering the large value of the distortion constant is discussed and interpreted as the manifestation of the floppy character of that molecular ion. A supplementary ab initio study was independently performed by

Mladenović (2014), entitled "The B11244 story: Rovibrational calculations for C_3H^+ and C_3H^- revisited." They concluded that numerically exact rovibrational computations are required in ab initio calculations and can be used to predict the spectroscopic molecular properties within the quartic internal coordinate force fields. Overcoming the perturbative approach obviously implies a great care in the choice of the basis functions and in the test of convergence, and requires a significant increase in computational time.

8.2.4 Laboratory Confirmation

Meanwhile, Brünken et al. (2014a) were able to synthesize the C_3H^+ molecular in a cryogenic ion trap apparatus employing a novel mass-selective action spectroscopy method based on light-induced reaction (LIR), bringing a definite end to the controversy. McCarthy et al. (2015), also motivated by the original astrophysical observations, performed cavity Fourier transform (FT) microwave spectroscopy of a supersonic molecular beam and found the lowest rotational transitions in a hydrocarbon discharge in exact frequency agreement with those predicted from the astronomical data. However, these laboratory measurements, pointing to an ion containing three carbon atoms and one hydrogen, could not conclusively distinguish between the anion and the cation candidate. The experimental results of Brünken et al. (2014a) allowed the carrier to be unambiguously settled.

8.3 Success and Limitations of Gas-phase Chemistry

The interstellar medium is not in thermodynamic equilibrium. That fact can be anticipated from the list of the molecular content as reported in Tables 8.1, 8.2, and 8.3, where reactive species and radicals can be found, along with more complex stable molecules. The low-density and low-temperature physical conditions of the medium, mainly constituted by hydrogen, in atomic or molecular form, do not allow chemical equilibrium to be achieved through collisions. Then, kinetic models are required to follow the chemical evolution of the various atomic and molecular components of the medium and the resulting chemical composition at steady state. Both the time evolution and steady-state solution are critically dependent on the physical conditions, such as: density; presence of a radiation field coming from the mean galactic star distribution or a nearby star; or the presence of cosmic rays, which are primarily high-energy protons and electrons, accelerated at the shock front powered by supernova explosions. A temperature can usually be defined that derives from the thermal balance between heating and cooling processes, which is usually attained in a few years, a short time in terms of astrophysical timescales. Its value is on the order of 70 K in diffuse clouds, whereas values down to $\sim$10 K are obtained in the cold dense clouds and prestellar dense core conditions. The first chemical models relevant for interstellar chemistry have been available for less than fifty years, and they have allowed the basic mechanisms at work in the gas phase to be settled, as reviewed in the piloting studies by Herbst & Klemperer (1973); Dalgarno & Black (1976), and Watson (1978).

Chemistry is also very dependent on the overall composition of the medium in the gas phase. The interstellar medium has strong reducing properties, as the principal element is hydrogen and the number of all species is given according to their abundance relative to hydrogen. This composition results from the processing of various nuclei in stars that eject the elements at the end of their evolution. While the elemental composition may be well-constrained in the solar nebula (Asplund et al. 2009), its value in the galactic or extragalactic environment is still a matter of study and debate. However, some general consensus is achieved for the diffuse gas environment, thanks to visible/UV absorption observations of the principal atomic and ionic constituents. Denser environments are less well constrained, as the gas-phase composition of heavy elements is decreased due to their condensable properties on dust particles in a still-hypothetical way. The elemental budget derived from the composition of interstellar ices seems to be incomplete, as pointed out in the review paper by Boogert et al. (2015), for several important elements such as carbon, oxygen, nitrogen, and sulfur.

Interstellar chemical models assume different elemental composition: the so-called low-metal and high-metal conditions have been introduced by Graedel et al. (1982), where the high-metal values refer to absorption observations towards the diffuse ζ Oph interstellar cloud and low-metal values were introduced for condensable species (S, Si, P, Metals) decreased by a factor of 100. It is remarkable that present-day model computations still keep that terminology. Table 8.5 reports the elemental abundances used in standard time-dependent (TD) interstellar models as originally introduced by Graedel et al. (1982) and in more recent studies of dark

Table 8.5. Elemental Abundances Relative to H_2 from Graedel et al. (1982) and Agúndez & Wakelam (2013)

Species	Abundances		
	High-metal, ref 1	Low-metal, ref 1	ref 2
H	1	—	
He	0.28	—	0.17
C^+	1.46(−4)	—	3.6(−4)
O	3.52(−4)	—	6.6(−4)
N	4.28(−5)	—	1.24(−4)
S^+	1.66(−5)	1.66(−7)	1.60(−7)
Si^+	1.66(−6)	1.66(−8)	1.60(−8)
P^+			6(−9)
Na^+			4(−9)
Mg^+			1.4(−8)
Fe^+			6(−9)
Cl^+			8(−9)
F			2(−8)

Notes. Parentheses refer to powers of 10.
Ref 1: Graedel et al. (1982).
Ref 2: Agúndez & Wakelam (2013).

clouds (Agúndez & Wakelam 2013), as well as the initial reservoir that is used for these TD models. The C/O carbon over oxygen ratio is sometimes introduced, following Bergin et al. (1997) and Terzieva & Herbst (1998a, 1998b), and varied over a 0.5–1.5 interval allowing the description of carbon-rich environments, such as the TMC1 molecular cloud, where large carbon chains have been detected (Gratier et al. 2016). It is also worth emphasizing that the low density of the medium implies that only binary gas-phase reactions may occur, which considerably simplifies the building of gas-phase chemical networks. An additional constraint in the cold interstellar environments is provided by energy conservation requirements, which suggest that exothermic reactions take place essentially.

8.3.1 Successful Achievements of Gas-phase Chemical Models

The Low-temperature Chemistry

Interatomic or intermolecular potentials between the two reactants essentially drive the efficiency of the gas-phase chemical reactions. Long-range interactions are determinant for very low-energy collisions and $1/R$ expansions of the interatomic or intermolecular electronic interaction, where R is the reaction coordinate, allow analytic expressions to be derived in terms of the electric properties of the two reactants (q = charge, d = dipole moment, α = polarizability, Q = quadrupole moment, …). Assuming such an intermolecular interaction, one can further consider the collision dynamics and derive the reactive collision cross section $\sigma(\nu)$ as a function of the relative velocity ν, and finally obtain the reaction rate coefficient after averaging the collision cross section over a Maxwellian distribution of collision velocities $k(\mathrm{T}) = \langle \nu\sigma \rangle$. Such studies have been performed in a classical mechanic approach by Gioumousis & Stevenson (1958) for ion–neutral molecule collisions and revisited recently by Georgievskii & Klippenstein (2005) in a quantum variational transition state theory for various types of interactions. Specifically, any isotropic ν_0/R^n long-range expansion of the intermolecular potential leads to an analytic expression of the classical reactive cross section, and consequently of the reaction rate coefficient. The general expression is

$$k(T) = (8\pi)^{1/2} \cdot \left(\frac{n-2}{2}\right)^{2/n} \cdot \Gamma(1 - 2/n) \cdot \mu^{-1/2} \cdot \nu_0^{2/n}(k_B T)^{1/2-2/n}, \tag{8.2}$$

where $\mu = m_1 m_2/(m_1 + m_2)$ is the reduced mass of the reactants and $\Gamma(z)$ is the Gamma function $\Gamma(z) = \int_0^\infty x^{z-1}e^{-x}dx$. These properties were acknowledged long ago for charge–neutral molecule interactions (Gioumousis & Stevenson 1958). In that particular case, the isotropic long-range interaction takes place between a charge and an induced dipole, and the reaction rate coefficient (as displayed in Table 8.6) corresponds to the well-known so-called Langevin value. The resulting expression is temperature invariant; its value is on the order of 10^{-9} cm^3 s^{-1} when the neutral molecule is molecular hydrogen, in very good agreement with experimental results. Another interesting case is obtained for the dispersion interaction C_6/R^6, where the temperature dependence is small and varies as $T^{1/6}$. The

Table 8.6. Long-range Interactions and their Corresponding Reaction Rate Coefficients Obtained in Variational Transition-state Theory (Georgievskii & Klippenstein 2005)

	Long-range expansion of the interatomic potential	Reaction rate coefficient Analytic expression	Value $cm^3\ s^{-1}$
Charge–dipole	$-\frac{qd\cos(\theta)}{R^2}$	$\sim\sqrt{2\pi}\,qd\mu^{-1/2}(k_BT)^{-1/2}$	A few 10^{-9}
Charge–quadrupole	$-\frac{qQ}{4R^3}(3\cos^2\theta - 1)$	$C^{(\pm)}\mu^{-1/2}\lvert qQ\rvert^{2/3}(k_BT)^{-1/6}$	$\sim 10^{-10}$
Charge–induced dipole	$-\alpha q^2/2R^4$	$2\ \pi(q^2\alpha/\mu)^{1/2}$	$\sim 10^{-9}$
Dipole–dipole	$\frac{d_1d_2}{R^3}\times$ $(\sin\theta_1\sin\theta_2\cos\phi - 2\cos\theta_1\cos\theta_2)$	$C\mu^{-1/2}(d_1d_2)^{2/3}(k_BT)^{-1/6}$	$\sim 10^{-10}$
Dipole–quadrupole	$\frac{3dQ}{4R^4}(3\cos\theta_d\cos^2\theta_Q - \cos\theta_d$ $-\sin\theta_d\sin 2\theta_Q\cos\phi)$	$C\mu^{-1/2}\lvert dQ\rvert^{1/2}$	$\sim 10^{-11}$
Dispersion (induced dipole–induced dipole)	C_6/R^6	$8.55\ \mu^{-1/2}C_6^{1/3}(k_BT)^{1/6}$	$\sim 10^{-11}$
Dipole–induced dipole	$-\frac{\alpha d^2}{2R^6}(1 + 3\cos^2\theta)$	$C\mu^{-1/2}(d^2\alpha)^{1/3}(k_BT)^{1/6}$	$\sim 10^{-10}$

Notes. The last column reports the orders of magnitude of the rate coefficients at 10 K. Expressions are reported in CGS units.
α: polarizability (L^3).
q: electric charge ($M^{1/2}L^{3/2}T^{-1}$).
d: dipole moment ($M^{1/2}L^{5/2}T^{-1}$).
Q: quadrupole moment ($M^{1/2}L^{7/2}T^{-1}$).
k_B is the Boltzmann constant. T is the temperature in Kelvin.

C_6 coefficient, called the van der Waals constant, is often approximated as $C_6 = 1.5\alpha_1\alpha_2E_1E_2/(E_1 + E_2)$, where α_i and E_i are respectively the polarizabilities and the ionization energies of the two reactants (Hirschfelder et al. 1964). The early astrochemical models have extensively used that property and successfully accounted for the presence of molecular ions such as HCO^+, N_2H^+. However, this formula does not hold anymore when the interaction involves some angular dependence, such as in charge-permanent dipole interaction (e.g., H_3^+ + HCN), charge–quadrupole, and dipole–dipole (e.g., OH + CH) interactions, so the collision treatment becomes more tricky and involves some additional approximations. Table 8.6 summarizes the expressions of the $1/R$ expansion terms of the intermolecular long-range potential, where R is the reaction coordinate and corresponding reaction rate coefficients. Some dimensionless parameters, C and $C^{\pm}$, introduced in the formulae, account for different approximations and are usually of the order of a few. They are displayed in Georgievskii & Klippenstein (2005). We refer the reader to that paper for more information, and report (in the third column of Table 8.6) the expected order of magnitude of the reaction rate coefficients for $T = 10$ K when H_2 is one of the reactant without dipole moment and OH for a reactant with dipole moment. All the reaction rate coefficients are seen to vary as $1/\sqrt{\mu}$, such that reactions involving heavy complex molecules are reduced accordingly.

The $1/T^{\beta}$ dependence ($\beta > 0$) arising for some cases is removed in more accurate treatments (Troe 1987, 1996), so that there is no divergence at very low temperatures. However, the formulae of Table 8.6 allow us to recover the order of magnitude of the reaction rate coefficients down to ~10K, as β is always $\leqslant 1/2$.

Refined expressions of reaction rate coefficients are derived for ion–dipole interactions, where the second term of the perturbation expansion of the intermolecular potential is introduced.

I want to mention a last remark concerning the possible occurrence of chemical tunneling at very low temperatures (Sims 2013), which results in some increase of the reaction rate coefficient at low temperature for reactions possessing a small barrier. One of the most spectacular examples concerns the $F + H_2$ reaction giving $HF + H$, which has extensively been studied both in the laboratory and theoretically up to very low temperatures (Tizniti et al. 2014). This reaction is actually of astrochemical relevance, as HF has been detected in diffuse environments. Its abundance is thus directly related to H_2, and Sonnentrucker et al. (2015) suggest using HF as a proxy of molecular hydrogen. These issues are discussed in other chapters of this book from both the theoretical and experimental points of view. However, such basic considerations have been made at the start of astrochemical models and allowed a general plausible description of the chemistry taking place in the dense interstellar clouds and star-forming regions.

The Role of State-to-state Chemistry

The significance of state-dependent chemistry in star-forming regions has been realized in different contexts. It is another signature of non-thermodynamical equilibrium conditions prevailing in such media. We consider below three examples that allow to emphasize the subtleties of the chemical reactivity mechanisms.

- Deuterium fractionation reactions

  The large amount of deuterated compounds found in comparison to their hydrogenated counterpart is a nice example of a case where taking into account detailed gas-phase mechanisms offers a straightforward way to explain the observations. The D/H elemental ratio is an important ingredient of cosmological models, and its value is related both to the helium abundance and the baryon density in Standard Big Bang Nucleosynthesis (SBBN) cosmological theory. Cyburt et al. (2016) summarize the present constraints derived from the recent cosmic microwave background (CMB) measurements obtained by the Planck mission. Indeed, the abundances of the four light nuclei (D, ^{3}He, ^{4}He, and ^{7}Li) can be expressed in terms of the baryon-to-photon ratio, or to the present baryon density $\Omega h^2 \equiv \omega_b$. The primordial D/H ratio is given as $\left(\frac{D}{H}\right)_p = (2.62 \pm 0.05) \times 10^{-5}$ from the latest determination in an absorbing system at $z = 3.572$ (Cooke et al. 2014; Riemer-Sørensen et al. 2017). After its formation during the Big Bang, the abundance of deuterium is decreasing monotonically because no nuclear process allows it to be reformed. The present-day elemental abundance is obtained from galactic observations

(Linsky et al. 2006) and the derived total (gas + dust) local D/H elemental ratio is larger than or equal to (2.31 ± 0.24) 10^{-5}. Even if that ratio remains somewhat uncertain, the ratio between deuterated molecules and their hydrogenated counterparts reaches much higher values, at least two to three orders of magnitude larger, as summarized in Roueff & Gerin (2003). Remarkably, kinetic considerations for low-temperature conditions allow the enrichment of deuterated molecules to be explained, as will be described below.

In the gas phase, under dense cloud conditions, the initial step of deuteration lies in the $H_3^+ + HD \rightleftharpoons H_2D^+ + H_2$ deuteron exchange reaction, which is followed by the deuteron transfer reaction of H_2D^+ to stable species, such as CO and N_2, giving DCO^+ and N_2D^+, which have been detected in a variety of interstellar regions. The forward deuteron exchange reaction between H_3^+ and HD is exothermic under low-temperature conditions, as the corresponding exothermicity is about 232 K when expressed in Kelvin, so the reverse reaction is almost forbidden. The equilibrium constant at thermal equilibrium of the $A\mathcal{L}^+ + B\mathcal{H} \rightleftharpoons A\mathcal{H}^+ + B\mathcal{L}$ exchange reaction between a "Light" $A\mathcal{L}^+$ ion ($H_2H^+ = H_3^+$) and a "Heavy" $B\mathcal{H}$ neutral reservoir (HD) is given by

$$K(T) = \frac{k_f}{k_r} = \left(\frac{m_{A\mathcal{H}^+} \cdot m_{B\mathcal{L}}}{m_{A\mathcal{L}^+} \cdot m_{B\mathcal{H}}}\right)^{3/2} \times \frac{q(A\mathcal{H}^+) \cdot q(B\mathcal{L})}{q(A\mathcal{L}^+) \cdot q(B\mathcal{H})} \times \exp(\Delta E_0/kT) \tag{8.3}$$

where ΔE_0 corresponds to the exoergicity of the exchange reaction. The q factors stand for the internal partition functions of the reactants and products involved in the reaction, which are functions of temperature T. They can be computed from the eigenvalues of electronic, rotational, and vibrational Hamiltonian under thermodynamical equilibrium, as described in various textbooks (e.g., Atkins & de Paula 2006). The mass-dependent terms arise from the translational motion partition functions.

In the case of the $H_3^+ + HD \rightleftharpoons H_2D^+ + H_2$ reaction, ΔE_0 is obtained from the difference in vibrational zero-point energies (Δ_{VZPE}) of the products and the reactants (the vibrational zero-point energies (VZPE) are reported in Table 8.7), plus the value of the rotational energy of the 1_1 para level of H_3^+, whose energy term is given by B + C, the sum of the two rotational constants of the corresponding symmetric top. The 0_0 level of H_3^+ is indeed forbidden, thanks to Fermi selection rules stating that the total wave function is antisymmetric in the exchange of the H nuclei.

Figure 8.3 displays the various energy levels involved in the reaction. As perceived early by Pagani et al. (1992), the exothermicity involved in the reaction may depend on the rotational population of the molecular hydrogen. Indeed, the $J = 1$ level of H_2, corresponding to the ground state ortho form of H_2, is at 170 K above the $J = 0$ para ground state. Then, even a small fraction of ortho-H_2[6] (Pagani et al. 2012) modifies the equilibrium reaction rate

[6] The ortho/para balance of H_2 is not discussed here, but it involves various possible mechanisms such as reactive collisions with H^+ (Dalgarno et al. 1973), H (Mandy & Martin 1993), and H_2 (Hugo et al. 2009), as well as ortho/para conversion on grains (Le Bourlot 2000).

Table 8.7. VZPE of H_2, H_3^+ and Their Deuterated Substitutes

Species	VZPE cm^{-1}	Δ(VZPE) K	References
H_2	2168.1	0	(1)
HD	1885.0	407.3	(1)
D_2	1543.4	898.9	(1)
H_3^+	4361.7	0	(2)
H_2D^+	3978.0	552.1	(2)
D_2H^+	3561.4	1151.5	(2)
D_3^+	3112.3	1797.7	(2)

Notes. References. (1): Herzberg (1989), (2): Ramanlal et al. (2003).

coefficient at low temperatures, as the corresponding exothermicity becomes 62 K, reducing significantly the energy barrier of the reverse reaction. This is also the case if the H_2D^+ product is in its ortho form, as the ground ortho level of H_2D^+ reacting with p-H_2 has an endothermicity of 146 K. Consideration of the various rotational levels involved in the reaction has been further introduced and quantitatively discussed by Hugo et al. (2009) in the frame of a complete scrambling approximation during the reaction, with the restriction that the total nuclear spin is conserved during the collision. The transitions between the lowest para and ortho levels of H_2D^+ have been detected as displayed in Figure 8.3 and are beautiful illustrations of the relevance of this state-to-state chemistry. The ortho emitted transition at 372 GHz has been found almost accidentally in the highly depleted L1544 cold pre-stellar core by Caselli et al. (2003) after the first detection by Stark et al. (1999) in NGC 1333 IRAS4A and subsequently toward different other sources (Vastel et al. 2006; Friesen et al. 2014). Another far-infrared ortho transition at $\lambda = 126.853\ \mu$m was detected in absorption in the line of sight toward SgrB2 in the galactic center by Cernicharo et al. (2007) with the *Infrared Space Observatory*. Such observations witness the gas-phase mechanisms involved in the formation of o-H_2D^+. The para-transition at 1.2 THz was found in absorption very recently toward the protostellar source IRAS 16293-2422 by Brünken et al. (2014b), thanks to the SOFIA telescope,[7] as the corresponding frequency range was not provided by the Herschel Observatory. The corresponding transitions are displayed on Figure 8.3. The dependence of the ortho-to-para ratio of H_2D^+ on its reaction with H_2 in its ortho and para forms was discussed by Gerlich et al. (2002) in the light of experimental ion trap studies performed at low temperature.

This chemical schema is further validated by the subsequent detection of D_2H^+, where the chemical sequence involves the following reaction of H_2D^+

[7] SOFIA is an 80/20 partnership of NASA and the German Aerospace Center.

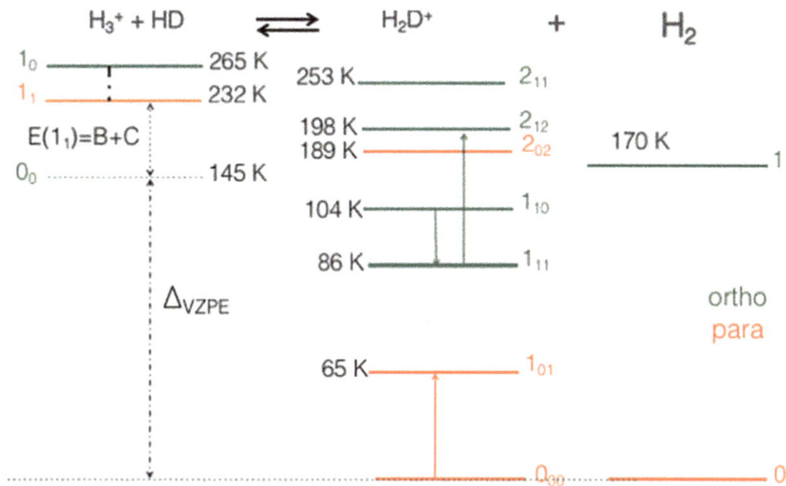

Figure 8.3. Levels involved in the H_3^+ + HD reaction. The astronomical detections reported in the text (Stark et al. 1999; Caselli et al. 2003; Cernicharo et al. 2007; Brünken et al. 2014b) are indicated with the corresponding arrows.

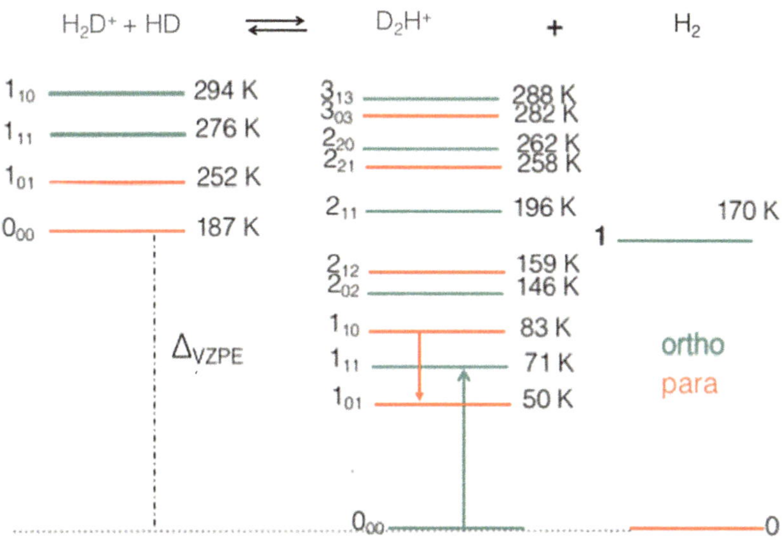

Figure 8.4. Levels involved in the H_2D^+ + HD reaction. The astronomically detected transitions reported in the text (Vastel et al. 2004; Parise et al. 2011; Harju et al. 2017) are indicated.

with HD, as shown on Figure 8.4. The para transition of D_2H^+ at 692 GHz was indeed detected very rapidly after the first detection of H_2D^+ in another cold dense cloud, L16293E (Vastel et al. 2004), and in another similar source by Parise et al. (2011). More than 10 years were necessary to finally detect the ortho D_2H^+ transition at 1.3 THz (Harju et al. 2017), thanks to the SOFIA telescope.

Inspired by these observations, Walmsley, Flower, and Pineau des Forêts (Walmsley et al. 2004) considered a fully depleted chemistry involving hydrogen and deuterated species in their lowest para and ortho levels. Such a network introduces the possibility of ortho/para transfer reactions of H_3^+ in reaction with H_2, in order to consistently solve all involved spin states of H_2, H_2^+, H_3^+. Such ortho/para transitions occur through proton exchange in reactive collisions that can involve different mechanisms, such as direct proton exchange and/or complete scrambling collisions, which have deserved several studies. Quack (1977) first pointed out the detailed symmetry selection rules for reactive collisions when fermions such as protons were involved. Whereas Walmsley et al. (2004) and Flower et al. (2004) introduced approximate rules for these exchange reactions, Hugo et al. (2009) and Sipilä et al. (2010) carefully reconsidered the symmetry restrictions within a microcanonical model in order to build up the different state-to-state proton exchange reactions. However, the relative balance between proton hopping or full scrambling of the protons is not fully resolved. The model results are moderately dependent on these fine tunings, which do not modify the general trends (Sipilä et al. 2010).

The sequence of deuteration enhancements via reaction with HD may continue and is displayed in Figure 8.5, resulting in the formation of D_3^+. As D_3^+ involves three deuterons that have a nuclear spin of 1, the Fermi principle on the total wave function does not hold anymore and $(2I + 1)^3 \equiv 27$ different nuclear spin wave functions are present. The corresponding rotational levels are spanned over three different symmetry groups: ortho (o) levels corresponding to a statistical weight of 10 have an even J rotational

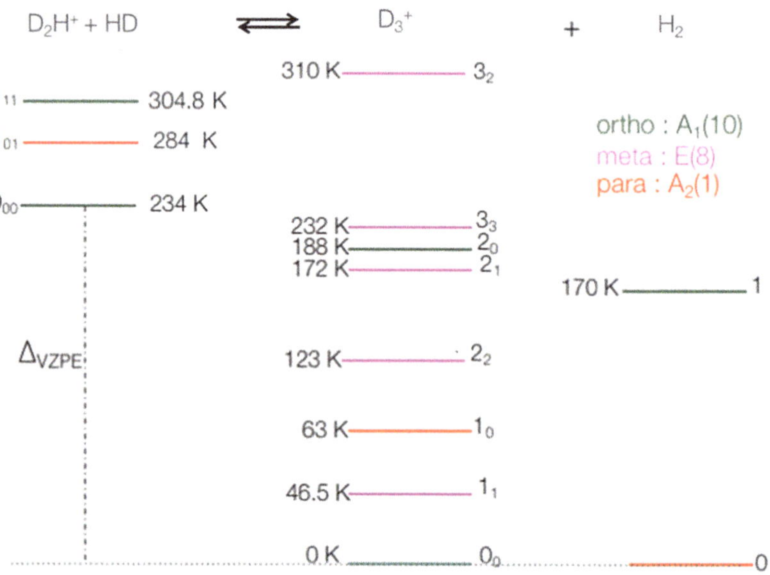

Figure 8.5. Levels involved in the D_2H^+ + HD reaction.

quantum number with $K = 0$ and belong to the A1 symmetry group; meta (m) levels have a non-zero K projection quantum number and belong to the E symmetry group with a statistical weight of 8 (involving then 16 spin states as E is doubly degenerate); and para (p) levels belong to the A1 symmetry group with a unit statistical weight and involve odd rotational levels with $K = 0$. As a general rule, the statistical weights are such that o > m > p. These symmetries and the corresponding level positions are shown in Figure 8.5.

Understanding the ortho-to-para ratio of H_2, H_2D^+, and D_2H^+ becomes an interesting issue that may help to constrain the chemical evolution time. When H_2 forms on-grain, one usually assumes that its ortho-to-para ratio is 3, corresponding to the high energy released in the formation process. The ortho-to-para ratio of H_2 decreases then as a function of time up to values of some 10^{-3}–10^{-4} at steady state (Pagani et al. 2011). The concomitant evolution of spin states of H_2D^+ and D_2H^+ involves mainly exchange reactions with o- and p-H_2 and depends also on the possible chemical selection rules in the reactive processes forming these molecular ions. Whereas some simplified expressions have been derived for the ortho-to-para ratio of H_2D^+ (Brünken et al. 2014a; Furuya et al. 2015), full chemical modeling involving ortho/para forms of hydrogen compounds is required to follow the time evolution of the ortho-to-para ratios H_2D^+ and D_2H^+, as shown in Figure 8.6 from the gas-phase

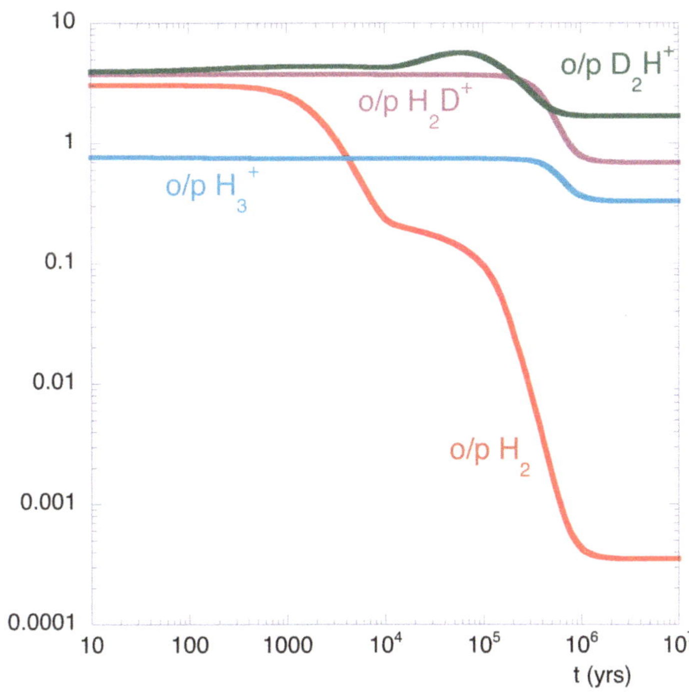

Figure 8.6. Time evolution of the ortho-to-para ratio of H_2, H_3^+, H_2D^+, and D_2H^+ for a gas-phase model with $n_H = 2 \times 10^5$ cm^{-3} and $T = 10$ K. O/H = 3×10^{-5}, C/O = 0.57, S/H = 6×10^{-7}, and $\zeta = 5 \times 10^{-17}$ s^{-1}.

chemical model used in Pagani et al. (2011). The actual values depend on the elemental abundances of C, O, N, S, and metals.

Spin conversion of H_2 may also occur on grain surfaces through the interaction of an adsorbed H_2 with an electron of the substrate surface leading to a transfer of angular momentum to the electron, as first suggested by Le Bourlot (2000) in the interstellar literature. The induced magnetic dipole may then interact with the nuclear spin and result in a change of modification. On graphite, the process is believed to occur with unpaired electrons at point defects in the crystalline structure (Ilisca 1992; Ilisca & Ghiglieno 2016, 2017). That effect is discussed quantitatively in a recent paper by Bovino et al. (2017) in the context of H_3^+ deuteration in fully depleted conditions, and could significantly modify the timescales involved in the formation of deuterated interstellar molecules. Further consideration of ortho/para chemistry of H_2 is also introduced for other deuterated species such as water (Furuya et al. 2015) and ammonia (Hily-Blant et al. 2018).

- The oxygen–proton charge exchange reaction: $O + H^+ \rightleftharpoons O^+ + H$

Accidental resonance may occur between the energies corresponding to the entrance and output channels of chemical reactions. That situation holds for the oxygen–proton charge exchange reaction, which is very prominent in interstellar conditions because it is the source of oxygen ions that subsequently react with molecular hydrogen in a sequence of exothermic reactions until the formation of the stable molecular ion H_3O^+, as shown in Figure 8.7. Such a reaction scheme beautifully describes the formation of the various oxygen hydrides that have been detected in the interstellar environment. The initial step is provided by the cosmic-ray ionization of hydrogen, where the cosmic rays are displayed as CR on Figure 8.7. Atomic ions may also recombine on grains or large polyaromatic molecules, denoted as g in the figure. It is then very gratifying that, with such a well-understood reaction scheme, one may reverse and use the observations of the OH^+ molecular ions to derive the primary cosmic-ray ionization rate of atomic hydrogen, as suggested by Gerin et al. (2010) after the detection of the submillimeter rotational transition of OH^+ by Herschel and followed by Indriolo et al. (2015). Further detections of OH^+ in the visible by Krełowski et al. (2010), Porras et al. (2014), Zhao et al. (2015), and Bacalla et al. (2018) are actually used for inferring cosmic ionization rate in diffuse cloud conditions. However, the charge exchange reactions occurring between H^+ and atomic oxygen may warrant some additional remarks. The charge transfer reaction rate coefficient is recorded through the $7.0 \times 10^{-10} \exp(-232/T)$ cm^{3}s^{-1}

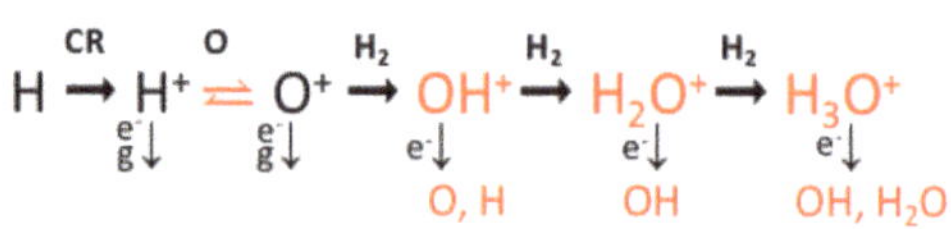

Figure 8.7. Chemical network displaying the formation routes to OH, H_2O, and the associated molecular ions in diffuse interstellar clouds. Detected molecular species are exhibited in red. See text for the description.

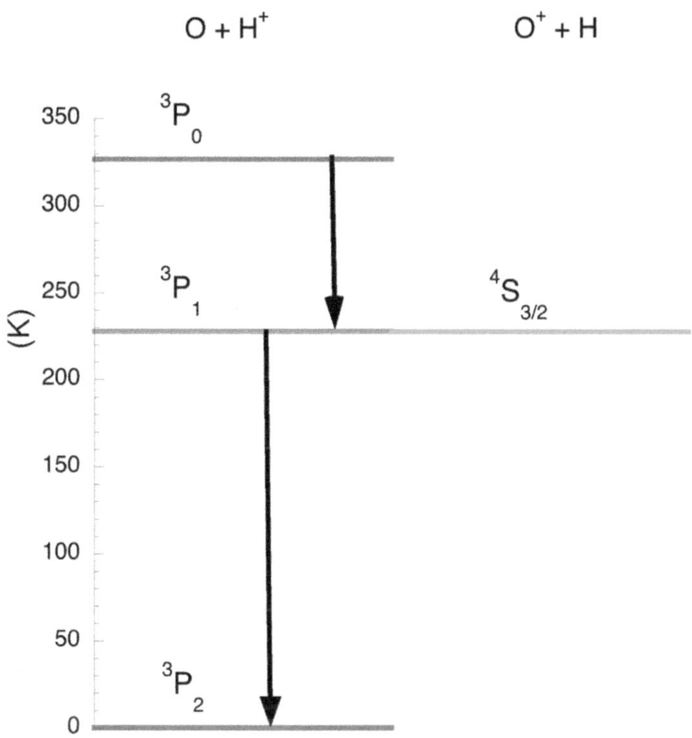

Figure 8.8. Energies involved in the entrance and exit channels of the O–H^+ charge transfer reaction. The origin of the energies is taken at the ground state $^3P_2$ of atomic O. The arrows refer to the magnetic dipole radiative transitions taking place within the atomic oxygen ground state.

expression in the KIDA chemistry database, whereas the UMIST database suggests $6.86 \times 10^{-10}(T/300)^{0.26}\exp(-224.3/T)$. The 232 (224.3) K value represents the energy defect when only the ground state $J = 2$ of atomic oxygen is considered, as displayed in Figure 8.8. We see that the $J = 1$ level of O in presence of H^+ is almost at the same energy as the O^+ ground-state level in the presence of atomic H. We have here an accidental resonance, where the fine-structure-level excitation of O by protons and charge exchange reaction of O^+ with H are strongly coupled. The radiative magnetic dipole transition between the $J = 1$ and $J = 2$ levels, which takes place at 63.1852 microns with a $8.91 \times 10^{-5}s^{-1}$ radiative emission probability, is detected in PDR and shocked regions, where the density and temperature are sufficiently large to excite the $J = 1$ level and atomic oxygen. In Table 8.8, we report the energy level positions as given in Kramida et al. (2015), which are expressed in K in Figure 8.8. A full coupled quantal collision treatment of the O–H^+ system was considered first by Chambaud et al. (1980), who also computed the relevant potential curves and radial coupling terms between the two $^3\Sigma$ potentials involved in the O + H^+ and O^+ + H channels. That work allowed both the fine structure excitation rates of atomic oxygen by protons and the charge exchange collisional rate coefficient to be obtained. This

Table 8.8. Energy Level Positions Involved in the O + H^+ Charge Transfer

	cm^{-1}	K
$O(^3P_2) + H^+$	0	0
$O(^3P_1) + H^+$	158.265	227.72
$O(^3P_0) + H^+$	226.977	326.59
$O^+(^4S_{3/2}) + H$	158.138	227.54

Notes: These values are taken from the NIST database (Kramida et al. 2015).

reaction was revisited by Stancil et al. (1999) through quantal, semiclassical, and distorted wave approximation calculations, and slightly different results were reported. However, Spirko et al. (2003) subsequently reported some inconsistencies in those calculations and recomputed the rate coefficients with the introduction of new potential curves computed in this same group (Spirko et al. 2000). Finally, the charge transfer reaction rates were found close to those computed in Chambaud et al. (1980), which may be explained by the similarity between the computational results of the $^3\Sigma$ potentials obtained in both calculations. It is remarkable that all theoretical calculations recover the experimental value of the thermal equilibrium rate at 300 K (Fehsenfeld & Ferguson 1972), which involves the thermal average of the populations of the fine structure states of atomic oxygen. Thus, some uncertainty remains at low temperature, and further theoretical calculations of the state-to-state reaction rate coefficients are welcome. In the diffuse environments where OH^+ is detected, atomic oxygen most likely is in its $^3P_2$ ground state, thanks to the efficient magnetic dipole radiative transition.

- $N^+ + H_2 \rightarrow NH^+ + H$

  This reaction is almost exoergic, but probably slightly endothermic. The reactive channel here is mixed both with fine structure excitation of the N^+ ion and rotational excitation of molecular hydrogen. The chemical ion–molecular hydrogen network applied to nitrogen chemistry is displayed in Figure 8.9. The starting point is the direct ionization of nitrogen through cosmic rays, as the charge transfer reaction between H^+ and N is highly endothermic and does not occur in interstellar environments. The uncertainty on the energy balance of the title reaction arises from the approximate knowledge of the NH^+ molecular ion dissociation energy. Building a thermodynamic cycle involving the different products and reactants involved, as shown in Figure 8.10, allows us to compute the energy released (in the case of an exothermic reaction) or required (in the case of an endothermic reaction) for the reaction to occur.

$$\Delta E + D_0(NH^+) + I_H - I_N - D_0(H_2) = 0. \tag{8.4}$$

A positive value of ΔE corresponds to an exothermic reaction. The determination of the threshold energy for that reaction to occur was performed via an ion-guided beam tandem spectroscopy experiment by Ervin & Armentrout (1987), where

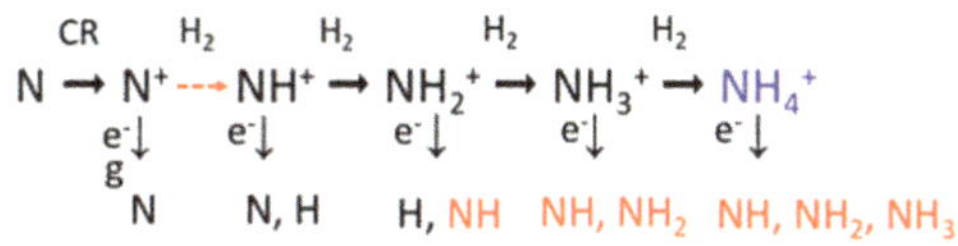

Figure 8.9. Chemical network displaying the formation routes of nitrogen hydrides, following cosmic-ray ionization of nitrogen. The detected molecular species are shown in red. The fully symmetric NH_4^+ molecular ion is shown in blue, as it can not be detected via rotational emission, but its presence is registered due to the detection of its deuterated form NH_3D^+.

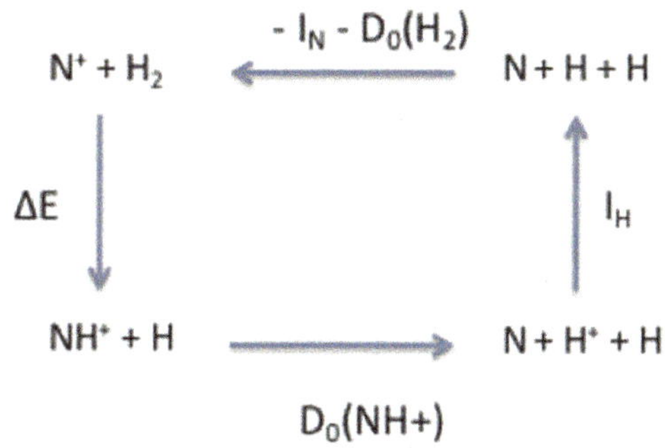

Figure 8.10. Thermodynamic cycle for the $N^+ + H_2$ reaction.

the energy of the incident ions is varied concomitantly with the analysis of the formation products. The reactions of N^+ ions with H_2, HD, and D_2 allowed a dissociation energy value of NH^+ to be derived: $D_0(NH^+)$ of 3.51 ± 0.03 eV. Taking that value and introducing the known values of the ionization energies of H ($I_H = 13.598434$ eV), N ($I_N = 14.53413$ eV) and dissociation energies of H_2 ($D_0(H_2) = 4.478072$ eV) leads to the conclusion that the reaction is almost exoergic and possibly slightly endothermic: $\Delta E = 0.03 \pm 0.03$ eV. The dissociation energy uncertainty is slightly larger than the energies of the ^3P_J fine structure levels $J = 1$ and $J = 2$, which are respectively 0.00604 and 0.01622 eV above the $J = 0$ ground level from the NIST database (Kramida et al. 2015), or the rotational energy of the $J = 1$ level of hydrogen, which is at 0.0147 eV above the $J = 0$ level. Thus, any possible excitation of the N^+ ion or H_2 molecule may influence the reaction kinetics. Theoretical considerations are helpful in understanding the mechanisms taking place in the reaction. The NH^+ molecular ion has an open shell structure with a regular X $^2\Pi_{1/2}$ ground electronic state. The fine structure splitting constant is about 81 cm^{-1}. An additional issue arises because the $a^4\Sigma^-$ excited electronic potential is less than 500 cm^{-1} above the $^2\Pi$ ground-state potential and induces significant perturbation in the rotational spectrum within the $\nu = 0$ and $\nu = 1$ vibrational levels (Colin 1968; Beloy et al. 2011). The schematic correlation between the entrance and exit channels electronic levels is displayed on Figure 8.11 and reproduced from (Zymak et al. 2013). The ion-guided beam experiment

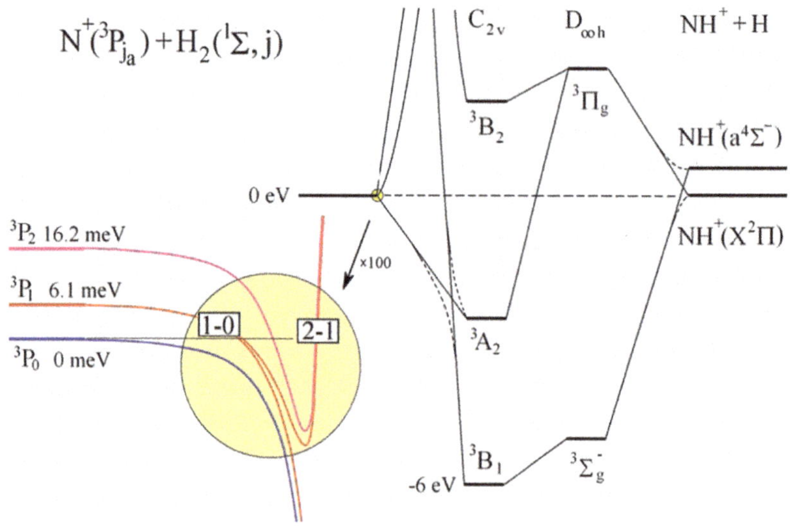

Figure 8.11. Energies involved in the $N^+ + H_2 \rightarrow NH^+ + H$ reaction. (Credit: Zymak et al. 2013.)

(Ervin & Armentrout 1987) was analyzed via a theoretical statistical approach by Grozdanov & McCarroll (2015), who analyzed the channel products NH^+ and ND^+ of the N^+ + HD reaction. That work focuses on the electronic potential energies involved in the reaction, which is independent of the isotopic composition of the reactants and products. Introducing the zero-point energy (ZPE) of the reactants and products, one can write the relation linking the energy difference between the dissociation energies at equilibrium D_e of H_2 and NH^+ and the experimentally measurable dissociation energies D_0, as well as the ZPEs of each molecule. The following relations are obtained for the case of N^+ reactions with H_2, HD, and D_2.

$$D_e^{H_2} - D_e^{NH^+} = D_0^{H_2} - D_0^{NH^+} + \mathrm{ZPE}(H_2) - \mathrm{ZPE}(NH^+)$$

$$= D_0^{HD} - D_0^{NH^+} + \mathrm{ZPE}(HD) - \mathrm{ZPE}(NH^+)$$

$$= D_0^{HD} - D_0^{ND^+} + \mathrm{ZPE}(HD) - \mathrm{ZPE}(ND^+)$$

$$= D_0^{D_2} - D_0^{ND^+} + \mathrm{ZPE}(D_2) - \mathrm{ZPE}(ND^+)$$

Grozdanov and McCarroll (2015) found a range of values between 101 and 106 meV for $D_e^{H_2} - D_e^{NH^+}$ from the analysis of the $N^+ + HD \rightarrow ND^+ + H$ reactive cross sections. This energy gap was then further restricted in a subsequent state-to-state study by Grozdanov et al. (2016), when comparing to the experimental reaction rate coefficients derived by Marquette et al. (1988) with normal and para-H_2. In that latest study, the endothermicity of the reaction was restricted to 101 meV, which allowed the various state-to-state cross sections and subsequent reaction rates with H_2, HD, and D_2 to be computed. The comparison of these latest results with experiments (Marquette et al. 1988) is moderate for H_2, as shown in Figure 8.12.

Moreover, a significant discrepancy is obtained for the N^++ HD reaction-rate coefficient, which is found to be much smaller (4.05×10^{-11} $cm^3\,s^{-1}$ for ground-state N^+ reacting with HD, and 6.3×10^{-11} $cm^3\,s^{-1}$ for thermalized reactants) than the experimental value of 1.4×10^{-10} $cm^3\,s^{-1}$ reported in Marquette et al. (1988) at 20 K. However, only a single point was studied in that experiment (J. B. Marquette 2014, private communication). The various energies computed in Grozdanov et al. (2016) lead to the conclusion that the reaction with HD should be more endothermic than the reaction with o-H_2. Nonetheless, the experimental value of the reaction rate coefficient of N^+ with o-H_2 is found to be respectively 5.3×10^{-11} $cm^3\,s^{-1}$ and 8.4×10^{-11} $cm^3\,s^{-1}$ at 20 K in the two experiments of Marquette et al. (1988) and Zymak et al. (2013). The larger value

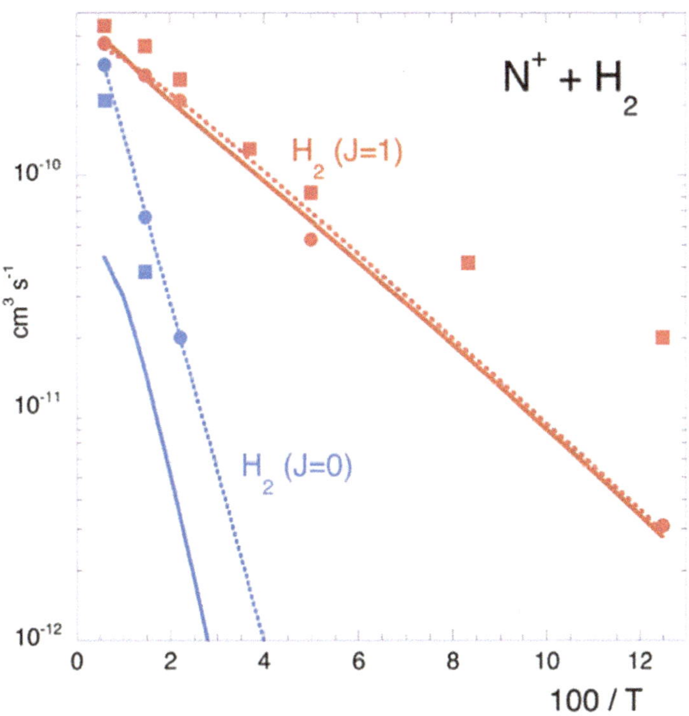

Figure 8.12. Comparison between computed rate coefficients for the reactions of $^{14}N_+$ (j = 0) with para- and ortho-H_2 (blue and red solid lines, respectively) with the analytic forms suggested by Dislaire et al. (2012; dotted lines). The experimental values of Marquette et al. (1988) are indicated as full circles, whereas the results of Zymak et al. (2013) are full squares. (Credit: Grozdanov et al. 2016, reproduced with permission of ESO.)

of 1.4×10^{-10} cm^3s^{-1} obtained at 20 K by Marquette et al. (1988) for the reaction with HD is therefore very unexpected, and more experiments with HD are obviously welcome. To conclude this section, let us mention that a preliminary experimental study of the reverse reaction $NH^+ + H \rightarrow N^+ + H_2$ is reported by Plašil et al. (2015), who find an activation barrier of about 150 K for that reverse reaction, with a reaction rate coefficient of a few 10^{-11} cm^3 s^{-1}, close to the experimental detection limit.

8.3.2 The Limitations of Gas-phase Chemistry

The presence of molecular ions in interstellar conditions provides significant support to the occurrence of the gas-phase ion–molecule reactions schema. Recent laboratory and theoretical studies have both revealed that neutral–neutral reactions could be rapid (a few 10^{-10} cm^3 s^{-1}) at low temperatures, opening a new route for molecular formation. We will not address here the problem of H_2, which cannot be formed efficiently through gas-phase reactions under current interstellar conditions. A recent review (Wakelam et al. 2017) summarizes the different viewpoints related to that important issue. Nevertheless, several molecular observations remain definitely unexplained through gas-phase chemistry, at least with our present chemical knowledge. A few examples are reported below.

Molecular Oxygen

Even a simple molecule like O_2 represents a chemical puzzle, as current dense cloud models predict a large abundance of that species, mainly due to the rapid formation channel through the $O + OH \rightarrow O_2 + H$ chemical reaction as measured in the laboratory (Howard & Smith 1980; Carty et al. 2006) and discussed in many theoretical studies (Quéméner et al. 2008; Lique et al. 2009; Teixidor & Varandas 2015). For example, early studies by Bergin et al. (1995) found a relative abundance of O_2 relative to H_2 between 10^{-4} and 10^{-5} in dense cloud conditions at 10 Myrs, depending on whether a pure gas-phase model is used or if depletion/desorption processes on grain surfaces are introduced, sharing the principal elemental oxygen content with atomic oxygen and the CO molecule. Electric dipole transitions are strictly forbidden for this symmetric molecule, but magnetic dipole rotational transitions can take place within the $^3\Sigma_g^-$ ground electronic state, as emphasized by Black & Smith (1984). However, observations from the ground have rather focused on the $^{18}O^{16}O$ isotopologue in order to avoid the atmospheric oxygen pollution (Black & Smith 1984) without any successful detection, and an upper limit derivation of the $[O_2]/[CO]$ ratio of ~0.1, smaller by an order of magnitude than the model expectations (Marechal et al. 1997). However, observations from space allow the main isotopologue to be searched, and the quest for O_2 was a major objective for the *Submillimeter Wave Astronomy Satellite* (*SWAS*) (Melnick et al. 2000) and ODIN (Nordh et al. 2003). A Herschel Open Time Key Project (Herschel Oxygen Project) was designed to investigate the issue of the unexpectedly low O_2 abundance in a number of interstellar sources through 140 h of observing time. To date, O_2 has been definitely detected in only two sources, namely ρ Oph A (Liseau et al. 2012) and

Orion (Goldsmith et al. 2011), at a level of ~5 × 10^{-8} and 10^{-6} of fractional abundance relative to molecular hydrogen, respectively. The question of why O_2 is such an elusive molecule in the ISM remains open. The role of oxygen depletion on dust allows the modeled O_2 abundance to be reduced (Hollenbach et al. 2009), but not at the level of the astrophysical observations. In addition, adsorption and diffusion energies of atomic oxygen have been recently measured in the laboratory on amorphous water surfaces by Minissale et al. (2016a), giving desorption energies much higher (E_{des} ~1410 K) and diffusion energies much lower (E_{dif} ~ 990 K) than previously assumed in the astrophysical models. Such trends have led to an increase of predicted gas phase O_2, in contradiction to observational results. The possibility that O_2 is a transient molecule is also suggested by Liseau et al. (2012), which could explain the paucity of interstellar detections.

Methanol

Methanol (CH_3OH) is observed ubiquitously in interstellar clouds, protoplanetary disks, and even extragalactic environments. Its fractional abundance relative to molecular hydrogen is a few times 10^{-9} in dark clouds and in extended portions of giant clouds such as the extended ridge in Orion. However, this value is increased dramatically in star-forming regions, such as the hot core and compact ridge sources in the direction of Orion KL, whereas a low fractional abundance is derived in the disc of TW Hya (Walsh et al. 2016). The microwave spectrum is very dense due to the different Hamiltonian interactions taking place (internal rotation or torsion, torsion–rotation interaction, and large effects of centrifugal distortion) and the global modeling of the resulting spectrum is a challenging task (Xu et al. 2008). A direct formation mechanism in the gas phase of protonated methanol was suggested by Herbst (1991), where radiative association of CH_3^+ with H_2O leads to $CH_3OH_2^+$, with a reaction rate coefficient of $5.5 \times (T/300)^{-1.7} \times 10^{-12}$ cm^3 s^{-1}, which then dissociatively recombines with electrons to form CH_3OH. However, this reaction was subsequently studied in an ion trap at low temperatures by Voulot et al. (2002), who found that the radiative association reaction does not proceed at the previously assumed reaction rate coefficient. In addition, experimental studies on dissociative recombination of protonated methanol performed with the CRYRING cryogenic storage ring (Geppert et al. 2006) showed that the saturated molecular ion tends to fragment into three bodies, CH_3, OH, and H, rather than just breaking one hydrogen bond to produce methanol. Thus, current pure gas-phase chemical models definitely fail to produce methanol at a satisfactory level, compared to observations. Additionally, the large amount of methanol found in star-forming regions does imply that another mechanism is at work. Garrod & Herbst (2006) question the pertinence of gas-phase models of interstellar chemistry through the case of methanol. They introduce a new desorption mechanism in a gas–grain code where the energy released in exothermic chemical reactions on grain surfaces may allow to eject adsorbed species back in the gas phase. Methanol is suggested to form in a sequence of hydrogenation reactions of CO occurring at the surface of the ice mantles. The last step, $CH_2OH + H \rightarrow CH_3OH$, has an exothermicity of 4.4 eV, whereas the desorption energy of methanol is about 0.18 eV, so some CH_3OH molecules formed on the ice mantle may contain some internal energy, allowing them to overcome the desorption barrier. A fraction of a few percent of desorbed methanol formed through this reaction can reproduce the observed amount of methanol in dense clouds. However, these findings are somewhat contradicted by a recent laboratory study by Minissale et al. (2016b) showing that adsorbed CO can go back to the gas phase under the action of two successive H atoms, and that the direct hydrogenation of CO to methanol could be too simplistic a description of the gas–grain chemical network.

8.3.3 Complex Organic Molecules

The presence of complex organic molecules (COMs) and their possible link to interstellar organisms was fiercely debated several decades ago (Davies et al. 1984) during the discussion of the carriers of diffuse interstellar bands (Hoyle & Wickramasinghe 1977), but gas-phase rotational spectra have demonstrated the unambiguous presence of complex organic molecular interstellar content, as displayed in Table 8.1. Two principal locations, showing the manifestation of complex organic molecules such as alcohols R-OH (R : CH_3, C_2H_5), acetone (CH_3COCH_3), methylamine (CH_3NH_2), vinyl cyanide (CH_2CHCN), and ethyl cyanide (CH_3CH_2CN) emerged in the 1980s, namely the Sg B2 giant molecular cloud close to the Galactic Center and the bright Orion Compact Ridge region (see Molecules in Space; Endres et al. 2016). Detailed spectroscopic studies have been performed and reported on these molecules, allowing a precise check of *all* spectral transitions over significant spectral ranges as reported in spectral databases (Müller et al. 2001, 2005; Pearson et al. 2010; Pickett et al. 2010; Endres et al. 2016). Herbst & van Dishoeck (2009) reviewed the observational status as well as the different chemical scenarios involved. Observations in hot cores such as the Orion Compact Ridge are interpreted with the help of multiphase models where additional physical conditions are introduced: during a cold-phase era, atoms and molecules accrete on grain surfaces. Surface chemical processes lead to the building of ices composed of water, carbon dioxide, and saturated COMs formed through association reactions between radicals. Examples are provided in Charnley & Rogers (2005), Garrod et al. (2006), Charnley et al. (2005), and Garrod & Herbst (2006). A warm-up phase is introduced afterwards, arising from the heating of the inner envelope by the protostar, during which the icy complex molecules evaporate and eject the complex molecules in the gas phase. The assumed timescale of this passive phase ranges from tens of thousands to one million years, and the temperature rise between 10 and 200 K scales linearly or quadratically. Finally, a hot-core phase corresponding to a temperature of 200 K allows various complex chemical reactions to occur in the gas phase. The rationale behind the chemical complexity found in the Galactic Center environment is not obvious, and additional studies are ongoing. More recently, interferometric observations have allowed COMs emission to be detected from two compact regions centered on the two components of the IRAS 19293-2422 binary system, the so-called corinos (Bottinelli et al. 2004), at a near-arcsecond resolution. Bacmann et al. (2012) obtained the first detection of COMs in a cold environment, the L1689B pre-stellar core, with the IRAM 30 m telescope. COMs have also been found in the PDR region of the Horsehead Nebula (Guzmán et al. 2014) and in the Orion Bar (Cuadrado et al. 2017). The presence of COMs is anticipated by Walsh et al. (2014) in protoplanetary disks. However, only methanol, methyl cyanide (CH_3CN), and cyanoacetylene (HC_3N) have been found up to now in those environments, thanks to ALMA interferometric studies. Dedicated large observational programs are being pursued to further constrain our knowledge of organic chemistry in solar-like star-forming regions (e.g., the SOLIS project; Ceccarelli et al. 2017). Very interestingly, such a project involves surface scientists as well as gas-phase physical chemists, in addition to observers, in order to insure the required link between the observations and the laboratory astrophysics studies.

The debate about the origin of complex organic molecules in space is far from being settled and additional work is expected, both from the observational side and theoretical/experimental facets. In conclusion, let us mention the theoretical suggestion that gas-phase neutral neutral reactions such as $NH_2 + H_2CO$ could be at work in the formation of the prebiotic formamide ($HCONH_2$) molecule (Skouteris et al. 2017). Other tentative attempts to find gas-phase pathways to COMs are considering the ion–molecular scheme, such as the reactions between formaldehyde, ammonia, or NH_2OH, and various molecular ions such as H_2COH^+, NH_4^+, NH_3OH^+, and $NH_2OH_2^+$. Theoretical calculations with state-of-the-art ab initio techniques are used to constrain the energetics of the reactions and the possible potential barriers. A direct dynamic approach is additionally introduced to provide information on the reactivity responsible for the formation of organic molecules without imposing any reaction channel (Spezia et al. 2016).

8.4 The Importance of Surface Chemistry

The difficulties encountered in the gas-phase modeling of particular molecules when comparing to observations are often interpreted as a failure of gas-phase chemistry and a hint of surface chemical processes. Nevertheless, the presence of dust particles is testified by different manifestations, including: the extinction of starlight; the dust continuum emission in the infrared, submillimeter, and millimeter wavelength ranges; and some broad infrared absorption features detected in front of bright stars, as recently reviewed by Boogert et al. (2015). Despite a reduced sensitivity compared to gas-phase spectroscopy, infrared signatures of icy components allow the main constituents of interstellar ices to be identified. H_2O is by far the main constituent. When evaluated with respect to water ice, CO_2 and CO are present at a level of 10–20%, whereas methanol ice represents about 5–10%, and ammonia and methane are in the 3–10% range. Other species are present as well, such as H_2CO, OCS, etc., but at a much smaller percentage and their identification is not fully secure. The launch of *JWST* will certainly allow a profound change in our knowledge of the icy universe and significantly improve the chemical inventory of interstellar ices. However, the available information, albeit limited, is still critical to include in astrochemical modeling studies in order to bring some constraints on the considered surface chemistry processes. In addition, summing the amount of gas phase + solid phase content of the different elements allows us to quantify our knowledge about the interstellar material and raise some issues, due to our incomplete census of the different constituents.

8.4.1 Bottom-up versus Top-down Chemistry

Different mechanisms may occur on the grain surfaces; these include accretion (i.e., adsorption), desorption from the surface, diffusion on the grain surface, and reaction between two adsorbed species. Desorption from the surface may be of thermal origin, if the ratio of binding energy to thermal energy is a factor of a few; it may also be triggered by UV photons, cosmic high-energy particles. Then, even gas-phase species, ejected from the grains, may witness the occurrence of solid-phase chemistry. This is the so-called top-down scheme, which is often opposed to the bottom-up approach considered in gas-phase chemical scenarios.

Such a top-down proposition has been developed by Zhen et al. (2014) to explain the presence of C_{60} fullerenes detected in a young planetary nebula by Cami et al. (2010) from the photochemical evolution of large polycyclic aromatic hydrocarbon molecules (PAHs). Laboratory studies performed on large PAHs show indeed that sequential losses of atomic H may lead to fully dehydrogenated PAHs, which may further be converted into cages from subsequent losses of C_2 units. Top-down chemistry may also be at work in producing smaller carbon chain molecules detached by photo-induced processes on large amorphous a-C(:H) nanoparticles, and be at the heart of formation and evolution of the Diffuse Interstellar Bands (DIBs) that have been observed for over a century and remain unassigned to this day (Jones 2016). An extensive literature is available, as reported at the occasion of the 297 IAU symposium (Cox et al. 2014). The subject is obviously beyond the scope of the present review. Could the observations help to resolve that issue as well? Guzmán et al. (2015) claim that the spatial resolution provided by interferometric observations of the C_3H^+ molecular ion permits the suggestion that top-down chemistry may indeed be at work in the Horsehead Nebula. In the first PDR layers where the 7.7 micron band of PAH emission is peaking, significant emission of C_2H, c-C_3H_2, and C_3H^+ are found to be significantly abundant and underpredicted by gas-phase PDR models. The early detection of eight visible emission transitions of CH^+ in the Red Rectangle Nebula by Balm & Jura (1992) was also thought to result from the destruction of grain mantles due to the harsh ultraviolet environment (Duley et al. 1992). The Red Rectangle Nebula is a spectacular bipolar nebula centered around the eccentric evolved HD44179 binary system. Among its rich spectral properties, one can refer to it as one of the brightest sources of mid-IR PAH bands. It is also well known for both Extended Red Emission (ERE) and Blue Luminescence (BL; Van Winckel 2014), making it an unique object. Other visible emissions of CH and CN (together with absorption lines) have been observed in that line of sight (Hobbs et al. 2004). However, no specific chemical modeling studies related to these emission lines have been reported so far. This issue warrants further investigation.

8.4.2 The Sulfur Problem

The elemental gas-phase abundance of sulfur compared to elemental hydrogen in present-day solar photosphere is $1.3 \pm 0.1 \times 10^{-5}$, as reviewed by Asplund et al. (2009) and 2.8×10^{-5} in the ζ Oph diffuse interstellar cloud (Savage & Sembach 1996). As a moderately volatile element ($T_C \sim 700$ K, per Lodders 2003), low metal elemental abundances are often assumed in dense cloud chemical models, as shown in Table 8.5. However, only very small amounts of sulfur ice compounds have been found so far (Boogert et al. 2015) and presently detected sulfur molecules do not account for the amount of available sulfur. One crucial uncertainty lies in the extent of neutral atomic sulfur, whose fine structure transitions at 25.249 and 56.311 μ are not detectable from the ground. This fact is sometimes referred as "the sulfur problem" in the literature. Vidal et al. (2017) recently revisited the different gas-phase chemical reactions occurring between sulfur containing molecules under interstellar conditions. They also analyzed the possible chemical processes taking place at the surface of interstellar grains. Forty-seven new sulfur gas and surface compounds, including S_3 to S_8 clusters, as well as $C_nH_pS_q$ and H_nCNS compounds, were introduced in the network. The general approach used in that study is to assume that surface reactions result from the interaction between physisorbed species, which are assumed to diffuse through thermal hopping only. Surface adsorbed H (labeled as *s*-H) is reacting with adsorbed sulfur (*s*-S), yielding adsorbed SH (*s*-SH). Additional hydrogenation may occur on the surface of the grain, yielding the saturated s-H_2S hydrogen sulfide. The presence of an activation barrier is checked through quantum chemical calculations involving gas-phase species. The model predicts that *s*-SH and *s*-H_2S become the main ice sulfur components, similarly to the results of the model developed by Furuya et al. (2015). However, no *s*-H_2S signatures have been detected in ice spectroscopic studies. The computed relative abundance of *s*-H_2S appears much larger than the upper limits toward W33 and IRAS18316-0602 reported in Jiménez-Escobar & Muñoz Caro (2011). The overall conclusion of that study appears somewhat optimistic, as the claim that sulfur elementary and chemical abundances are reconciled has been called into questioned by new observations, as will be reported shortly below.

Another approach to tackle the problem of missing sulfur was put forward by Martín-Doménech et al. (2016), who suggest that doubly sulfuretted molecules could be formed on ice mantles and subsequently desorbed in the gas phase, as studied in the laboratory by Jiménez-Escobar & Muñoz Caro (2011). However, the observations performed towards IRAS 16293-2422 led only to upper limit values for H_2S_2, S_2H, and S_2. The presence of four weak transitions of S_2H in the WHISPER survey at both the PDR and Core positions of the Horsehead Nebula had been overlooked in previous studies (Gerin et al. 2009; Goicoechea et al. 2009; Pety et al. 2012; Guzmán et al. 2012; Gratier et al. 2013; Guzmán et al. 2013, 2014, 2015). They were fortuitously found in the spectral archives and confirmed through discretionary director time at the IRAM 30 m telescope by Fuente et al. (2017), thanks to a project dedicated to Gas-phase Elementary abundances in Molecular cloudS (GEMS),[8]

[8] Large program accepted at IRAM: http://www.oan.es/gems/.

following the link between ionization and sulfur molecules discussed in Fuente et al. (2016). The identification appears secure, with two different rotation doublets within the X $^2A''$ ground electronic state, involving N: 6 → 5 and N: 7 → 6 rotational transitions corresponding to low excitation conditions. No other S_2H are available in the survey, but additional transitions could be searched at lower frequencies to better constrain the excitation diagram and column density determinations. The derived fractional abundance is about 1.0×10^{-10} and 1.3×10^{-10} in the core and PDR regions, respectively. Nevertheless, understanding the chemical processes involved in the formation/destruction balance of S_2H is far from straightforward. Photodesorption efficiencies of H_2S_2 and S_2H have been found on the order of 10^{-5}, so such a mechanism is not expected to be at the origin of S_2H. Further studies will be pursued in order to better constrain the respective roles of gas and dust in the sulfur compound chemistry. Finally, let us mention the very recent detection of the thioformyl radical HCS and its HSC metastable isomer by Agúndez et al. (2018) in the L483 molecular cloud. Even if these species represent a small fraction of the available sulfur, understanding the full set of detected sulfur species is an important goal that is yet to be completed.

8.5 Conclusions

We have given some examples of the fertile interplay between astrophysical spectroscopic observations and fundamental microscopic atomic and molecular processes. Both gas-phase and surface chemistry contribute to the chemical balance of observed interstellar molecular species. More studies are required in both fields, which will allow us to constrain not only the formation/destruction mechanisms but also the physical conditions where these reactions are taking place. The details of the gas-phase reactivity involving possible state-selective reactions may and should be introduced in chemical astrophysical models. The implementation of surface processes in such models is not straightforward and has not been discussed here. However, significant efforts are ongoing in this area, as described in the review paper by Cuppen et al. (2017). The challenge is considerable, but the beautiful observations currently available, as well as those anticipated from new instrumentation such as ALMA, IRAM, and *JWST* deserve such joined attempts.

Acknowledgements

This work was supported by the Programme National "Physique et Chimie du Milieu Interstellaire" (PCMI) of CNRS/INSU with INC/INP, co-funded by CEA and CNES. Helpful comments from the referee are gratefully acknowledged.

References

Adams, W. S. 1941, ApJ, 93, 11

Agúndez, M., Marcelino, N., Cernicharo, J., & Tafalla, M. 2018, A&A, 611, L1

Agúndez, M., & Wakelam, V. 2013, ChRv, 113, 8710

Asplund, M., Grevesse, N., Sauval, A. J., & Scott, P. 2009, ARA&A, 47, 481

Atkins, P. W., & de Paula, J. 2006, Physical Chemistry (8th ed; Oxford: Oxford Univ. Press)

Bacalla, X., Linnartz, H., & Cox, N. L., et al. 2018, A&A, in press (arXiv:1811.08662)
Bacmann, A., Taquet, V., Faure, A., Kahane, C., & Ceccarelli, C. 2012, A&A, 541, L12
Balm, S. P., & Jura, M. 1992, A&A, 261, L25
Bates, D. R., & Spitzer, L. Jr. 1951, ApJ, 113, 441
Beloy, K., Kozlov, M. G., Borschevsky, A., et al. 2011, PhRvA, 83, 062514
Bergin, E. A., Goldsmith, P. F., Snell, R. L., & Langer, W. D. 1997, ApJ, 482, 285
Bergin, E. A., Langer, W. D., & Goldsmith, P. F. 1995, ApJ, 441, 222
Bertaux, J.-L., & Lallement, R. 2017, MNRAS, 469, S646
Black, J. H., & Smith, P. L. 1984, ApJ, 277, 562
Boogert, A. C. A., Gerakines, P. A., & Whittet, D. C. B. 2015, ARA&A, 53, 541
Botschwina, P., Stein, C., Sebald, P., Schröder, B., & Oswald, R. 2014, ApJ, 787, 72
Bottinelli, S., Ceccarelli, C., Neri, R., et al. 2004, ApJL, 617, L69
Boulanger, F., Ensslin, T., Fletcher, A., et al. 2018, JCAP, 2018, 049
Bovino, S., Grassi, T., Schleicher, D. R. G., & Caselli, P. 2017, ApJL, 849, L25
Brünken, S., Kluge, L., Stoffels, A., Asvany, O., & Schlemmer, S. 2014a, ApJL, 783, L4
Brünken, S., Sipilä, O., Chambers, E. T., et al. 2014b, Natur, 516, 219
Calmonte, U., Altwegg, K., Balsiger, H., et al. 2016, MNRAS, 462, S253
Cami, J., Bernard-Salas, J., Peeters, E., & Malek, S. E. 2010, Sci, 329, 1180
Carty, D., Goddard, A., Köhler, S. P. K., Sims, I. R., & Smith, I. W. M. 2006, JPCA, 110, 3101
Caselli, P., van der Tak, F. F. S., Ceccarelli, C., & Bacmann, A. 2003, A&A, 403, L37
Ceccarelli, C., Caselli, P., Bockelée-Morvan, D., et al. 2014, in Protostars and Planets VI (Tuscon AZ: Univ. Arizona Press), 859
Ceccarelli, C., Caselli, P., Fontani, F., et al. 2017, ApJ, 850, 176
Cernicharo, J., Polehampton, E., & Goicoechea, J. R. 2007, ApJL, 657, L21
Chambaud, G., Levy, B., Millie, P., et al. 1980, JPhB, 13, 4205
Charnley, S. B., & Rodgers, S. D. 2005, in IAU Symp. 231, Astrochemistry: Recent Successes and Current Challenges, ed. D. C. Lis, G. A. Blake, & E. Herbst (Cambridge: Cambridge Univ. Press), 237
Colin, R. 1968, CaJPh, 46, 61
Cooke, R. J., Pettini, M., Jorgenson, R. A., Murphy, M. T., & Steidel, C. C. 2014, ApJ, 781, 31
Cox, N. L. J. & Cami, J. (ed.) 2014, in IAU Symp. 297, The Diffuse Interstellar Bands (Cambridge: Cambridge Univ. Press), 412
Cuadrado, S., Goicoechea, J. R., Cernicharo, J., et al. 2017, A&A, 603, A124
Cuadrado, S., Goicoechea, J. R., Pilleri, P., et al. 2015, A&A, 575, A82
Cuppen, H. M., Walsh, C., Lamberts, T., et al. 2017, SSRv, 212, 1
Cyburt, R. H., Fields, B. D., Olive, K. A., & Yeh, T.-H. 2016, RvMP, 88, 015004
Dalgarno, A., & Black, J. H. 1976, RPPh, 39, 573
Dalgarno, A., Black, J. H., & Weisheit, J. C. 1973, ApL, 14, 77
Davies, R. E., Delluva, A. M., & Koch, R. H. 1984, Natur, 311, 748
Dislaire, V., Hily-Blant, P., Faure, A., et al. 2012, A&A, 537, A20
Douglas, A. E., & Herzberg, G. 1941, ApJ, 94, 381
Drozdovskaya, M. N., van Dishoeck, E. F., Jørgensen, J. K., et al. 2018, MNRAS, 476, 4949
Duley, W. W., Hartquist, T. W., Sternberg, A., Wagenblast, R., & Williams, D. A. 1992, MNRAS, 255, 463
Dulieu, F., Minissale, M., & Bockelée-Morvan, D. 2017, A&A, 597, A56
Dunham, T. Jr. 1937, PASP, 49, 26

Endres, C. P., Schlemmer, S., Schilke, P., Stutzki, J., & Müller, H. S. P. 2016, JMoSp, 327, 95
Ervin, K. M., & Armentrout, P. B. 1987, JChPh, 86, 2659
Fehsenfeld, F. C., & Ferguson, E. E. 1972, JChPh, 56, 3066
Flower, D. R., Pineau des Forêts, G., & Walmsley, C. M. 2004, A&A, 427, 887
Fortenberry, R. C. 2015, JPCA, 119, 9941
Friesen, R. K., Di Francesco, J., Bourke, T. L., et al. 2014, ApJ, 797, 27
Fuente, A., Cernicharo, J., Roueff, E., et al. 2016, A&A, 593, A94
Fuente, A., Goicoechea, J. R., & Pety, J., et al. 2017, ApJL, 846, L49
Furuya, K., Aikawa, Y., Hincelin, U., et al. 2015, A&A, 584, A124
Garrod, R. T., & Herbst, E. 2006, A&A, 457, 927
Georgievskii, Y., & Klippenstein, S. J. 2005, JChPh, 122, 194103
Geppert, W. D., Hamberg, M., Thomas, R. D., et al. 2006, FaDi, 133, 177
Gerin, M., de Luca, M., Black, J., et al. 2010, A&A, 518, L110
Gerin, M., Goicoechea, J. R., Pety, J., & Hily-Blant, P. 2009, A&A, 494, 977
Gerin, M., Neufeld, D. A., & Goicoechea, J. R. 2016, ARA&A, 54, 181
Gerlich, D., Herbst, E., & Roueff, E. 2002, P&SS, 50, 1275
Gioumousis, G., & Stevenson, D. P. 1958, JChPh, 29, 294
Goicoechea, J. R., Pety, J., Gerin, M., Hily-Blant, P., & Le Bourlot, J. 2009, A&A, 498, 771
Goldsmith, P. F., Liseau, R., Bell, T. A., et al. 2011, ApJ, 737, 96
Graedel, T. E., Langer, W. D., & Frerking, M. A. 1982, ApJS, 48, 321
Gratier, P., Majumdar, L., Ohishi, M., et al. 2016, ApJS, 225, 25
Gratier, P., Pety, J., Guzmán, V., et al. 2013, A&A, 557, A101
Grozdanov, T. P., & McCarroll, R. 2015, JPCA, 119, 5988
Grozdanov, T. P., McCarroll, R., & Roueff, E. 2016, A&A, 589, A105
Guzmán, V., Pety, J., Gratier, P., et al. 2012, A&A, 543, L1
Guzmán, V. V., Goicoechea, J. R., Pety, J., et al. 2013, A&A, 560, A73
Guzmán, V. V., Pety, J., Goicoechea, J. R., et al. 2015, ApJL, 800, L33
Guzmán, V. V., Pety, J., Gratier, P., et al. 2014, FaDi, 168, 103
Harju, J., Sipilä, O., Brünken, S., et al. 2017, ApJ, 840, 63
Herbst, E. 1991, in ASP Conf. Ser. 16, Atoms, Ions and Molecules: New Results in Spectral Line Astrophysics, ed. A. D. Haschick, & P. T. P. Ho (San Francisco, CA: ASP), 313
Herbst, E., & Klemperer, W. 1973, ApJ, 185, 505
Herbst, E., & van Dishoeck, E. F. 2009, ARA&A, 47, 427
Herzberg, G. 1989, Molecular Spectra and Molecular Structure (2nd ed; Malabar, FL: Krieger)
Hily-Blant, P., Faure, A., Rist, C., Pineau des Forêts, G., & Flower, D. R. 2018, MNRAS, 477, 4454
Hirschfelder, J. O., Curtiss, C. F., & Bird, R. B. 1964, Molecular Theory of Gases and Liquids (New York: Wiley)
Hobbs, L. M., Thorburn, J. A., Oka, T., et al. 2004, ApJ, 615, 947
Hollenbach, D., Kaufman, M. J., Bergin, E. A., & Melnick, G. J. 2009, ApJ, 690, 1497
Howard, M. J., & Smith, I. W. M. 1980, CPL, 69, 40
Hoyle, F., & Wickramasinghe, N. C. 1977, Natur, 266, 241
Huang, X., Fortenberry, R. C., & Lee, T. J. 2013, ApJL, 768, L25
Hugo, E., Asvany, O., & Schlemmer, S. 2009, JChPh, 130, 164302
Ilisca, E. 1992, PrSS, 41, 217
Ilisca, E., & Ghiglieno, F. 2016, RSOS, 3, 160042

Ilisca, E., & Ghiglieno, F. 2017, CPL, 667, 233
Indriolo, N., Neufeld, D. A., Gerin, M., et al. 2015, ApJ, 800, 40
Jiménez-Escobar, A., & Muñoz Caro, G. M. 2011, A&A, 536, A91
Jones, A. P. 2016, RSOS, 3, 160223
Kaifu, N., Ohishi, M., Kawaguchi, K., et al. 2004, PASJ, 56, 69
Keeney, B. A., Stern, S. A., A'Hearn, M. F., et al. 2017, MNRAS, 469, S158
Kramida, A., Ralchenko, Yu., Reader, J., & NIST ASD Team 2015, NIST Atomic Spectra Database (version 5.3), Available: T_E http://physics.nist.gov/asd (2017 May 22), (Gaithersburg, MD: National Institute of Standards and Technology)
Krełowski, J., Beletsky, Y., & Galazutdinov, G. A. 2010, ApJL, 719, L20
Le Bourlot, J. 2000, A&A, 360, 656
Linsky, J. L., Draine, B. T., Moos, H. W., et al. 2006, ApJ, 647, 1106
Lique, F., Jorfi, M., Honvault, P., et al. 2009, JChPh, 131, 221104
Liseau, R., Goldsmith, P. F., Larsson, B., et al. 2012, A&A, 541, A73
Lodders, K. 2003, ApJ, 591, 1220
Lucas, R., & Liszt, H. S. 1993, A&A, 276, L33
Mandy, M. E., & Martin, P. G. 1993, ApJS, 86, 199
Marechal, P., Viala, Y. P., & Pagani, L. 1997, A&A, 328, 617
Marquette, J. B., Rebrion, C., & Rowe, B. R. 1988, JChPh, 89, 2041
Martín-Doménech, R., Jiménez-Serra, I., Muñoz Caro, G. M., et al. 2016, A&A, 585, A112
McCarthy, M. C., Crabtree, K. N., Martin-Drumel, M.-A., et al. 2015, ApJS, 217, 10
McGuire, B. A., Carroll, P. B., Gratier, P., et al. 2014a, ApJ, 783, 36
McGuire, B. A., & Carroll, P. B. 2013, ApJ, 774, 56
McGuire, B. A., Carroll, P. B., Sanders, J. L., et al. 2014b, MNRAS, 442, 2901
Melnick, G. J., Stauffer, J. R., Ashby, M. L. N., et al. 2000, ApJL, 539, L77
Minissale, M., Congiu, E., & Dulieu, F. 2016a, A&A, 585, A146
Minissale, M., Moudens, A., Baouche, S., Chaabouni, H., & Dulieu, F. 2016b, MNRAS, 458, 2953
Mladenović, M. 2014, JChPh, 141, 224304
Müller, H. S. P., Schlöder, F., Stutzki, J., & Winnewisser, G. 2005, JMoSt, 742, 215
Müller, H. S. P., Thorwirth, S., Roth, D. A., & Winnewisser, G. 2001, A&A, 370, L49
Nordh, H. L., von Schéele, F., Frisk, U., et al. 2003, A&A, 402, L21
Pagani, L., Lesaffre, P., Roueff, E., et al. 2012, RSPTA, 370, 5200
Pagani, L., Roueff, E., & Lesaffre, P. 2011, ApJL, 739, L35
Pagani, L., Salez, M., & Wannier, P. G. 1992, A&A, 258, 479
Parise, B., Belloche, A., Du, F., Güsten, R., & Menten, K. M. 2011, A&A, 526, A31
Pearson, J. C., Müller, H. S. P., Pickett, H. M., Cohen, E. A., & Drouin, B. J. 2010, JQSRT, 111, 1614
Pety, J., Gratier, P., Guzmán, V., et al. 2012, A&A, 548, A68
Pickett, H. M., Poynter, I. R. L., Cohen, E. A., et al. 2010, JQSRT, 111, 1617
Plašil, R., Roučka, Š., Mulin, D., et al. 2015, J. Phys. Conf. Ser., 635, 022024
Polo, S. R. 1957, CaJPh, 35, 880
Porras, A. J., Federman, S. R., Welty, D. E., & Ritchey, A. M. 2014, ApJL, 781, L8
Pulliam, R. L., McGuire, B. A., & Remijan, A. J. 2012, ApJ, 751, 1
Quack, M. 1977, MolPh, 34, 477
Quéméner, G., Balakrishnan, N., & Kendrick, B. K. 2008, JChPh, 129, 224309

Radom, L., Hahiharan, P. C., Pople, J. A., & Schleyer, P. R. 1976, JACS, 98, 10
Ramanlal, J., Polyansky, O. L., & Tennyson, J. 2003, A&A, 406, 383
Riemer-Sørensen, S., Kotuš, S., Webb, J. K., et al. 2017, MNRAS, 468, 3239
Roueff, E., & Gerin, M. 2003, SSRv, 106, 61
Savage, B. D., & Sembach, K. R. 1996, ARA&A, 34, 279
Sims, I. R. 2013, NatCh, 5, 734
Sipilä, O., Hugo, E., Harju, J., et al. 2010, A&A, 509, A98
Skouteris, D., Vazart, F., Ceccarelli, C., et al. 2017, MNRAS, 468, L1
Snow, T. P., & McCall, B. J. 2006, ARA&A, 44, 367
Solomon, P. M., & Klemperer, W. 1972, ApJ, 178, 389
Sonnentrucker, P., Wolfire, M., Neufeld, D. A., et al. 2015, ApJ, 806, 49
Spezia, R., Jeanvoine, Y., Hase, W. L., Song, K., & Largo, A. 2016, ApJ, 826, 107
Spirko, J. A., Mallis, J. T., & Hickman, A. P. 2000, JPhB, 33, 2395
Spirko, J. A., Zirbel, J. J., & Hickman, A. P. 2003, JPhB, 36, 1645
Stancil, P. C., Schultz, D. R., Kimura, M., et al. 1999, A&AS, 140, 225
Stark, R., van der Tak, F. F. S., & van Dishoeck, E. F. 1999, ApJL, 521, L67
Teixidor, M. M., & Varandas, A. J. C. 2015, JChPh, 142, 014309
Terzieva, R., & Herbst, E. 1998a, ApJ, 509, 932
Terzieva, R., & Herbst, E. 1998b, ApJ, 501, 207
Tizniti, M., Le Picard, S. D., Lique, F., et al. 2014, NatCh, 6, 141
Troe, J. 1987, JChPh, 87, 2773
Troe, J. 1996, JChPh, 105, 6249
van Dishoeck, E. F., & Black, J. H. 1986, ApJS, 62, 109
Van Winckel, H. 2014, in IAU Symp. 297, The Diffuse Interstellar Bands, ed. J. Cami, & N. L. J. Cox (Cambridge: Cambridge Univ. Press), 180
Vastel, C., Caselli, P., Ceccarelli, C., et al. 2006, ApJ, 645, 1198
Vastel, C., Phillips, T. G., & Yoshida, H. 2004, ApJL, 606, L127
Vidal, T. H. G., Loison, J.-C., Jaziri, A. Y., et al. 2017, MNRAS, 469, 435
Voulot, D., Luca, A., & Gerlich, D. 2002, WDS'02 Proc. Contributed Papers, Part II, Vol. 2 (Prague: Matfyzpress), 294
Wakelam, V., Bron, E., Cazaux, S., et al. 2017, MolAs, 9, 1
Walmsley, C. M., Flower, D. R., & Pineau des Forêts, G. 2004, A&A, 418, 1035
Walsh, C., Loomis, R. A., Öberg, K. I., et al. 2016, ApJL, 823, L10
Walsh, C., Millar, T. J., Nomura, H., et al. 2014, A&A, 563, A33
Watson, W. D. 1978, ARA&A, 16, 585
Wilson, S., & Green, S. 1980, ApJ, 240, 968
Wilson, S., & Green, S. 1982, ApJ, 253, 989
Xu, L.-H., Fisher, J., Lees, R. M., et al. 2008, JMoSp, 251, 305
Zhao, D., Galazutdinov, G. A., Linnartz, H., & Krełowski, J. 2015, ApJL, 805, L12
Zhen, J., Castellanos, P., Paardekooper, D. M., Linnartz, H., & Tielens, A. G. G. M. 2014, ApJL, 797, L30
Zymak, I., Hejduk, M., Mulin, D., et al. 2013, ApJ, 768, 86

Lightning Source UK Ltd.
Milton Keynes UK
UKHW051424130619
344242UK00004B/50/P

9 780750 314268